D0947056

KARL IMHOFF'S

HANDBOOK OF URBAN DRAINAGE AND WASTEWATER DISPOSAL

KARL IMHOFF'S

HANDBOOK OF URBAN DRAINAGE AND WASTEWATER DISPOSAL

VLADIMIR NOVOTNY, *Editor*
Marquette University
Milwaukee, Wisconsin

KLAUS R. IMHOFF
Ruhrverband, Essen
Federal Republic of Germany

MEINT OLTHOF
Duncan, Lagnese, and Associates
Pittsburgh, Pennsylvania

PETER A. KRENKEL
University of Nevada
Reno, Nevada

WILEY

A Wiley-Interscience Publication

JOHN WILEY & SONS

New York ▪ Chichester ▪ Brisbane ▪ Toronto ▪ Singapore

Library of Congress Cataloging in Publication Data:

Handbook of urban drainage and wastewater disposal.

"A Wiley-Interscience publication."
Based on the latest German edition of Karl and Klaus R. Imhoff's
Taschenbuch der Stadtentwässerung.
Bibliography: p.
1. Urban runoff—Management—Handbooks, manuals, etc.
2. Sewage disposal—Handbooks, manuals, etc.
I. Novotny, Vladimir, 1938- . II. Imhoff, Karl,
1876–1965. Taschenbuch der Stadtentwässerung.
TD657.H36 1989 628'.2 88-20593
ISBN 0-471-81037-1

Printed in the United States of America

10 9 8 7 6 5 4 3 2

Dedicated to Karl Imhoff (1876–1965),
a pioneer of modern environmental engineering.

Preface

The foundation for this book was laid in 1906 when Karl Imhoff, a young engineer in charge of a newly established sewage disposal agency in the industrialized Emscher area of Germany, prepared an informative brief manual on urban sewerage and treatment of wastewater. The primary objective of the sewage disposal practices in the early 1900s was to keep the sewage and receiving waters fresh and without objectionable odor. The emphasis was on constructing open, naturally aerated channels and odorless treatment plants that consisted mainly of Imhoff tanks (a two-story settling tank with a digestion compartment) in conjunction with drying beds for sludge disposal.

In 1913, a new agency was founded that was given the task of keeping the river Ruhr reasonably clean so it could be used as a source of drinking water for most of the Ruhr area and the industrialized Emscher zone, in which a large portion of German coal, steel, and other heavy industry was, and still is, located. As a result of these developments, sections on biological treatment and later on self-purification of streams were added to the book.

In 1926, Karl Imhoff met Gordon M. Fair, a prominent engineer who later became a distinguished professor of sanitary engineering at Harvard University in Cambridge, Massachusetts. Fair translated the 5th edition of Karl Imhoff's book into English, which was then published in the United States in 1929. Subsequent translations and adaptations of the book appeared in 12 other countries. The work of Karl Imhoff became known throughout the world and he is considered a founder of modern European sanitary engineering.

In 1938, Imhoff and Fair began to work on a new joint U.S. adaptation. The book was completely reworked and published in 1940 (Imhoff–Fair: *Sewage Treatment*, John Wiley & Sons) with the second edition appearing in 1956. Especially important were Fair's pioneering contributions on self-purification in receiving waters, theory of decomposition of organics in sewage and sludges, and sludge handling. These sections were then incorporated in the subsequent German editions. Also in 1956, the British adaptation of the book was published (Imhoff–Müller–Thistlethwayte: *Disposal of Sewage and Other Water-borne Wastes*, Butterworth). In summary, in addition to the three U.S. editions, the book was published in Germany 26 times (last in 1985 by R. Oldenbourg Publishing Company in Munich). It appeared five times in France, four times in Poland, and at least twice in several other countries.

Karl Imhoff, founder of this book, died on September 28, 1965 after completing work on the 21st edition. The subsequent German editions have been prepared by his son Klaus.

In 1983, the present authors formed a team that assumed the task of preparing a new edition of a practical handbook of urban stormwater management and wastewater treatment and disposal, following as closely as possible the tradition of Karl Imhoff and Gordon M. Fair. This book is based on the latest German edition; however, it is completely reworked to reflect new U.S. technology, practices, and conditions. A section on urban nonpoint sources and abatement of pollution from urban runoff was added along with current analysis of the effects of effluent disposal into receiving waters, and a description of the newest technology in wastewater treatment. Numerous practical design examples have been included. Also, additional English language references have been included, replacing some less accessible references in the German version.

Although the book reflects the latest U.S. and European state-of-the-art technology, the authors also realized that the book will be used as a handy reference by English reading audiences in many countries, including those that are now rapidly advancing and developing. Thus, in the tradition of the book founders, the authors made an effort to prepare a practical reference "pocketbook" for engineers, city and government officials, and students.

In the preface to the first edition of the book, Karl Imhoff expressed the philosophy and practicality of the handbook by saying that "the first rule of all engineering computations is to make them to conform with the task for which they are used. The more detailed and unnecessarily complicated the computation becomes, the greater is the likelihood of error." Therefore, the emphasis is on simple graphical and computational procedures that will provide quick and reliable overview answers. Descriptions of various processes covered in the book are brief, with references included in the text for a further detailed study. The wisdom of Karl Imhoff's philosophy is even more evident today when computer software packages are being used in designing urban drainage and treatment systems. The engineer or designer who is using the software must not be a blind user. The simple and quick procedures delineated in this book can be used to make a quick check of the correctness of the computerized design.

The book uses almost exclusively the SI (metric) system of units. However, as the old English system still persists in the United States, convenient conversions to the English system have been included throughout the text and a conversion table is provided in Chapter 15.

In conclusion, the authors would like to thank all their colleagues and friends in the United States and Germany who contributed directly or indirectly to this work. Special appreciation is also expressed to the German publisher of the Karl Imhoff book, Oldenbourg Publishing Company, for permission to freely use the materials from the German editions. Several American and European journals were also used as a source of information from which the *Journal of Water Pollution Control Federation* and *Journal of Environmental Engineering Division* of the ASCE are the most frequently quoted ones. In addition to the references following each chapter, a list of pertinent journals and information sources is included in Chapter 16.

Everett Smethurst and David Eckroth were the editors of Wiley Interscience who were in charge of publishing the book.

New drawings and artwork were done by James Studeny and Shuhai Zheng.

Finally, the authors would like to thank the reviewers of the book for their valuable comments and contributions.

Vladimir Novotny

Marquette University
Milwaukee, Wisconsin
Fall 1988

Contents

PART II

SEWAGE AND WASTEWATER TREATMENT

PART III

PROTECTION OF RECEIVING WATERS

KARL IMHOFF'S

HANDBOOK OF URBAN DRAINAGE AND WASTEWATER DISPOSAL

PART I

URBAN DRAINAGE

CHAPTER 1

Fundamentals of Urban Drainage

1.1 INTRODUCTION

A complete well-functioning urban drainage (sewerage) and treatment system is the most effective solution to the sewage and urban runoff problem. Such systems allow utilization of the most modern methods for household and commercial wastewater disposal, including flushing toilets and in-sink garbage grinders. The need for cesspools, septic tanks, and other on-site sewage disposal systems is therefore eliminated. A complete urban storm drainage and waste collection system can be optimally and economically developed when protection of receiving waters and their ability to assimilate wastewaters are included in the design. In some countries, less efficient and cheaper treatment can be included temporarily until, finally, more expensive treatment plants are affordable.

Urban Drainage Systems

Urban drainage systems consist of a network of sewers and drainage channels (minor system) and flood runoff paths (major system). The minor system includes streets, sewers, and open channels, either natural or constructed. The minor system is almost always planned and designed to safely convey runoff from a storm with a specified recurrence interval, usually 5–10 years. A larger recurrence interval has been considered an "Act of God" by the courts.

The major system is utilized when the capacity of the minor system is exceeded. It is composed of whatever paths have been provided for the runoff to flow to a receiving stream.

The flood drainage systems are included in the flood insurance studies required in the United States by the Office of Insurance and Hazard Mitigation of the Federal Emergency Management Agency.

Preliminary Investigations and Plans.[1–3] A planning report must be prepared that would delineate (1) the drainage area characteristics (topography, drainage patterns, surface conditions, location of streets and right-of-way, existing drainage systems to which the proposed system may be connected, flow rates, and quality of receiving waters), (2) future development (population trends, land use, present and projected quantity and quality of sewage and stormwater runoff, location and profiles of main trunks and interceptors, location of treatment plants, stormwater and combined sewer overflows, and effluent outfalls), (3) political information (political boundaries, service agreements, water quality standards, wastewater control ordinances, subdivision ordinances and building codes), (4) financial (availability of federal and state assistance, taxes, proposed users fees, and all other fundamental data necessary to establish a feasible financing program).

Maps and supporting data bases for preliminary design of urban drainage systems can be obtained in the United States from the National Oceanic and Atmospheric Administration (NOAA) (local and regional climatic data), the U.S. Geological Survey (topographic maps and information on geology and hydrology as well as water quality of major streams), National Aeronautic and Space Administration (NASA) (satellite photographs), the U.S. Soil Conservation Service (soil information), the U.S. Environmental Protection Agency (water and air quality data, storm runoff and receiving water body models, water quality data), various state agencies, city engineers, and local planning agencies. In many cases, data are stored in computerized data base banks from which information can be retrieved by a computer terminal.

Field Surveys. The use of mathematical models for planning and design requires special surveys that are then used in the design and calibration of the models. Local rainfall and runoff data should be evaluated to estimate the design storms and their frequency as well as the magnitude and probability of minimum low-flow conditions in the receiving waters.

Subsurface explorations should establish the type of soils and geologic formations in the areas affected by construction. The borings should extend to a depth of 1.5 m (5 ft) below the bottom of sewer lines or foundation of the major construction works.

In the United States, some parts of the urban drainage system development may qualify for federal grants. In such cases, the drainage plan must be included in a regional pollution abatement plan prepared by a state designated planning agency.[4,5]

Detailed Contract Drawings, Design Plans, and Specifications. These include plans, profiles, and detailed drawings of all construction works. The overview map is usually prepared on a scale of 1:5 000 to 1:25 000 (50 m to 1 cm to 250 m to 1 cm, roughly equivalent to 500 ft to 1 in. to 2500 ft per 1 in.). Such maps should show contour lines. City maps are usually on a scale of 1:5 000 to 1:1 000 (50 m to 1 cm to 10 m to 1 cm, roughly equivalent to 500 ft to 1 in. to 100 ft to 1 in.). Profiles have the horizontal scale similar to that of the maps; the vertical scale is exaggerated to 1:100 (10 m to 1 cm, roughly equivalent to 100 ft to 1 in.) or less. Depth of basements, groundwater levels and conditions, and soil characteristics should be reported. Detailed contract plans showing house connections should be on a scale of 1:1 000 to 1:500. The drawings should also include profiles of all conduits, plans and details for manholes, treatment plans and other appurtenances, complete results of soil and subsurface condition tests, complete hydraulic computations of sewer networks, and detailed descriptions.

The contract drawings should be prepared only for existing streets. Future expansions should be elaborated only approximately. As a rule, sewage in sewers must be maintained aerobic (fresh) and deposition of solids in sewers ought to be avoided.

Street Conveyance. The drainage patterns and storm inlets must be designed to minimize interference of flooding with traffic. The American Association of State, Highway and Transportation Officials (AASHTO) criteria state that for curbed streets, stormwater inlets should be spaced so that not more than half of a through traffic lane would be flooded during the 10-year recurrence interval storm.[6]

1.2 SEWER SYSTEMS

Separated Sewers consist of *storm sewer networks* that carry mainly surface runoff from streets, roofs, parking lots, and other surfaces toward the nearest receiving water body or a man-made channel, and *sanitary sewers* that carry household and commercial sewage and industrial wastewater toward a treatment plant. Urban drainage by separate systems is more expensive than a combined sewers system since it uses two parallel networks of conduits.

Sanitary sewers carry a mixture of sewage, industrial wastewater, and clean water from groundwater infiltration, cross-connections between stormwater drainage, and basement and foundation drainage connections. The clean water inputs into sanitary sewers may be of the same magnitude or even greater than the sewage and industrial inputs. Clean water infiltration into sanitary sewers should be avoided wherever possible. Excessive clean water inputs into sanitary sewers will result in treatment plant hydraulic overloads and thus bypasses and overflows of untreated sewage into receiving waters.

The most convenient unit of expressing infiltration rates is cubic meters per day per centimeter of sewer diameter per kilometer of sewer. Emerson[7] has found that infiltration rates in wet ground ranged between 3 and 560 $m^3/day^{-1}/km^{-1}$ of pipe.

Sanitary and storm sewers are designed commonly as gravity flow conduits. Pressure or vacuum sewers are still rare.

Combined Sewers are used for collection and conveyance of both surface runoff and sanitary sewage and industrial wastes in one conduit. Combined sewer systems are common in older U.S. and European urban centers. There has been a trend in the United States to replace combined sewers by separate sewer systems, however, recently this trend appears to have leveled off.

The average biochemical oxygen demand (BOD_5) in the overflows from the stormwater systems is about 30 mg/liter. The average BOD_5 in combined sewer overflows is between 60 and 120 mg/liter.[8] Because a large part of the flow in the combined sewer systems is treated in the treatment plant, the overall load of organic pollution to receiving waters is about the same from both systems.

Open Channel Drainage includes swales, roadside ditches, and drainage channels. The channels should be lined either with man-made lining (asphalt, concrete, gabions, mats) or natural grasses and sod. Grassed, maintained waterways are an excellent and cheap alternative to underground storm sewers, especially in suburban zones. They enhance infiltration and attenuation of pollutants.[9]

Sewer Types and Selection Parameters

Sewers can be divided into building sewers, lateral sewers, branch sewers, main (trunk) sewers, interceptors, and outfall (relief) sewers.

Building Connections connect the building plumbing to the nearest lateral or branch public sewer line. The building connections should be equipped by backwater gates if there is a danger of sewer backup and basement flooding. This device consists of a gate or float or a hydraulically regulated check valve that opens only in the direction toward the sewer. Building sewer lines from upper elevations should bypass the basement. The minimum recommended size of building connections is 100 mm (4 in.) for single family homes and the minimum slope should be 2%.

Lateral and Branch Sewers are used to collect wastewater from building connections and convey it to a main sewer. A lateral has no other sewer tributary except building connections. The minimum diameter of lateral and branch sewers should be 200 mm (8 in.).

Main (Trunk) Sewers accept flow from several branch or lateral sewers and convey it to a treatment plant or to an intercepting sewer. In older combined sewer systems trunk sewers convey the mixture of sewage and stormwater runoff to a flow divider whereby the excess flow is diverted into a relief sewer and then to the nearest watercourse. Regulations in some countries (e.g., the United States and Germany) now require interception and treatment of combined sewer overflows.

Interceptors are large sewers that were originally designed to accept dry weather flow from a number of trunk sewers and carry the flow to a treatment facility. In more recent systems large interceptors are used for storage of combined sewer overflows (in-line storage). The in-line storage can be supplemented by off-line surface or underground basins that store the excess mixture of sewage and stormwater runoff or the stormwater runoff alone before it enters the sewer system during a storm event. The stored water is then released for subsequent treatment and disposal.

The layout of main sewers and interceptors should follow the natural slope of the terrain. Separate interceptors for upland and low lying areas should be considered, since in low areas or when pumping of sewage is required installing safe stormwater overflows may not be possible.

Longitudinal Profiles (Sewer Slopes). The minimum slope of sewers should result in flow velocities large enough to prevent deposition of solids (self-cleaning velocity) and keep sewage fresh. Commonly, the slopes are calculated assuming a minimum velocity for the dry weather (sewage) flow of 0.6 m/s (2 fps) when sewers are flowing half-full. Considering that the debris and solids in stormwater will normally have a higher specific gravity than sewage-suspended solids, a minimum velocity of 0.9 m/s (3 fps) is recommended for full flow (wet weather) of combined and storm sewers. European practices permit minimum velocities for sand free, dry weather sewage flow of 0.4 m/s (1.3 fps). In the United States, the minimum velocities are often specified by local or state building codes.

The maximal velocity for concrete pipes and common sewage–stormwater mixture or stormwater should be 3 m/s (10 fps). When grit is not a problem maximum velocities can be increased up to 6 m/s (20 fps) or 8 m/s (27 fps). At velocities exceeding 12 m/s (40 fps) abrasion by cavitation becomes a problem.[10]

After selecting the minimum and maximum velocity the slope of the sewer for a given diameter or cross section can be determined from the nomographs in Appendix 1 (Graphs 1 to 5). For circular pipes with diameters up to 1 m see Figure 1.1. For larger diameters the critical slope can be computed using a simple formula proposed by Schütz[11] or

$$S_{crit} = 0.001 \times D^{-1.3}$$

where S_{crit} = dimensionless critical minimal slope (m/m or ft/ft), and
D = the cross-sectional diameter in meters.

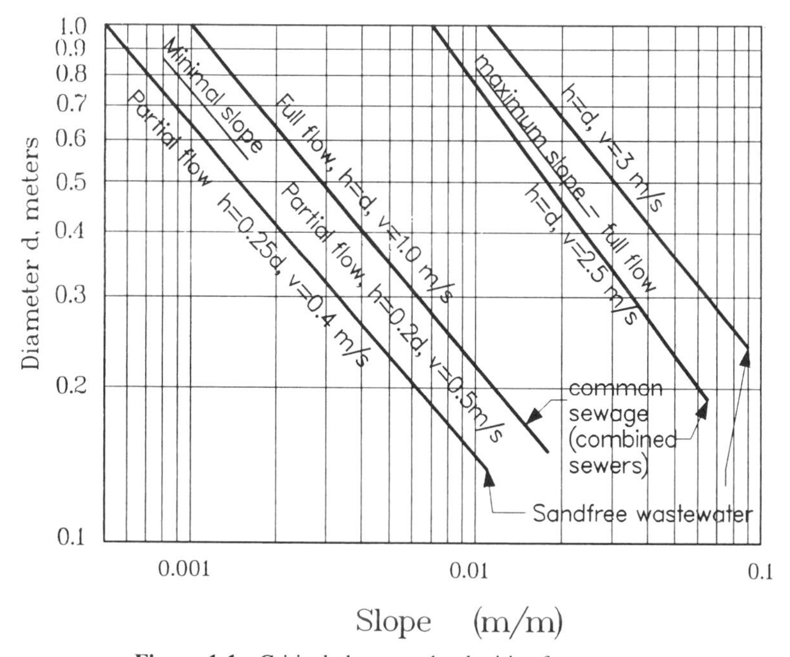

Figure 1.1 Critical slopes and velocities for sewers.

Minimum velocities should be based on a reasonable average flow rate during early stages of operation. If the design flows based on projected future expansion are too high, deposition may occur in the first years of operation.

Sewers should not be designed for flow levels more than one-half full during normal dry weather flow periods.

Example What is the minimum slope for a concrete circular sewer with a diameter of $d = 0.4$ m carrying sewage containing grit (sand). From Figure 1.1 following the curve for common sewage (combined sewers), the slope is 0.004. This slope will result in a flow velocity of $v = 1$ m/s for full flow and $v = 0.5$ m/s for partial flow ($h = 0.2\ d$). This can be confirmed by Nomographs 1 or 5 in the Appendix. For sand-free sewage the corresponding slope is 0.002, which will yield a velocity at a partially full flow ($h = 0.25\ d$) of approximately $v = 0.4$ m/s.

Hydraulic Characteristics. Typical sanitary sewers are most often designed as gravity sewers flowing partially full. Pressure and vacuum sanitary sewers are sometimes built in areas unsuitable for gravity sewers. These systems are usually small and should not carry groundwater infiltration and stormwater runoff. Storm and combined sewers almost always have gravity flow.

Sewer Materials.[2,10,12–15] Common sewer materials include concrete (plain or reinforced), vitrified clay, asbestos-cement, iron and steel (cast iron, ductile iron, corrugated steel), and polyvinyl chloride (PVC). Small and medium sized sewers are assembled mostly from concrete or vitrified clay pipes. In general, both materials have similar economy and, when installed properly, infiltration or exfiltration should be minimal. For larger flows, cast-in-place concrete sewers are used when standard concrete pipes are not available.

Practices differ for storm sewers and combined or sanitary sewers in which flow is more damaging to the sewer materials. In sanitary or combined sewers concrete pipe can be subjected to corrosion by acidic groundwater, wastewater, and hydrogen sulfide, when present. A special lining or coating mostly made of asphalt, coal tar, or epoxies can be applied to sewers to reduce the damage by corrosion. Such sewers should also be made from a more corrosion-resistant cement (Type II-Portland instead of Type I). Concrete pipe should be impermeable, which is accomplished by the proper composition of concrete and by good workmanship and quality control. The ability of clay pipes to better resist corrosion gives them a distinct advantage over other common sewer materials when acids and/or hydrogen sulfide are present. These pipes also have better resistance to scour.

Corrosion by acidic groundwater can be remedied by lowering the groundwater table (which may be difficult and costly), by enrichment of concrete, or by lining the pipe with a corrosion-resistant mortar or material. The septic condition of sewage that causes the formation of hydrogen sulfide may be the result of low velocities at insufficient slopes of the sewers, low flows, or high temperatures. These conditions can be alleviated by ventilation and/or by flushing. The reader is referred to an article by Pomeroy[16] for additional reading.

For pipe connections most manufacturers employ belt-and-spigot or tongue-

and-groove joints with a rubber (or a rubber-like elastomer) sealant. Tight joints are necessary if infiltration is to be kept low. For the magnitudes of infiltration see the paper by Santry.[17]

Asbestos-cement pipes used to provide an alternative to concrete.[2,13] They are light weight and easy to handle and with rubber-ring joints, infiltration is minimal. The pipes might be subjected to corrosion when acids or hydrogen sulfide are present. It should be noted that asbestos dust when inhaled may cause cancer. Therefore precautions and proper and safe installation procedures must be employed and exposed asbestos pipes or the use of asbestos in indoor plumbing must be avoided.

Pipes of PVC and cast iron are primarily used for indoor plumbing and building connections.

Fabricated steel pipes[15] made of galvanized corrugated steel and iron are manufactured in various shapes and sizes, large size segments being field bolted from plates. The exterior and interior of the pipes may be lined with a protective coating to reduce corrosion attack. The hydraulic roughness of these pipes is about twice that of concrete pipes. The manufacturers also provide various appurtenances such as pipe connections and manholes. Corrugated steel pipes are used primarily for storm runoff drainage and for culverts.

Cross-sections. Circular pipes made of concrete, asbestos-cement, or clay are used for cross sections between 100 mm (4 in.) and 60 cm (25 in.). Above 60 cm, the sewers are made from reinforced or prestressed concrete. Larger intercepting sewers also may have noncircular cross sections, such as eliptical (egg), horseshoe, basket-handle, or pentagonal. Various shapes of sewers and their hydraulic characteristics are shown in Figures 2.7 to 2.12 (see Section 2.2). Corrugated steel pipes can be assembled to very large diameters. Deep tunnels for combined sewer overflows have been bored in rock formations and may require grouting or lining.

Sewer Maintenance.[3,18,19] Flushing is required, particularly in sanitary and combined sewers, where solids may accumulate during dry weather periods. The accumulated solids are then carried by subsequent rainwater inflows through the combined sewer overflows or sewer bypasses into receiving waters. The source of water for flushing can be external (fire hydrant or tank truck) or sewage impounded in sewers and manholes by gates.[20]

Sewers are also cleaned by mechanical devices (bag, power shovel, mole, kite) that are dragged by a pulley through the sewer section between two manholes. Inspection of sewers can be done by a visual survey or by a TV camera.[21]

Ventilation of sanitary and combined sewers is accomplished mainly through building connections to house plumbing vent stacks; therefore, no hermetic closures should be installed on main house sewers. Odor traps should be placed on each individual plumbing connection. Perforated manhole covers should be used only on combined sewers since influx of stormwater into sanitary sewers must be avoided. Ventilation of storm and combined sewers is accomplished through the stormwater inlets. Ventilation is necessary, particularly when sewer slopes change. Forced ventilation may be needed for long enclosed sections.

Sewer Appurtenances

Sewer appurtenances are components of the sewerage system other than pipes.

Manholes should be installed at every change of slope, direction of sewer lines, change of sewer size, and sewer connection. On straight smaller sized sewer sections the maximum spacing of manholes should be 50–120 m (150–400 ft). Intervals greater than 120 m are acceptable only for larger sewers. The bottoms of connecting sewers of the same size in a manhole should be on the same level to avoid formation of deposits.

Drop manholes are installed when the elevation difference between the inlet and outlet sewers exceeds 0.5 m. Changes in direction or in size of the sewers will result in a headloss in the manhole.[6,22,23]

Street Inlets are installed as curb-opening inlets, gutter grate inlets, combination inlets, or slotted drain inlets. Odor seals should be avoided since the inlets provide an effective ventilation of sewers. Grate inlets may become clogged by street debris and require regular maintenance and cleaning. Curb openings are not suitable for steep streets as it is difficult to direct the surface runoff into the inlet.[6,24,25] Catch-basins for grit removal should be used when runoff is expected to carry large amounts of grit and subsequent sewer slopes do not guarantee a self-cleaning velocity.[26]

Depressed Sewers (inverted siphons) (Fig. 1.2) carry flow under depressed highways, streams, and other obstructions. For combined sewer systems both European and U.S. practices recommend two- or multiple-barrel siphons with a minimum velocity of 1 m/s (3.3 fps) to be achieved during dry weather conditions at least once a day. Additional barrels are connected with the primary barrel by a regulating weir or a small dam. German practice recommends installing a gate for flushing. Vertical risers should be avoided; the slope of the rising pipe should be less than 1 to 2 (vertical to horizontal) to permit flushing of grit deposits from the siphons. For small flows one barrel siphons may be adequate, but regular flushing must be provided.

Above Ground Sewers require prevention against freezing and leakage.

Pressure Sewers[27] from pumping (lift) stations have design requirements similar to depressed sewers. Nomographs 1 to 3 and 5 in the Appendix can be used for the design calculations with the headloss per length of the pipe substituted as a slope. Equivalent lengths of the fittings (valves, bends, etc.) should be added to the length of the pipe. The design velocity should be about 1 m/s. At long sections it is very difficult to keep sewage fresh.

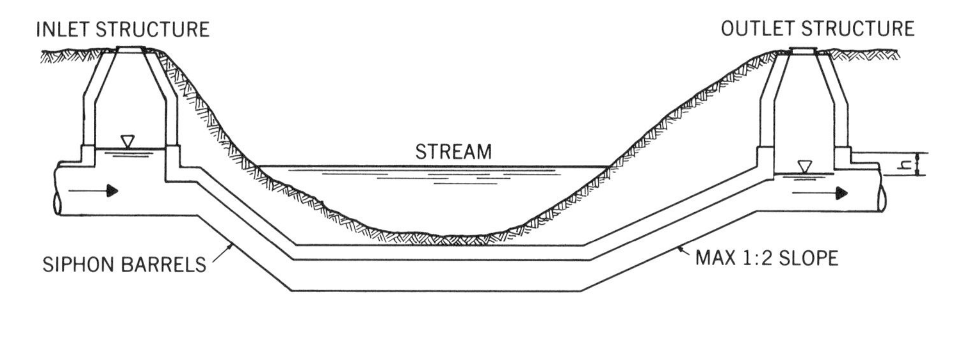

Figure 1.2 Inverted siphon.

Pumping (Lift) Stations.[28–32] The efficiency of pumping stations depends on the drainage requirements. The stations should have at least two pumps for safety reasons. Coarse screens or bar racks with 5–9 cm (2–3.6 in.) openings are not always necessary since grit should pass through the pumps. Centrifugal pumps are most common but they require a wet well (sump) to collect water.

The necessary volume of the wet well, V (in m^3), depends on the inflow, Q (in liters/s), pump flow capacity, Q_p (liters/s), and switch-on frequency, Z (hr^{-1}). The maximum practical wet well volume for one pump when inflow equals one-half of the pump capacity is

$$V(m^3) = \frac{0.9Q_p}{Z}$$

For a motor power rating of less than 7.5 kW the switch-on frequency $Z \leq 15$, between 7.5 and 30 kW $Z \leq 12$, and over 30 kW $Z \leq 10$. For submersible pumps the corresponding Z should be doubled.

For a pump with a capacity of 100 liters/s and 12 pumping cycles per hour the practical wet well volume is $0.9 \times 100/12 = 7.5\ m^3$.

For several pumps in a pumping station, the wet well volume is computed only for the largest capacity pump; however, the switch-on levels for the additional pumps are set at proportional heights (the smallest pump has the lowest switch-on elevation, after a 10 cm rise a larger pump is switched on, etc.). Inlets to the dry weather flow (sewage) pumps should be located at the lowest point of the wet well with the bottom sloped toward the inlet to prevent sludge accumulation.

For pumping dry weather flows, radial horizontal centrifugal pumps are commonly selected. The pumps are located in an adjoining dry well with the top of the pump casting below the low water level.[30] Axial pumps located in a vertical shaft are generally used for pumping large volumes of stormwater or wet weather flow and should not be used for pumping sewage. Submersible pumps are economical for smaller flows, whereas screw pumps are efficient for large flows and small hydraulic heads—up to 5 m (16 ft). The screw pumps do not require wet wells and can be installed outdoors. They are located in an open trough made either of concrete or steel casting. Screening or grit removal is not necessary.

Pneumatic ejectors and air-lift pumps can be installed when the flows are small (less than 80 liters/s). Pneumatic ejectors work on a principle of compressed air ejecting water from a closed tank. The flow of water and air is controlled by check valves, however, their efficiency is relatively low. In air-lift pumps, air is injected at the bottom of an up-draft tube and the water–air mixture is lifted by atmospheric pressure. The air requirement is approximately twice the wastewater flow, and the air inlet should be located deep below the water level.

Pumping stations for treatment plants should maintain uniform delivery to the plant. The flow is divided between several pumps that automatically turn on and off at different wastewater levels. For larger volumes, screw pumps are advantageous because they can adapt to the delivery flow rate.

All electric installations must be located in a dry space. Entry to the pump shaft should be outside the pumping station.

The energy requirement for the pump (for pumping water) is

$$K = \frac{9.81HQ}{\eta_A} \qquad \text{(kW)}$$

where H is the total dynamic head for the pump in meters, Q is the flow in m^3/s, and η_A is a joint efficiency of the pump and motor under normal operating conditions. Typical efficiencies for centrifugal pumps range from $\eta_A = 0.3–0.7$, for screw pumps $\eta_A = 0.6–0.7$.

Diversion (Bypasses) of Stormwater.[33] In combined sewer systems the overflows of excess rainwater are directed to the nearest watercourse. Past U.S. and European practices based the design of overflow facilities on the permissible dilution ratio of rainwater with the dry weather sewage flow (for example 1 part of sewage per 7 parts of rainwater). The treatment plant bypasses were designed for 1 to 4 to 1 to 6 dilution of sewage by rainwater. The dilution volume was related to the peak hourly flow rate of sewage that was estimated, for example, as 1/10 to 1/14 of the average daily sewage volume.

According to recent European practices, it is more appropriate to base the permissible overflow estimates on critical rainfall intensity in liters/(s × ha) (1 liter/(s-ha) = 0.014 in./hr), which reflects the rate at which the overflow begins.[34] The "critical storm intensity" in Switzerland is 15–30 liters/(s-ha) (0.2–0.4 in./hr); in Germany it is 7 to 15 liters/(s-ha) (0.1–0.2 in./hr).

The relationship between the critical rainfall intensity and the dilution ratio is as follows:

$$r_{\text{crit}}[\text{liters}/(\text{s-ha})] = (m - 1)\,\frac{Q(\text{liters}/\text{s})}{A(\text{ha})C_{\text{m}}}$$

where Q is the dry weather flow from a contributing area A, and C_{m} is the average runoff coefficient for the area. As a result of water losses, infiltration, and the magnitude of the runoff coefficient, the ratio Q/AC_{m} fluctuates between 1 and 3 liters/(s-ha). A rainfall with this intensity would result in flow that would equal the dry weather sewage flow. For example, assume Q/AC_{m} as 2 liters/(s * ha) and one to five dilution, then the overflow begins at $r_{\text{crit}} = (5 - 1) \times 2 = 8$ liters/(s-ha) (0.11 in./hr). For further analysis, a modified rainfall frequency–duration curve can be used (Fig. 1.3). Its delineation can be obtained for various geographic locations from a known rainfall frequency–duration curve by setting the total annual rainfall time as 100% on the horizontal axis. For example Zürich (Switzerland) has on the average 1 048 hr of rainfall per year and Essen (FRG) has 750 hr. For $r_{\text{crit}} = 8$ liters/(s-ha) in Essen there would be approximately $0.08 \times 750 = 60$ hr of overflows per year. Curve b on Figure 1.3 presents a corresponding cumulative rainfall depth. At $r_{\text{crit}} = 8$ liters/(s-ha), 78% of the total annual rainfall would remain in the sewer system and 22% would overflow in a receiving water body. Doubling the critical rainfall rate would result in reducing the overflow rainfall volume to 11%. In this way, the size and economy of sewer systems can be related to a corresponding permissible overflow volume.

In the United States, the number of overflows per year (overflow frequency) is

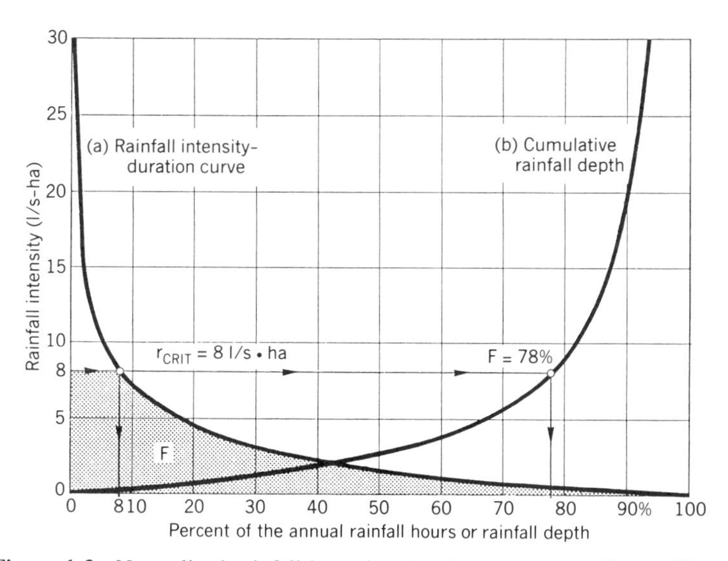

Figure 1.3 Normalized rainfall intensity–duration curve according to Hörler.

often regulated by state pollution control agencies. The permissible frequency of overflows may vary from 0.5 to 2 per year to more than 10 per year. The permissible frequency of overflows should be established by a waste assimilative study of the receiving water body and water quality modeling (see Section 14.5).

The overflow frequency or volume can also be estimated from the overall pollution removal requirement of the entire drainage system. For example in Germany, as a goal, the combined sewer abatement projects ought to consider that 90% of total waste load should receive biologic (secondary) treatment, which would imply that approximately 65% of the mixed flow in the sewers during a storm ought to be biologically treated.[35] This can be accomplished provided that the drainage area is not too large, the sewage volumes or storage volume requirements are not excessive, and the overflows are not initiated at low flow rates Q_{crit}. Similar regulations are now being considered in the United States and other countries.

Simple Computation of Combined Sewer Overflows

In order to compute Q_{crit}, the critical rainfall rate should first be determined from Figure 1.4. In the figure, r_{crit} is related to the low flow conditions of the receiving water body (MNQ) and to the treatment capacity available for the drainage area (Q_s). If the corresponding connected impervious area (paved streets, roofs, driveways) of the drainage basin is A_{red}(ha) then the critical overflow rate is

$$Q_{rcrit} = r_{crit} Z A_{red} \qquad \text{(liters/s)}$$

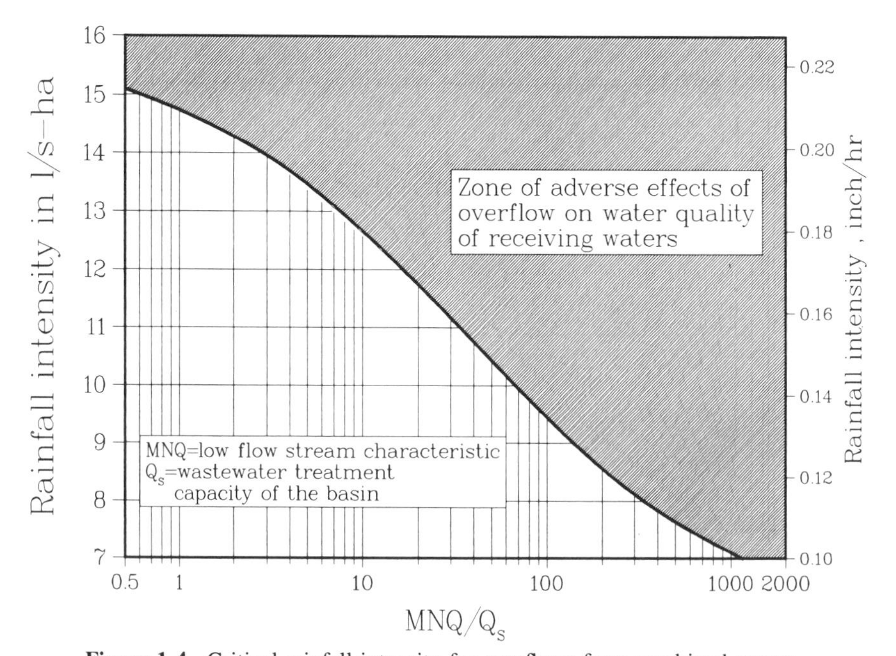

Figure 1.4 Critical rainfall intensity for overflows from combined sewers.

where $Z = 1 - (t_f/200)$ is a reduction coefficient for runoff that accounts for the fact that the peak flow, peak flow volume, and pollution loading decrease as the overland flow time (time of concentration), t_f, increases. The minimum value for Z is 0.5. t_f is in minutes.

Then

$$Q_{crit} = Q_t + Q_{rcrit} + Q_r \qquad \text{(liters/sec)}$$

where Q_t is the dry weather flow (= sewage flow Q_s + clean water inflow, Q_f). Q_r is the residual flow in the sewer from the upstream section if overflows in succession are considered.

A Combined Sewer Flow Regulator (flow divider) generally consists of an overflow weir and an orifice outlet. The diameter of the orifice, D, should be greater or equal to 0.2 m (0.6 ft). Flow dividers can be designed with side weirs, baffled side weirs, leaping weirs, or relief siphons.[3,33] A typical flow regulator-divider is shown on Figure 1.5.

Flow regulators with an elevated side weir have the crest of the weir located higher than the normal water level of the inflow at Q_{crit}. If the incoming velocity in the inlet sewer at Q_t is greater than or equal to 0.5 m/s (1.6 fps), then the crest of the weir should be set as high as backwater effects will permit or at least 0.6 d_0. The crest should be above the high water level of the receiving water body. If rapid (supercritical) flow conditions are expected in the inlet, the energy of the flow should be dissipated in a stilling section with a corresponding drop of the bottom, or the crest of the weir should be elevated to the top of the inlet sewer. Side (lateral) flow into flow regulators should be avoided.

The orifice outlet must safely convey the flow Q_t into the outlet sewer without backwater effects. The bottom elevation drop, Δs, along the weir is estimated from a condition that the energy head on the orifice (h_{Eu}) during the dry weather flow should be below that in the inflow (h_{E0}). Δs is commonly set between 3 and 5 cm (1.2–2 in.). The weir crest must be located at least 0.05 m (2 in.) above the top of the orifice outlet.

In a hydraulic computation, the elevation of the weir crest (heights s_0 and s_u

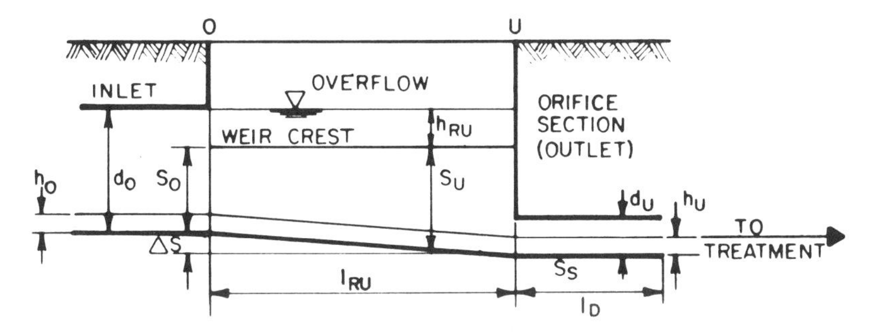

Figure 1.5 Typical flow divider in combined sewer systems.

above the bottom, respectively) and the bottom slope, S_s (not more than 0.3%), will be determined first. The diameter of the orifice, d_u, can be selected smaller than would be necessary for free flow conditions at Q_{crit}. At Q_t the flow velocity shall be >0.5 m/s. The necessary length of the orifice section computed for a sharp edged orifice inlet becomes

$$l_D = \frac{s_u - md_u - 0.07v_u^2}{S_e - S_s}$$

where v_u is the flow velocity and S_e is the slope of the energy line at flow Q_{crit}. m approximately equals one but becomes smaller if the Froude number $F = v_u/\sqrt{gd_u} > 1.0$. When $F = 2.5$, m decreases to 0.5. g is gravity acceleration (9.81 m/s²).

In some cases the orifice nozzle is not completely full, which causes the overflow to begin sooner. It should be checked whether such conditions may occur.

The length of the weir, l_{RU}, for fully overflowing water can be in a first approximation selected as

$$l_{RU} = \frac{4Q_{max}}{1000d_0}$$

The depth of water above the weir is then

$$h_{RU} = \left(\frac{3Q_{RU}}{2C \times 1000 l_{RU}\mu\sqrt{2g}}\right)^{2/3} \qquad \text{(meters)}$$

in which $Q_{RU} = Q_{max} - Q_{crit}$. For a perfect overflow the discharge coefficient $C = 1.0$ and for partial overflow C is smaller. μ can be roughly estimated as 0.6 for one-sided overflow and 0.5 for two-sided overflow. $\sqrt{2g} = 4.43$.

Example: An overflow has a contributing drainage area (approximately equaling the impervious surface area that is directly connected to the sewers) of $A_{red} = 16.5$ ha. The dry weather wastewater flow, $Q_t = 20$ liters/s, and the maximum flow in the sewer, $Q_{max} = 1650$ liters/s, and time of concentration (= inlet time + flow time in the sewer to the flow divider), $t_f = 13$ min. The inflow sewer has $d_0 = 1.20$ m, slope $S_s = 0.2\%$ (2 m/km), hydraulic roughness $k_B = 1.5$, and capacity flow $Q_v = 1\,700$ liters/s and $v_v = 1.5$ m/s (for known d_0 and S_s, Q_v and v_v can be obtained from Nomograph 2 in the Appendix). The overflow starts at $r_{crit} = 15$ liters/(s-ha).

From Figure 2.7, which shows the flow, depth, and velocity relationships in partially flowing circular sewers, find for $Q_{max}/Q_v = 1\,650/1\,700 = 0.97$, the depth of partial flow $h_0 = 1.06$ m and velocity $v_0 = 1.56$ m/s. Similarly for $Q_t = 20$ liters/s find $h_0 = 0.09$ m and $v_0 = 0.54$ m/s. The energy head is obtained from the formula $h_{E0} = h_0 + v_0^2/(2g) = 0.09 + 0.54^2/(2 \times 9.81) = 0.1$ m.

The sill (weir crest height) $s_0 \geq 0.6 \times 1.2 = 0.72$ m. Select 0.75 m (this is possible since, at Q_t, the velocity in the inflow sewer $v_0 \geq 0.5$ m/s). Also

$$Q_{rcrit} = 15 \times \left(1 - \frac{13}{200}\right) \times 16.5 = 231 \text{ liters/s}$$

$$Q_{crit} = 231 + 20 = 251 \text{ liters/s}$$

Orifice design: Select $d_u = 0.4$ m, $S_s = 0.2\%$. With the selected hydraulic roughness $k_B = 1.5$ mm the $Q_v = 114$ liters/s and $v_u = 0.91$ m/s. At Q_t the depth is $h_u = 0.11$ m and velocity $v_u = 0.69$ m/s (use Fig. 2.7 and the nomographs in the Appendix for the estimation), hence, the energy head is $h_{Eu} = 0.11 + 0.69^2/2g = 0.13$ m.

Then the following values of parameters can be estimated:

$$\Delta s = h_{Eu} - h_{E0} = 0.13 - 0.10 = 0.03 \text{ m} \qquad \text{select } 0.05 \text{ m}$$

$$s_u = s_0 + \Delta s = 0.75 + 0.05 = 0.80 \text{ m}$$

At Q_{crit} the velocity $v_u = 2$ m/s and $S_{Eu} = 0.9\%$. With

$$F = \frac{v_u}{\sqrt{gd_u}} = \frac{2.0}{\sqrt{9.81 \times 0.4}} = 1.0$$

the coefficient m can be considered as being 1.0.

For a sharp-edged orifice, the length of the orifice section is

$$l_D = \frac{0.8 - 1.0 \times 0.4 - 0.07 \times 2^2}{0.009 - 0.002} = 17.15 \text{ m}$$

The overflow weir characteristics:

$$l_{Ru} = \frac{4 \times 1650}{1000 \times 1.2} = 5.5 \text{ m}$$

$$Q_{Ru} = Q_{max} - Q_{crit} = 1650 - 251 = 1399 \text{ liters/s}$$

$$h_{Ru} = \left(\frac{3 \times 1399}{2 \times 1000 \times 5.5 \times 0.6\sqrt{2g}}\right)^{2/3} = 0.27 \text{ m}$$

In reality, the solution is a trial-and-error type since the selected crest height and the dimensions of the orifice will not always lead to the desired solution. The dimensioning process must then be repeated with different initially selected values. Additional examples of possible solutions can be found in Metcalf and Eddy[3] and in a publication by Benefield et al.[36]

When either the allowable capacity of the downstream sewer network or the capacity of the treatment plant is not sufficient, or if the waste assimilative capacity of the receiving water body is limited, the stormwater overflow can be connected to a detention–retention basin that acts as storage and, when necessary, as a treatment unit. For various schemes of combined sewer and stormwater water overflow control schemes see Chapter 3 and publications by APWA[6] and by Lager et al.[37]

REFERENCES

1. *Recommended Standards for Sewage Works.* Health Education Service, Albany, NY, 1978.
2. *Design and Construction of Sanitary and Storm Sewers,* ASCE-WPCF Manual No. 9. Am. Soc. Civ. Eng., New York, 1982.
3. Metcalf & Eddy, Inc. *Wastewater Engineering: Collection and Pumping of Wastewater.* McGraw-Hill, New York, 1981.
4. *The Water Pollution Control Act Amendments* (*The Clean Water Act*), 1972 (PL 92-500), 1977 (PL 95-217), 1987 (PL 100-4). U.S. Congress, Washington, DC.
5. *The National Environmental Policy Act of 1969,* PL 91-196. U.S. Congress, Washington, DC.
6. *Urban Stormwater Management,* Rep. No. 49. Am. Public Works Assoc., Chicago, IL, 1981.
7. C. A. Emerson, Jr., *Sewage Works J.* **5,** 988 (1933).
8. V. Novotny and G. Chesters. *Handbook of Nonpoint Pollution: Sources and Management.* Van Nostrand-Reinhold, New York, 1981.
9. P. A. Oakland, *An Evaluation of Urban Stormwater Pollution Removal Through Grassed Swale Treatment,* Proc. Int. Symp. Urban Hydrol., Hydraul., and Sediment Control. University of Kentucky, Lexington, 1983.
10. *Concrete Pipe Handbook.* Am. Concrete Pipe Assoc., Vienna, VA, 1980.
11. M. Schütz, *Korresp. Abwasser* **32,** 415–419 (1985).
12. *Clay Pipe Engineering Manual.* Nat. Clay Pipe Inst., Crystal Lake, IL, 1968.
13. K. Hünerberg and H. Tessendorff, *Handbuch für Asbestzementrohre* (*Handbook of Asbestos-cement Pipes*). Springer-Verlag, Berlin and New York, 1977.
14. *Modern Sewer Design.* Amer. Iron & Steel Inst., Washington, DC, 1980.
15. *Handbook of Steel Drainage & Highway Construction Products.* Amer. Iron & Steel Inst., Washington, DC, 1971.
16. R. Pomeroy, *Water Sewage Works* **107,** 400–403 (1960).
17. I. W. Santry, Jr., *J. Water Pollut. Control Fed.* **36,** 1256 (1964).
18. *Handbook for Sewer Evaluation and Rehabilitation.* Rep. No. 430/9-75-021. U.S. Environ. Prot. Agency, Washington, DC, 1975.
19. *Sewer System Evaluation, Rehabilitation and New Construction: A Manual of Practice.* Rep. No. 600/2-77-017d. U.S. Environ. Prot. Agency, Washington, DC, 1977.
20. J. B. Gifford. *Public Works* **105** (2), 93, (1974).
21. "Making Optimum Use of TV Sewer Inspection and Grouting," *Public Works* **107** (10), 86 (1976).
22. D. A. Howard and A. J. Saul. *Energy Loss Coefficient at Manholes,* Proc. 3rd Int. Conf. Urban Storm Drainage. IAHR-IAWPRC, Göteborg, Sweden, 1984.

23. G. Lindwall, *Headlosses at Surcharged Manholes with a Main Pipe and 90° Lateral,* Proc. 3rd Int. Conf. Urban Storm Drainage. IAHR-IAWPRC, Göteborg, Sweden, 1984.
24. U.S. Federal Highway Administration, *Urban Highway Storm Drainage Model,* Vol. 3, FHWA/RD-83-043. U.S. FHA, Washington, DC, 1983.
25. D. C. Woo, *Inlet in Stormwater Modeling,* Proc. 3rd Int. Conf. Urban Storm Drainage. IAHR-IAWPRC, Göteborg, Sweden, 1984.
26. J. A. Lager, W. G. Smith, and G. Tchobanoglous, *Catch-basins Technology Overview and Assessment,* Rep. No. 600/2-77-051. U.S. Environ. Prot. Agency, Cincinnati, OH, 1977.
27. D. Trasher, *Design and Use of Pressure Sewer Systems.* Lewis Publ., Chelsea, MI, 1987.
28. A. Hörler, *Lehr-und Handbuch der Abwassertechnik* (*Text-and Handbook of Wastewater Technology*). Wilhelm Ernst & Sohn, Berlin, 1973.
29. R. Walker, *Pump Selection.* Ann Arbor Sci. Publ., Ann Arbor, MI, 1980.
30. Task Force on Pumping Stations, *Design of Wastewater and Stormwater Pumping Stations.* Water Pollut. Control Fed., Washington, DC, 1981.
31. R. E. Barlett, *Pumping Stations for Water and Sewage.* Appl. Sci. Publ., London, 1974.
32. Submersible Water Pump Association, *Submersible Sewage Pumping Systems Handbook.* Lewis Publ., Chelsea, MI, 1986.
33. Am. Public Works Assoc., *Combined Sewer Regulator Overflow Facilities,* FWQA Pub. No. 11022/DMV07/70, Federal Water Qual. Administration, Washington, DC, 1970.
34. W. Munz, *GWF, das Gas-und Wasserfach: Wasser/Abwasser* **113,** 525 (1972).
35. *Richtlinien für die Bemessung und Gestaltung von Regenentlastungen in Mischwasserkanalen* (*Guidelines for Dimensioning and Construction of Stormwater Overflows in Combined Sewers*), ATV-Arbeitsblatt A128, St. Augustine, FRG, 1977.
36. L. D. Benefield, J. F. Judkins, Jr., and A. D. Parr, *Treatment Plant Hydraulics for Environmental Engineers.* Prentice Hall, Englewood Cliffs, NJ, 1984.
37. J. A. Lager, W. G. Smith, W. G. Lynard, R. M. Finn, and E. J. Finnemore, *Urban Stormwater Management and Technology: Update and User's Guide,* Rep. No. 600/8-77-014. U.S. Environ. Prot. Agency, Cincinnati, OH, 1977.

CHAPTER 2

Calculation of Drainage Systems

2.1 FLOW VOLUMES

Population Forecasting. Urban drainage systems are prepared for conditions 20–25 years in the future, sometimes 50 years. Engineering forecasting of the future population and hydrologic conditions is an integral part of the project. Population forecasts are basically an extension of past trends. The population of European cities has been increasing annually by 2% for smaller urban centers, 1% for larger urban centers, and up to 5% for some industrial cities. In the United States, the population growth of urban centers varies considerably depending on the general location (Sunbelt vs. Northeast and Midwest), type of industry, and other demographic factors. Urban centers of many developing countries have been anticipating very large population increases. This variability and unforeseeable demographic factors makes population forecasting often questionable; in addition, future increases of unit flows should be anticipated. Engineers of many coastal cities should also anticipate other factors such as atmospheric "greenhouse" effects that may change future hydrologic conditions.[1]

Techniques for engineering population forecasting have been covered in the ASCE-WPCF Manual No. 9[2] or in a publication by McJunkin.[3] The best model recommended by demographers is so-called population growth S-curve (Fig. 2.1). This curve has three stages: first, a geometric increase stage in which the annual population increase is expressed as a fraction of the present population, second, an arithmetic population progression during which the annual population increase remains more or less constant, and third, a geometric decrease phase during which the annual population increase is progressively decreasing until it reaches a saturation limit (no growth). These models can be best established by plotting the population on a semilog or arithmetic paper versus census years (plot the logarithm of the difference of the population from the saturation population vs. the census years if the urban area anticipates decreasing rate of increase). The maximum population densities in urban areas are given in Table 2.1.

Sewage Flows. When a conduit of a separate sewer system carries only sanitary sewage without rainfall inputs, the computation of the sewage flows is quite simple. European practice assumes an average per capita sewage flow of 200 liters/cap-day and the peak hourly flow as 1/14 of the total daily sewage volume. A 100%

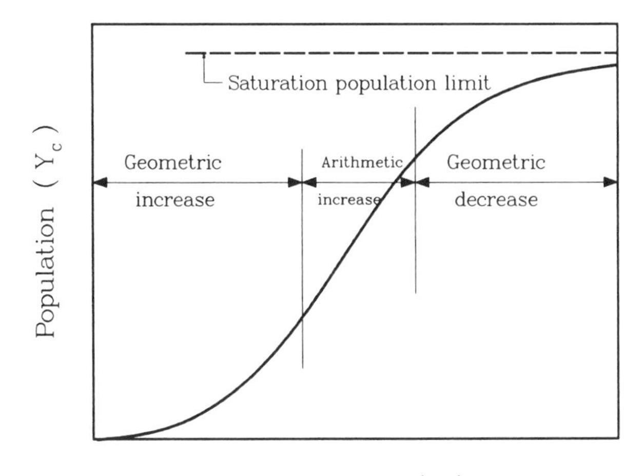

Figure 2.1 Population increase S-curve.

increase of the design flow over the peak flow is recommended to account for infiltration. In the United States the average daily per capita domestic wastewater flows vary considerably from one urban area to another. The ASCE-WPCF Manual No. 9[2] provides a list of average daily sewage flows and the design flows for a number of U.S. cities. A typical U.S. per capita sewage flow is between 250 and 450 liters/cap-day (60–120 gpcd) with the design sewage peak flow rates between 1.5 and 10 times the average daily sewage flow rate. Sewage flow rates from commercial establishments and sanitary flows from industrial and public facilities must be added to the domestic sewage volumes. Smaller sewage flow rates can be expected in developing countries.[4]

For planning purposes, it may be more convenient to express sewage flows per unit area rather than per capita. If, for example, per capita sewage flow is 300 liters/cap-day, the population density is 100 persons/ha, and the ratio of the peak flow rate to the average daily flow is 1/14; then allowing a 100% increase for infiltration, the design unit flow rate is 2 × 300 (liters/cap-day) × 100

TABLE 2.1 Population Density Ranges (U.S. Conditions)

Land Use	Persons per Hectare[a]
Residential	
Single family dwellings	15–75
Multiple family dwellings	75–250
Apartments	250–25 000
Commercial	40–75
Industrial	12–40

[a] To convert from persons/hectare to persons/acre divide by 2.45.

(persons/ha)/(14 × 60 × 60) = 1.2 liter/(s-ha). Using this method, it is easier to add industrial and commercial wastewater contributions.

Stormwater Flows

For the design of combined and storm sewer systems the sewage (dry weather) flows have very little effect. The sewage flow in combined sewers is evaluated only when partial filling and dry weather velocities in the sewers are considered. Modern design procedures use mathematical hydrologic models for determining the flows in large sections or in the entire sewer system. Many models (see the next section for a list of available hydrologic models) can be run on small personal computers. For small projects, the Rational Formula is still used.

The Rational Formula, which is known in Great Britain as the Lloyd–Davies formula, relates the peak flow rate, Q_p, in a sewer or an inlet to the rain intensity as

$$Q_p = CIA \qquad \text{(liters/min or cfs)}$$

where C is a runoff coefficient depending on the characteristics of the drainage area, I is the average rainfall intensity during a specified time interval called the time of concentration (I in mm/min or in./hr), and A is the contributing drainage area (m^2 or acres).

The time of concentration is the time required for the surface runoff to travel from the remotest part of the drainage area to the point of consideration. It consists of the overland flow time (inlet time) and sewer or channel flow time. The time of flow in sewers may be estimated closely from the hydraulic properties of the conduit. The inlet time will vary with surface characteristics, slope, and length of path of the surface flow, as well as the rainfall intensity and duration. In general, the higher the rainfall intensity the shorter is the inlet time. Reported inlet times in sewered watersheds vary from 5 to 30 min. In well-developed areas with closely spaced storm inlets, an inlet time of 5 min is common. In well-developed districts with relatively flat slopes, an inlet time of 10–15 min is common, whereas in flat residential districts with widely spaced street inlets, inlet times of 20–30 min are customary.

The runoff coefficient, C, of the Rational Formula is a ratio of the peak flow to the rainfall intensity of a constant rainfall of prolonged duration. This is not hydrologically correct since runoff is a residual of precipitation after losses (infiltration and surface storage) are subtracted. Thus, the coefficients given in Table 2.2 are only an approximation:

A good discussion on the nature of the Rational Formula and its coefficients was presented by Hall[5] who states that the runoff coefficient should be considered as an "impermeability" factor, which is more logical than to consider the coefficient as a proportionality factor. An obvious simplifying assumption is to use a ratio of paved (impervious) connected surface to the total area as an estimate of the runoff coefficient.

TABLE 2.2 Runoff Coefficients for the Rational Formula[2]

Description of the Area	Runoff Coefficient
Business	
Downtown	0.7 to 0.95
Neighborhood	0.5 to 0.70
Residential	
Single family	0.3 to 0.5
Multiunits—detached	0.4 to 0.6
Multiunits—attached	0.6 to 0.75
Residential—suburban	0.25 to 0.4
Apartment	0.50 to 0.70
Industrial	
Light	0.50 to 0.80
Heavy	0.60 to 0.90
Parks, cemeteries	0.10 to 0.25
Playgrounds	0.20 to 0.35
Unimproved	0.10 to 0.30
Parks and gardens	0.0 to 0.1

Runoff Coefficients Related to Type of Surface

Surface Character	Runoff Coefficient
Pavement	
Asphalt and concrete	0.70 to 0.95
Brick	0.70 to 0.85
Roofs	0.75 to 0.95
Lawns—sandy soil	
Flat, 2% or less	0.05 to 0.10
Average, 2–7%	0.10 to 0.15
Steep, greater than 7%	0.15 to 0.20
Lawns—tight soils	
Flat, 2% or less	0.13 to 0.17
Average, 2–7%	0.18 to 0.22
Steep, greater than 7%	0.25 to 0.35

If the drainage area is not homogeneous, the area must be first divided into homogeneous segments (roofs, streets, lawns) with partial areas of A_1, A_2, A_3, etc. The average runoff coefficient, C, can then be computed using the coefficients for the partial surface areas, C_1, C_2, C_3, etc. as follows:

$$C = \frac{A_1 C_1 + A_2 C_2 + \cdots}{A_1 + A_2 + A_3 \cdots}$$

An accurate estimation of the runoff coefficient is the most important task of the entire computation as can be seen when considering the large differences of the values given herein. The values of the runoff coefficients given in the table correspond to rather flat catchments. For steeper catchments, larger runoff coefficients should be selected. Also, the ranges of the reported runoff coefficients correspond to summer conditions.

The rainfall intensity for the Rational Formula is obtained from rainfall intensity–duration–frequency curves established from past meteorological observations. These curves express a relation between the intensity of a rainfall in cm/hr (in./hr) or liters/(s-ha), its duration (rainfalls of longer duration have lower average intensity than short-duration catastrophic storms), and the statistical probability of the storm occurrence expressed as the recurrence interval, T_r, in years or storm frequency $n = 1/T_r$.

In 1922, Karl Imhoff noted that the intensity–duration–frequency curves for most German cities were similar. Therefore, he normalized them by dividing the rainfall intensity values of the curve by the intensity of a standard once per year storm with a specified duration. Following this concept, storm intensity–duration curves were prepared for a standard once-per-year storm at different U.S. locations and for Essen, Germany. When these curves were normalized by the 1-hr duration once-per-year rainfall a similar pattern has emerged as shown in Figure 2.2a. The magnitude of the standard once-per-year 1-hr duration rainfall can be obtained from isopluvial maps prepared in the United States by the National Weather Service (Fig. 2.3). In Germany, geographic variations in storm intensities are not as great as in the United States. The intensities of the standard once-per-year 1 hr storm range there from 32 liters/(s-ha) (11.5 mm/hr) in the northern flat regions to about 43 liters/(s-ha) (16.2 mm/hr) in southern alpine regions.

In Figure 2.2a, two distinct relations of the intensities to duration can be detected inasmuch as the storm characteristics in the Pacific coastal areas of the United States (including Pacific coasts of Alaska and Hawaii) differ from the rest of the conterminous United States. The U.S. Soil Conservation Service classified the former type of precipitation as a Type I storm and the latter as a Type II storm, respectively. It is also interesting to note that most of Europe follows Type II, and hence a normalized intensity–duration curve for Essen is similar to those for Chicago or Miami. The magnitudes of the once-per-year storm for Esssen are smaller than those typical for the eastern United States.

Figure 2.2b can be used for extrapolating the intensity obtained for a once-per-year storm to storms with a higher recurrence interval.

Example: For a drainage network, determine the intensity of a 5-year design storm if the main sewer length is 1 500 m, the velocity in the sewer is 1 m/s, and the inlet time is 5 min, hence the time of concentration is $t_t = l/v + t_{\text{inlet}} = 1\,500/(1 \times 60) + 5 = 30$ min.

If the design location is in Chicago, IL read the intensity of the standard 1-year–1-hr storm from the isopluvial map in Figure 2.3, where $r_1^1 = 32.5$ mm/hr $= 32.5$ mm/hr $\times\ 2.777$ {[liters/(s-ha)]/(mm/hr)} $= 90.27$ liters/(s-ha). If the location was, for example, Essen, Germany the magnitude of the design 1-year–1-hr storm would be $r_1^1 = 34$ liters/(s-ha). To convert this 1-hr–1-year intensity storm to the desired 5 year–30 min design storm, read first the magnitude of the

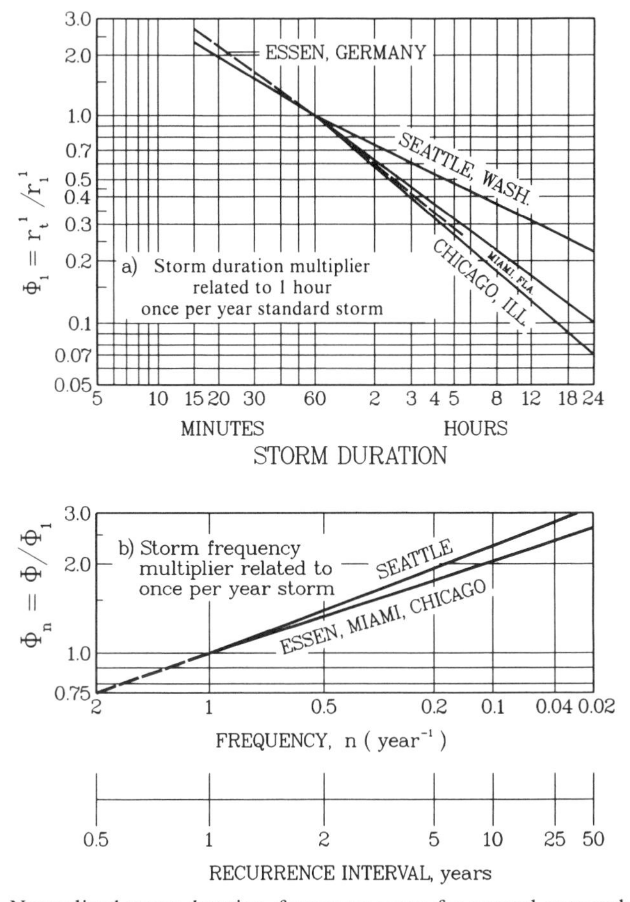

Figure 2.2 Normalized storm duration–frequency curve for several geographic locations.

duration multiplier, ϕ_1, from the upper portion of Figure 2.2. Herein, $\phi_1 = 1.4$ for Chicago (full line) and 1.7 for Essen (dashed line). Then from the lower portion of Figure 2.2 read the frequency multiplier, ϕ_n, which for both locations is 1.7. Hence, the average intensity of the design once-in-5 years–30-min duration storm is

$$\text{Chicago} \quad I = \phi_1 \phi_n r_1^1 = 1.4 \times 1.8 \times 32.5 \text{ mm/hr}$$

$$= 81.9 \text{ mm/hr} \times 2.78 \, [\text{liters/(s-ha)}]/(\text{mm/hr})$$

$$= 227.43 \text{ liters/(s-ha)}$$

$$\text{Essen} \quad I = 1.7 \times 1.8 \times 34 = 104 \text{ liters/(s-ha)}$$

When applying the Rational Formula, it is necessary to select the design storm for which the frequency–duration curve is available. Design storms can be prepared from rainfall maps published regularly by the National Weather Service (formerly

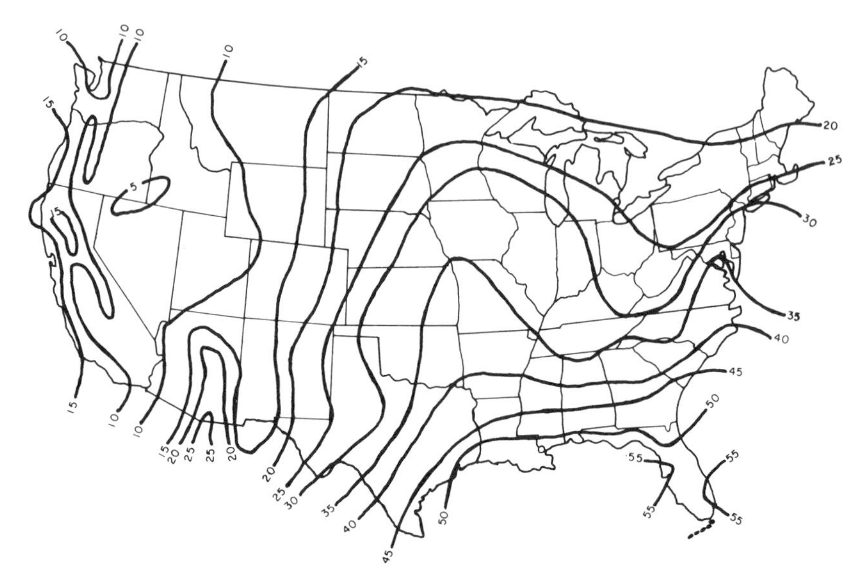

Figure 2.3 Isopluvial map of the United States for once-per-year–1-hr duration rainfall in mm (to convert from mm/hr to liters/(s × ha) multiply by 2.78; 1 in. = 25.4 mm).

U.S. Weather Bureau) of the National Oceanic and Atmospheric Administration (NOAA). This service also prepares rainfall intensity–duration curves for most U.S. urban centers. If the information is not available, it is necessary to prepare the curves from the local storm data and interpolate.

Example: Estimate the peak runoff for a 10-year precipitation falling on a mixed land use watershed. The watershed area is 300 ha with the following land use distribution:

Open space—lawn 2–7% slope	120 ha
Residential—single family	120 ha
Commercial—neighborhood business	60 ha

Using Table 2.2 to identify the individual runoff coefficients, the weighted runoff coefficient becomes

$$C = \frac{0.2 \times 120 + 0.5 \times 120 + 0.7 \times 60}{300} = 0.42$$

Assume an inlet time of 25 min and 5 min for sewer flow time. The time of concentration is then $t_c = 30$ min.

Again using Chicago as an example, the standard 1-year–1-hr storm can be read from Figure 2.3 as being 32.5 mm/hr and from Figure 2.2 the appropriate duration and frequency multipliers for a 10-year–30-min storm are $\phi_l = 1.4$ and $\phi_n = 2.1$, respectively. Then

$$I = 2.1 \times 1.4 \times 32.5 = 95.55 \text{ mm/hr} = 1.56 \text{ mm/min}$$

and the flow is

$$\begin{aligned} Q_p &= CAI = 0.42 \times 300 \text{ (ha)} \times 10\,000 \text{ (m}^2\text{/ha)} \times 1.56 \text{ (mm/min)} \\ &= 1\,965\,600 \text{ (liters/min)} \times \left\{(\text{m}^3/\text{s})/[60\,000 \text{ (liters/min)}]\right\} \\ &= 32.76 \text{ m}^3/\text{sec} \end{aligned}$$

Watkins[6] recommended that the use of the Rational Formula be confined to the design of drainage systems for small areas in which the diameter of the largest sewer is unlikely to exceed 0.6 m. For small watersheds with about 1–2 km of the flow length and with a uniform runoff coefficient the computation can proceed as shown on the table heading shown in Table 2.3.

TABLE 2.3 Computational Table for Hand Calculations

1	2	3	4	5	6	7	8
		Stretch from Manhole			Contributing area		
Area Number	Street Name	Upstr.	Downst.	Sewer Length L	Partial A_p	Total ΣA_p	Specific Flow Yield of Sewage q_s
#		#	#	m	ha	ha	liters/(s − ha)

9	10	11	12	13	14	15
		Flow				
Rainfall Intensity r_X = liters/s × ha n	Stormwater Flow Rate	Sewage		Stormwater		Total
Runoff coefficient C	$q_r = Cr_x$	$A_p q_s$	Q_s	$A_p q_r$	Q_r	$Q = Q_s + Q_r$
	liters/(s − ha)	liters/s	liters/s	liters/s	liters/s	liters/s

16	17	18	19	20
		Capacity		
Slope of the Sewer S_s	Cross-Section Shape Size	Flow Q_c	Velocity v_c	Comments
%	cm	liters/s	m/s	

Anderl[7] suggested a formula that combined the Rational Formula with the capacitive–storage–subtraction approach. The equation was proposed as

$$R = (P - I)C - \frac{C}{a}[1 - e^{-a(P-I)}]$$

where R = total runoff (mm)
P = total precipitation (mm)
I = initial subtraction (mm)
a = an empirical factor

The British Transport and Road Research Laboratory (TRRL) Hydrograph Method. This method was developed in Great Britain[6] and has been subsequently introduced in the United States[8] and other countries. The TRRL method requires relatively little additional effort compared to the Rational Formula. The basic assumptions connected with the applications of the TRRL method are

1. Most of the runoff originates from the directly connected impervious surface area that should be more than 15% of the catchment.
2. The recurrence interval of the selected design storm is less than 20 years.
3. The catchment area is less than 13 km^2.

The process of constructing a hydrograph from a design storm involves the following steps that are illustrated in Figure 2.4:

1. On the map of the catchment define the areas that are directly connected to the drainage system. Exclude all pervious areas and impervious areas that overflow on adjacent pervious surfaces.
2. Develop the area/time of flow diagram by subdividing the contributing areas according to their time of flow from the segment area to the drainage outlet. The diagram is then constructed by linear addition of the individual areas contributing to the flow rate after the start of the rainfall (Fig. 2.4a).
3. Divide the design storm according to the time steps of the area/time curve. This time increment is usually 1–5 min (Fig. 2.4b). About 2.5 mm can be optionally subtracted from the rainfall hyetograph to account for the surface storage.
4. Calculate the hydrograph by summing up the product of the flow time ordinate from the area/time curve and the corresponding rainfall intensity. Hence

$$R_1 = i_1 a_1$$

$$R_2 = i_2 a_1 + i_1 a_2$$

$$R_3 = i_3 a_1 + i_2 a_2 + i_1 a_3 + \cdots$$

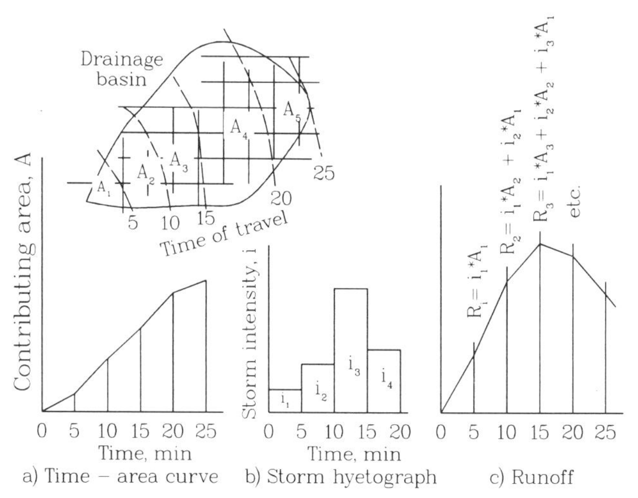

Figure 2.4 British Transport and Road Research Laboratory (TRRL) hydrograph concept.

The overland flow time necessary for the development of the area/time curve can be estimated from known or computed overland flow velocity or by using concepts associated with ascertaining the times of concentration.[5,9,10]

Several computer urban runoff models have incorporated the TRRL method. Of note is ILLUDAS.[11]

The SCS Runoff Model. The U.S. Soil Conservation Service (SCS) has developed a procedure by which volume and peak of the runoff can be estimated for a 24-hr design storm.[12,13] This method can be used for both urban and nonurban small watersheds. As specified previously, the SCS has developed its relationships for two basic storm patterns—Type I, typical for Hawaii, Alaska, the coastal side of the Sierra Nevada and Cascade Mountains in California, Oregon, and Washington, and Type II intended for the rest of the United States, Puerto Rico, and the Virgin Islands. The magnitude of the 24-hr storm can again be obtained from the isopluvial maps of the region or determined approximately from the magnitude of the standard 1-year–1-hr rainfall (Fig. 2.2) and reading the duration multiplier ϕ_1 for 24 hr and multiplying it by 24 hr to get the 24 hr volume.

The method also requires the determination of the runoff curve number, CN, for the drainage basin, which is a function of the soil and surface characteristics and land use (Table 2.4). The division of soils into hydrologic groups follows the SCS soil classification. Soil characteristics that are associated with each group are as follows:

Group A—deep sand, deep loess, aggregated silts.

Group B—shallow loess, sandy loam.

Group C—clay loams, shallow sandy loam, soils low in organic content, and soils usually high in clay.

TABLE 2.4 Runoff Curve Numbers for Hydrologic[12] Soil-Cover Complexes (Antecedent Moisture Condition)

Land Use Description/Treatment/Hydrologic Condition			Hydrologic Soil Group A	B	C	D
Residential[a]						
Average lot size	Average impervious[b] (%)					
1/8 acre or less	65		77	85	90	92
1/4 acre	38		61	75	83	87
1/3 acre	30		57	72	81	86
1/2 acre	25		54	70	80	85
1 acre	20		51	68	79	84
Paved parking lots, roofs, driveways, etc.[c]			98	98	98	98
Streets and roads:						
Paved with curbs and storm sewers[c]			98	98	98	98
Gravel			76	85	89	91
Dirt			72	82	87	89
Commercial and business areas (85% impervious)			89	92	94	95
Industrial districts (72% impervious)			81	88	91	93
Open spaces, lawns, parks, golf courses, cemeteries, etc.						
Good condition: grass cover on 75% or more of the area			39	61	74	80
Fair condition: grass cover on 50% to 75% of the area			49	69	79	84
Fallow	Straight row	—	77	86	91	94
Row crops	Straight row	Poor	72	81	88	91
	Straight row	Good	67	78	85	89
	Contoured	Poor	70	79	84	88
	Contoured	Good	65	75	82	86
	Contoured and terraced	Poor	66	74	80	82
	Contoured and terraced	Good	62	71	78	81
Small grain	Straight row	Poor	65	76	84	88
		Good	63	75	83	87
	Contoured	Poor	63	74	82	85
		Good	61	73	81	84
	Contoured and terraced	Poor	61	72	79	82
		Good	59	70	78	81
Close-seeded	Straight row	Poor	66	77	85	89
legumes[d] or	Straight row	Good	58	72	81	85
rotational	Contoured	Poor	64	75	83	85
meadow	Contoured	Good	55	69	78	83
	Contoured and terraced	Poor	63	73	80	83
	Contoured and terraced	Good	51	67	76	80
Pasture or range		Poor	68	79	86	89
		Fair	49	69	79	84
		Good	39	61	74	80
	Contoured	Poor	47	67	81	88
	Contoured	Fair	25	59	75	83
	Contoured	Good	6	35	70	79
Meadow		Good	30	58	71	78
Woods or Forest		Poor	45	66	77	83
land		Fair	36	60	73	79
		Good	25	55	70	77
Farmsteads		—	59	74	82	86

[a] Curve numbers are computed assuming the runoff from the house and driveway is directed toward the street with a minimum of roof water directed to lawns where additional infiltration could occur.
[b] The remaining pervious areas (lawn) are considered to be in good pasture condition for these curve numbers.
[c] In some warmer climates of the country a curve number of 95 may be used.
[d] Close-drilled or broadcast.
After Soil Conservation Service.[12]

Group D—soils that swell significantly when wet, heavy plastic clays, and certain saline soils.

Soil classification can be obtained from soil maps prepared by the SCS for many counties throughout the United States or from soil surveys.

The runoff curve number in Table 2.4 corresponds to average soil moisture conditions (antecedent soil moisture condition II). If the soils are dry or very wet the runoff curve number must be adjusted according to Table 2.5. The volume of

TABLE 2.5 Correction of the Runoff Curve Number for Wet and Dry Antecedent Soil Moisture Conditions.

	Antecedent Soil Moisture Conditions (AMC)	
	Total 5-Day Antecedent	Rainfall
AMC	Dormant Season	Growing Season
I	Less than 12 mm	Less than 35 mm
II	12 to 28 mm	35 to 55 mm
III	Over 28 mm	Over 55 mm

	Runoff Curve Number Correction	
CN for	Corresponding CN for Condition	
Condition II	I	III
100	100	100
95	87	99
90	78	98
85	70	97
80	63	94
75	57	91
70	51	87
65	45	83
60	40	79
55	35	75
50	31	70
45	27	65
40	23	60
35	19	55
30	15	50
25	12	45
20	9	39
15	7	33
10	4	26
5	2	17
0	0	0

After Soil Conservation Service.[12]

the runoff (excess rainfall) for a 24-hr design storm can be then obtained from the formula

$$Q = \frac{(P - 0.2S)^2}{P + 0.8S}$$

where P and Q are precipitation and runoff volumes in mm, respectively, $S = (25\,400/\mathrm{CN} - 254)$ is the storage characteristics of the basin in mm, and CN is the runoff curve number for the basin.

The relation of the surface runoff volume, Q, to the 24-hr precipitation, P, is also shown in Figure 2.5.

The peak runoff flow can be estimated by transforming the storm hyetograph into runoff either by using the SCS Unit Hydrograph or with the aid of a chart that relates the peak flow to the time of concentration for the watershed, t_c, and to the volume of the runoff, Q (Fig. 2.6).

The relationship in Figure 2.6 can be satisfactorily used only for large 24-hr rainfalls and higher runoff curve numbers. In all other cases, the excess rain should be transformed into runoff using the SCS Unit Hydrograph.

The time of concentration, t_c, in the SCS runoff procedure can be estimated from

$$t_c = \frac{l^{0.8}(S + 25)^{0.7}}{4\,234\gamma^{0.5}} \quad \text{(hr)}$$

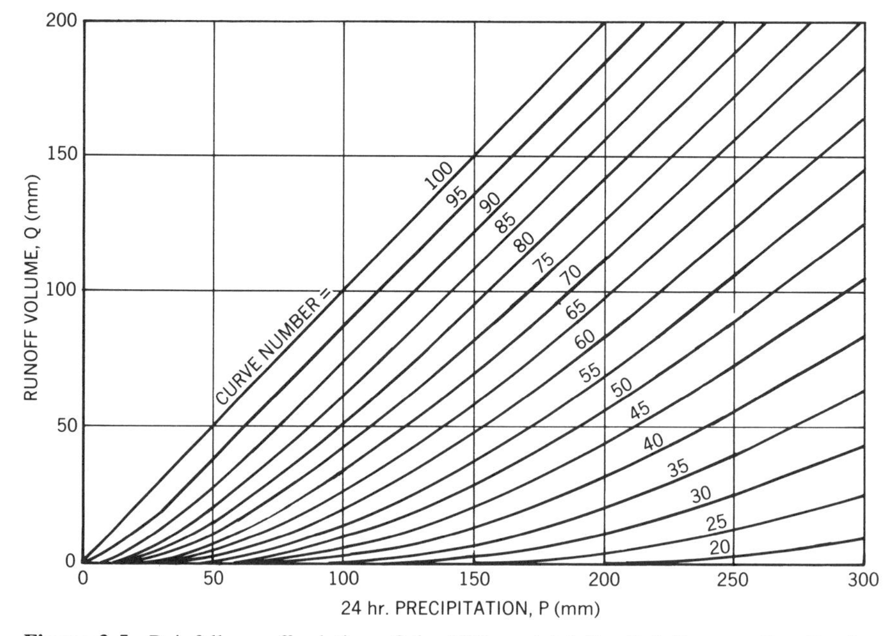

Figure 2.5 Rainfall–runoff relation of the SCS model (after Soil Conservation Service, U.S. Dept. of Agriculture): 1 in. = 25.4 mm.

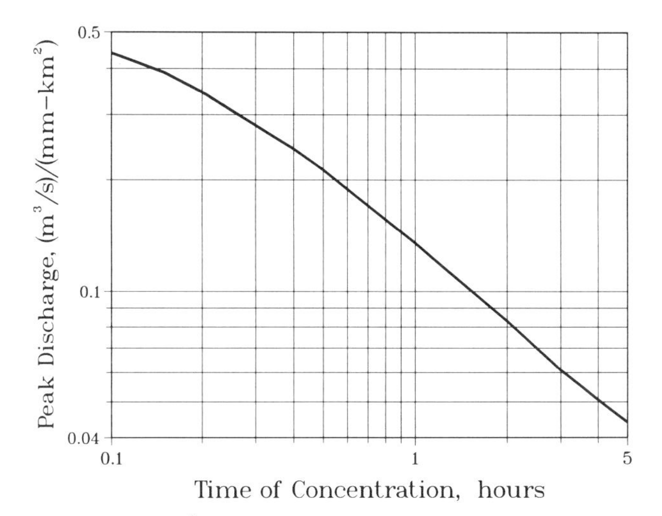

Figure 2.6 Peak runoff in m^3/s per mm of runoff volume versus time of concentration, t_c, for 24-hr Type II rainfall according to SCS procedure.

where l is the hydraulic length of the overland flow in meters, S is the storage in millimeters defined previously, and γ is the slope of the watershed toward the drainage in percent. The time of concentration can be also estimated as $t_c = l/v$, where v is the overload flow velocity in meters/hour.

In urban or urbanizing watersheds that are partially impervious the time of concentration would be shorter than that estimated by the preceding equation, since impervious surfaces provide more efficient flow patterns than pervious land surfaces. The time of concentration must then be adjusted by a corrective multiplier as follows:

$$t_{\text{c-urban}} = \text{LF} \times t_c$$

where

$$\begin{aligned}\text{LF} = 1 - \text{PIM}(&-0.006789 + 0.000335\ \text{CN} \\ &- 0.0000004298\ \text{CN}^2 - 0.00000002185\ \text{CN}^3)\end{aligned}$$

in which PIM is the percentage impervious area within the watershed.

Example: Determine the 24-hr runoff volume for a 10-year 24-hr precipitation of 125 mm. Assume antecedent soil moisture conditions (ASMC) II. The catchment has the following land use distribution and soils:

Area Fraction	Land Use	Soil Group
0.4	Open space—good conditions	C
0.4	Residential—38% impervious	B
0.2	Commercial	B

From Table 2.4, the following CN numbers are selected for the three land uses: 74, 75, and 92, respectively. Then the composite CN is

$$\mathrm{CN} = 0.4 \times 74 + 0.4 \times 75 + 0.2 \times 92 = 78$$

The basin storage constant $S = 25\,400/78 - 254 = 72$ mm and the runoff (excess rain) volume becomes

$$Q = \frac{(125 - 0.2 \times 72)^2}{125 + 0.8 \times 72} = 67 \text{ mm}$$

Example: For the watershed defined in the previous example, estimate the peak flow of the runoff. The watershed size is 300 ha (3 km^2), the average overland flow length is 800 m, and the slope is 3%. The watershed is 15% impervious.

The time of concentration becomes

$$t_c = \frac{800^{0.8}(72 + 25)^{0.7}}{4234 \times 3^{0.5}} = 0.7 \text{ hr}$$

This time should be corrected to account for the imperviousness of the watershed. Hence,

$$\begin{aligned} \mathrm{LF} = 1 - 15(&-0.006789 + 0.000335 \times 72 \\ &- 0.0000004298 \times 72^2 - 0.00000002185 \times 72^3) = 0.896 \end{aligned}$$

Then

$$t_c = 0.896 \times 0.7 = 0.62 \text{ hr}$$

From Figure 2.5, for the 24-hr Type II storm and the time of concentration of 0.62 hr the unit peak flow is

$$q_p = 0.19 \text{ m}^3 \text{ s}^{-1} \text{ km}^{-2} \text{ mm}^{-1}$$

Using the runoff (excess rain) volume of 67 mm determined in the preceding example, the peak runoff with a 10-year recurrence interval becomes

$$\begin{aligned} Q_p &= q_p A Q = 0.19 \left(\text{m}^3/(\text{s-km}^2\text{-mm})\right) \times 3 \left(\text{km}^2\right) \times 67 \text{ mm} \\ &= 38.9 \text{ m}^3/\text{s} \end{aligned}$$

Hydrologic Computer Models

A number of urban hydrologic models were developed during the 1970s. These models range from computerized procedures not much different from the simple methods discussed in the preceding section to very complex and detailed urban runoff models.

The Wallingford Hydrograph Method.[14] This method consists of several computerized procedures developed in the United Kingdom that include the following:

Method/Model	Function
A. Wallingford Modified Rational Method	To size pipes and/or calculate discharges using a Modified Rational Formula.
B. Wallingford Hydrograph method	To size pipes and/or simulate discharge hydrographs for observed or design rainfall events.
C. Wallingford Optimizing method	To design pipe diameters, depths, and gradients for a minimum construction cost using the Modified Rational Formula.
D. Wallingford Simulation model	To simulate time-dependent flows for observed or design rainfalls.

Illinois Urban Drainage Area Simulator[11] uses the TRRL method for estimating the runoff volumes and flow rates and provides for an optimal selection of pipe sizes of the sewer segments.

Similar in nature is the *SVK-System* developed in Denmark.[15]

These models can be considered as computerized design procedures. Other models are in the category of simulation models.

The Storm Water Management Model (SWMM)[16,17] is a complex model of urban hydrology that is structured into "blocks" along the lines of the physical processes involved in the runoff formation in the urban areas. The RUNOFF blocks generate the runoff hydrographs and their associated pollutant loadings. The TRANSPORT blocks route both the hydrographs and the pollutant histograms through the channels and sewerage systems. The STORAGE/TREATMENT block simulates the performance of devices (e.g., detention basins) that have the capability for storage and removal of pollutants. The RECEIV block then finally simulates the response of a receiving water body to the inputs of sewage and combined sewer overflows from the drainage systems.

Originally, the SWMM was a single event model. Later the model was modified and can accept a sequence of storms as the input.[18]

The application of SWMM requires a division of the drainage area into homogeneous land segments. The model can be used both for planning and design analysis of urban drainage systems and is being maintained and continuously updated by the U.S. Environmental Protection Agency (Athens, GA).

2.2 DETERMINATION OF CROSS SECTIONS

A short form of the now more than 100-year-old Kutter's formula was used earlier by German designers of sewer networks. The formula is

$$\text{velocity (m/s) } v = \frac{100\sqrt{R}}{b + \sqrt{R}} \sqrt{RS} \tag{1}$$

where R = hydraulic radius (m)
S = slope of the water surface
$b = 0.35$ = coefficient of roughness

$$\text{Flow} \quad Q = Av = \text{cross-sectional area} \times \text{velocity}$$

$$R(\text{m}) = A/O = \text{area/wetted perimeter } (\text{m}^2/\text{m})$$

The formula assumes full flow in the cross section. This form of Kutter's formula is relatively simple to use but it has certain drawbacks. Therefore, the Darcy–Weisbach formula was recommended by the German Association for Wastewater Technology (ATV) for the computation of the full flow in the sewers in combination with a complicated but physically correct Prandtl–Colebrook formula for the friction factor. The Darcy–Weisbach formula is

$$S = f \frac{1}{D} \frac{v^2}{2g} \tag{2}$$

or

$$v = \left(\frac{2g}{f}\right)^{0.5} D^{0.5} S^{0.5}$$

where S = slope
f = friction factor
D = diameter (m)
v = velocity (m/s)
g = gravity acceleration = 9.81 (m/s^2)

The formula for the friction factor was given by Prandtl–Colebrook as

$$\frac{1}{\sqrt{f}} = -2 \log \left(\frac{2.51}{\text{Re}\sqrt{f}} + \frac{k}{3.71D} \right) \tag{3}$$

where f = friction factor
Re = Reynolds number = vD/ν
D = diameter (m)
v = flow velocity (m/s)
ν = kinematic viscosity (m^2/s)
k = absolute roughness (m)

For computation for sewers use $\nu = 1.31 \times 10^{-6}$ m^2/s, which corresponds to the viscosity of clean water at 10°C.

Beside the roughness of the pipe materials, the roughness term, k, in the formula is also affected by other factors such as rough open joints, poor alignment and grade caused by settling, deposits in sewers, side flows from laterals, and other disrupting factors in the sewers. For these reasons the ATV has recommended a uniform value of $k = 1.5$ mm for sewers regardless of the material. Ackers and Pitt[19] confirmed these findings and found that the coefficient of roughness is more a function of sewer alignment than a characteristic of the sewer material. In their survey, aging of the sewer material or deposition of a thin (<2 mm) layer of slime did not affect the hydraulic roughness.

Nomographs 1 to 3 in the Appendix provide an overall representation of the relationships given by the Darcy–Weisbach and Prandtl–Colebrook formulas. Nomographs 1 and 2 are for circular pipes and Nomograph 3 expresses the Prandtl–Colebrook equation for a standard egg-shaped sewer. All three nomographs use a uniform roughness factor $k = 1.5$ mm. Nomograph 4 can then be used to adjust flows and velocity to another roughness.

Example: What is the magnitude of the flow and velocity in a circular concrete pipe with a diameter of 1 m and slope of 0.71% = 7.1 m/km.

In Nomograph 2 for $D = 1$ m and $S = 7.1$ m/km read $Q = 2$ m^3/s and $v = 2.5$ m/s.

Example: In a straight well-aligned concrete sewer pipe the roughness was determined as $k = 1$ mm. Determine the flow and velocity for $D = 1$ m and $S = 7.1$ m/km.

From Nomograph 2 ($k = 1.5$ mm) read $Q' = 2$ m^3/s and $v' = 2.5$ m/s. Then from Nomograph 4 for $D/k = 1/1.5$ m/mm and $k = 1$ mm read $\Delta Q/Q = 5\%$. Hence, $Q = 1.05Q' = 1.05 \times 2 = 2.1$ m^3/s and, similarly, $v = 1.05v' = 1.05 \times 2.5 = 2.63$ m/s.

Example: What is the necessary width (d) of a regular egg-shaped sewer ($h = 1.5\ d$) with a slope $S = 0.2\%$ (2 m/km) that should carry a flow of $Q = 1.2$ m^3/s, assuming full flow conditions.

In Nomograph 3, the crossing point of the horizontal ordinate $Q = 1.2$ and vertical ordinate $S = 2$ m/km lies near the line denoting a cross section of 0.9/1.35. The width is then 0.9 m.

In the United States, the *Manning formula* has been widely used both for sewer and open channel design. The Manning formula is

$$v = \frac{1}{n} R^{2/3} S^{1/2} \tag{4}$$

where v = velocity (m/s)
n = Manning roughness factor
$R = D/4$ = hydraulic radius (m)
S = slope (dimensionless)
D = diameter of the pipe (m)

This formula is almost identical to the Gauckler–Strickler equation known in Europe. n values for sewer design range from 0.013 to 0.015 for concrete sewers. Approximately the same roughness coefficients can be used for other sewer materials (vitrified clay, PVC, and cast iron). For corrugated steel pipes use n = 0.022–0.026. A relationship between the Darcy–Weisbach friction factor, f, and Manning's n would give (for D in meters)

$$f = \frac{124.6n^2}{\sqrt[3]{D}} \tag{5}$$

A Manning coefficient of n = 0.013 corresponds approximately to the roughness factor k = 1.5 mm. Nomograph 5 in the Appendix is provided for solutions of the Manning formula for a full flow in a circular pipe.

By substituting $D = 4R$, where R is the hydraulic radius of the conduit, Nomographs 1, 2, 4, and 5 can be used for solutions of any open channel flow problem, including that in sewers with atypical shapes.

The slope in the formulas should correspond to the slope of the water surface. In general, for free surface flow and longer sections this would be approximately equal to the slope of the pipe. For pressure pipes and pressure flows, the slope of the hydraulic grade line must be used, which is the headloss (pressure loss) between the beginning and the end of the section divided by the length of the pipe, including the equivalent lengths for fittings installed on the pipe. At the entrance of flow from a quiescent water body into the pipe the velocity head $h = v^2/(2 \times 9.81)$ must be subtracted.

Figures 2.7 and 2.8 show the hydraulic element curves for partial flows in 4 different sewer shapes. From these curves, one can obtain the relationships of the flow (Q) and velocity (v) for the partial flow to those under full flow conditions. The appropriate values for Q and v can be estimated by the Prandtl–Colebrook or Manning equations or with the aid of Nomographs 1 to 5.

As shown in the ASCE-WPCF Manual No. 9[2] the friction coefficients n and f are not independent of the depth and the channel or sewer slope. This may affect the shape of the hydraulic element curves by as much as 15% for flow and 20% for velocity, respectively. Therefore, the constant n and f assumption may underestimate the flow and velocity.

Example: Identical to the preceding example of the flow in an egg-shaped sewer, except the cross section is filled to 70% of the height of the egg-shaped profile.

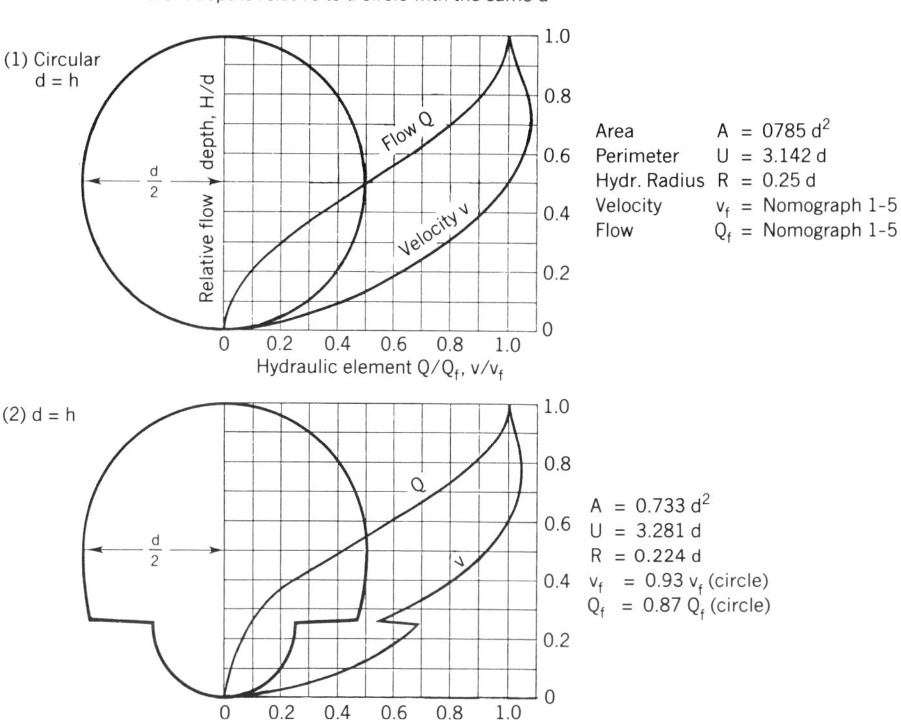

Figure 2.7 Hydraulic element curves: circular and semicircular cross sections.

From the hydraulic element curve for the standard egg-shaped cross section [Fig. 2.8(4)] the flow (Q) for the 0.7 filling is 0.74 of the flow when flowing full. From Nomograph 3 read the dimensions of the cross section, however, not for $Q = 1.28$ but for $Q = 1.28/0.74 = 1.7\ \text{m}^3/\text{s}$. The necessary cross section is then 1.00/1.5 and the corresponding full flow velocity is 1.5 m/s. On the hydraulic element curve [Fig. 2.8(4)], the velocity in the cross section flowing 70% full is 1.04 times that at full flow. Then the actual velocity is $1.04 \times 1.5 = 1.56$ m/s.

Example: What are the depth and velocity of partial flow in an egg-shaped conduit, 0.9/1.35 m, for a flow-rate of $Q = 0.2\ \text{m}^3/\text{s}$ and slope of 0.2% (2 m/km)?

The conduit at full flow (Nomograph 3) will carry $1.28\ \text{m}^3/\text{s}$ with a velocity of $v = 1.4$ m/s. The partial to full flow ratio is $0.2/1.28 = 0.16$. According to the hydraulic element curve [Fig. 2.8(4)] the depth of the partial flow at $Q = 0.16$ equals 0.29 of the cross-sectional height, hence $0.29 \times 1.35 = 0.39$ m and the velocity [Fig. 2.8(4)] at 0.29 partial filling is 0.76 times that at full flow, or $0.76 \times 1.4 = 1.1$ m/s.

In addition, Figures 2.7 and 2.8 also contain coefficients for each individual

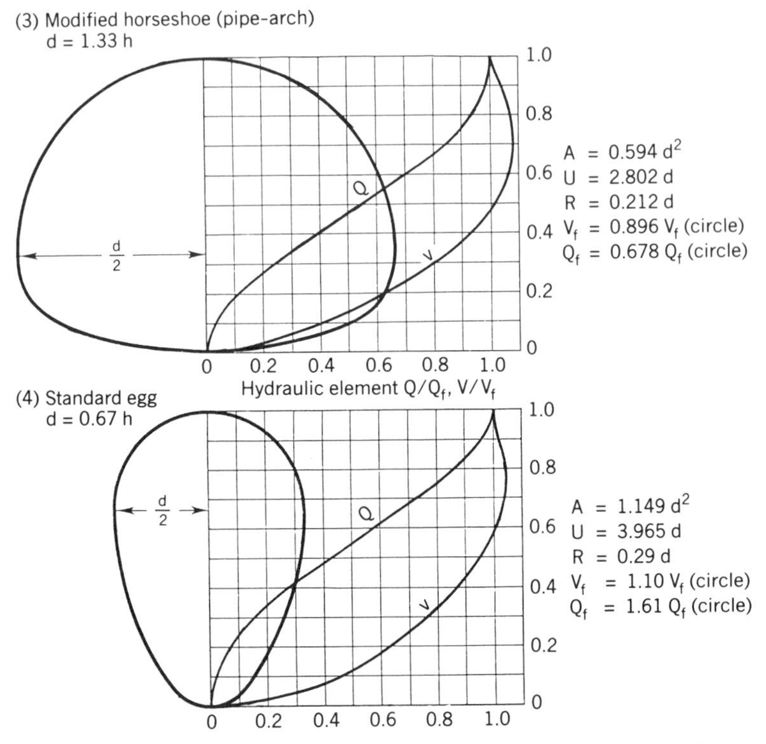

Figure 2.8 Hydraulic element curves: horseshoe (pipe-arch) and standard egg cross sections.

cross section that relate the values in the figure to those of a circle with the diameter equaling the width of the profile. This enables computation of selected cross sections using Nomographs 1 to 5 that were prepared for circular cross sections. The roughness is the same as that given above the nomographs.

Example: A pipe-arch conduit [Fig. 2.8(3)] made of corrugated steel (Manning $n = 0.024$) is dimensioned for storm flow of $Q = 2\ \text{m}^3/\text{s}$. The slope of the conduit is 0.002 (2 m/km).

According to Figure 2.8(3), the flow in the pipe-arch conduit is $Q = 0.678Q_1$ (circle), or the flow in this conduit equals 0.678 times that for a circular sewer with a diameter equaling the span of the arch of the pipe. Then in Nomograph 5 (Manning equation) for $Q = 2\ (\text{m}^3/\text{s})/0.678 = 2.95\ \text{m}^3/\text{s}$, $n = 0.024$ and $S = 0.002$; the corresponding diameter is $d = 1.6$ m. This will be the same as the span of the arch. The rise of the arch (height) is then $1.6/1.33 = 1.2$ m.

Of note is also a simple form of computation that was published in the first German edition of this book in 1907. If one compares the Manning (Gauckler–Strickler) formula for any cross section with that for a circular cross section it can be observed that for the same roughness and slope, the ratio of velocities is related to the 2/3 root of the ratio of the hydraulic radii. If one selects, for example, the hydraulic radius of an egg-shaped cross section presented in Figure 2.8(4) and relates that to the hydraulic radius of a circle ($0.25d$), a constant relationship of velocities can be obtained as

$$v_{egg} = (0.29/0.25)^{2/3} \times v_{circle} = 1.1 v_{circle}$$

When the coefficient of 1.1 is known, it is then possible to use Nomographs 1 to 3 for circular cross sections without any further complications. This is not necessary for a standard egg shape since Nomograph 3 provides the solution. For any other shapes this procedure will yield significant simplifications.

Figures 2.9 to 2.12 can be used for dimensioning of triangular and rectangular conduits. Open concrete-lined conduits shown in Figure 2.10 were introduced in Germany before World War I.[20] They have a triangular bottom section that has been found particularly advantageous for carrying sewage. Compared to other cross sections, the triangular sections have a larger capacity and they can be assembled from prefabricated parts, even under wet conditions. Today, open channel conduits for wastewater are used primarily within the boundaries of a treatment plant.

Example: A triangular open channel shown in Figure 2.12 is lined with concrete. The channel has a dimension $d = 2$ m and slope $S = 0.002$. What is the flow if the depth is $h = 1.2d = 2.4$ m.

From Figure 2.12, the flow ratio to that of a circular cross section is $Q = 1.7Q_1$ (circle) and, similarly, $v = 1.26v_1$ (circle). From Nomographs 2 or 5 ($k = 1.5$ mm or $n = 0.13$) the flow for a circular pipe with a diameter $d = 2$ m is $Q_1 = 6.7\ m^3/s$ and the velocity $v_1 = 2.1$ m/s. Hence, the flow and velocity in the triangular open channel are

$$Q = 1.7 \times 6.7 = 11.4\ m^4/s \quad \text{and } v = 1.26 \times 2.1 = 2.64\ m/s$$

Figures 2.11 and 2.12 contain the most common existing cross sections for open channels. The nomographs were prepared using the Manning equation. For triangular sections with concrete lining use $n = 0.015$ and for channels with banks lined with sod use $n = 0.025$. For grassed waterways Ree and Palmer[21] found that the roughness coefficient n in the Manning equation (k in the Gauckler–Strickler equation) is not a constant but is related to the product of the hydraulic radius and velocity and to the type of the vegetal cover (Fig. 2.13). The retardance category, curve A in Figure 2.13, corresponds to an excellent grass cover 1 m tall, curve B corresponds to good grass cover 30–40 cm tall, curve C to a good grass cover 15–30 cm tall, and the D retardance category is for mowed good grass cover 5–10 cm tall. The nonerosive maximal velocity ranges from 0.75 m/s for small grain annual grasses to about 1.5 m/s for a good grass cover.

At high slopes, the excessive flow velocity of floods can be slowed down by "flow or energy breakers," which are lateral rafters submerged across the channel.

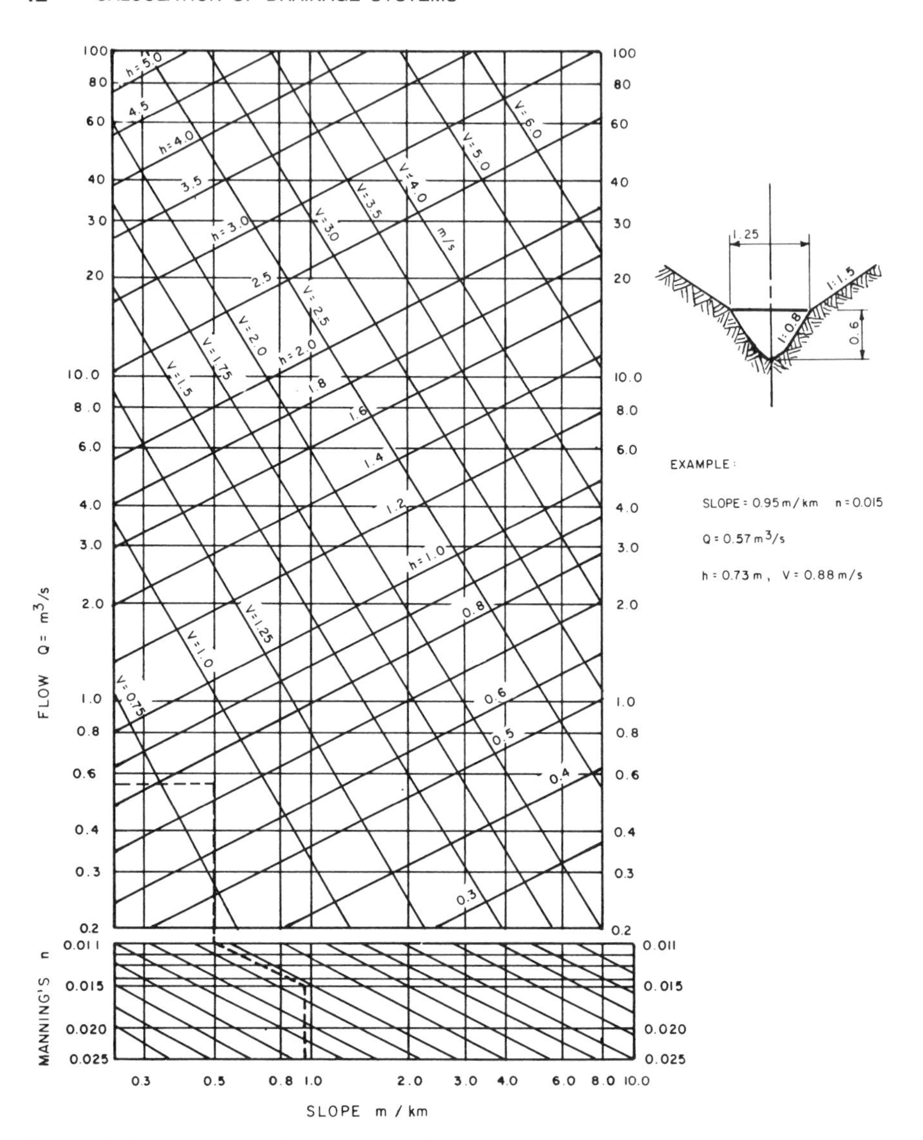

Figure 2.9 Nomograph for solution of triangular open channels using the Manning equation.

Example: Determine the nonerosive velocity and dimensions of a triangular grass-lined channel (mowed Kentucky blue-grass) given that $Q = 2\ \text{m}^3/\text{s}$, the slope of the channel is 1%, and the bank slope is 1 to 3.

The retardance factor for the moved Kentucky blue grass is D. The range of the Mannings roughness factor for this retardance factor (Fig. 2.13) is approximately 0.03–0.1; select $n = 0.04$.

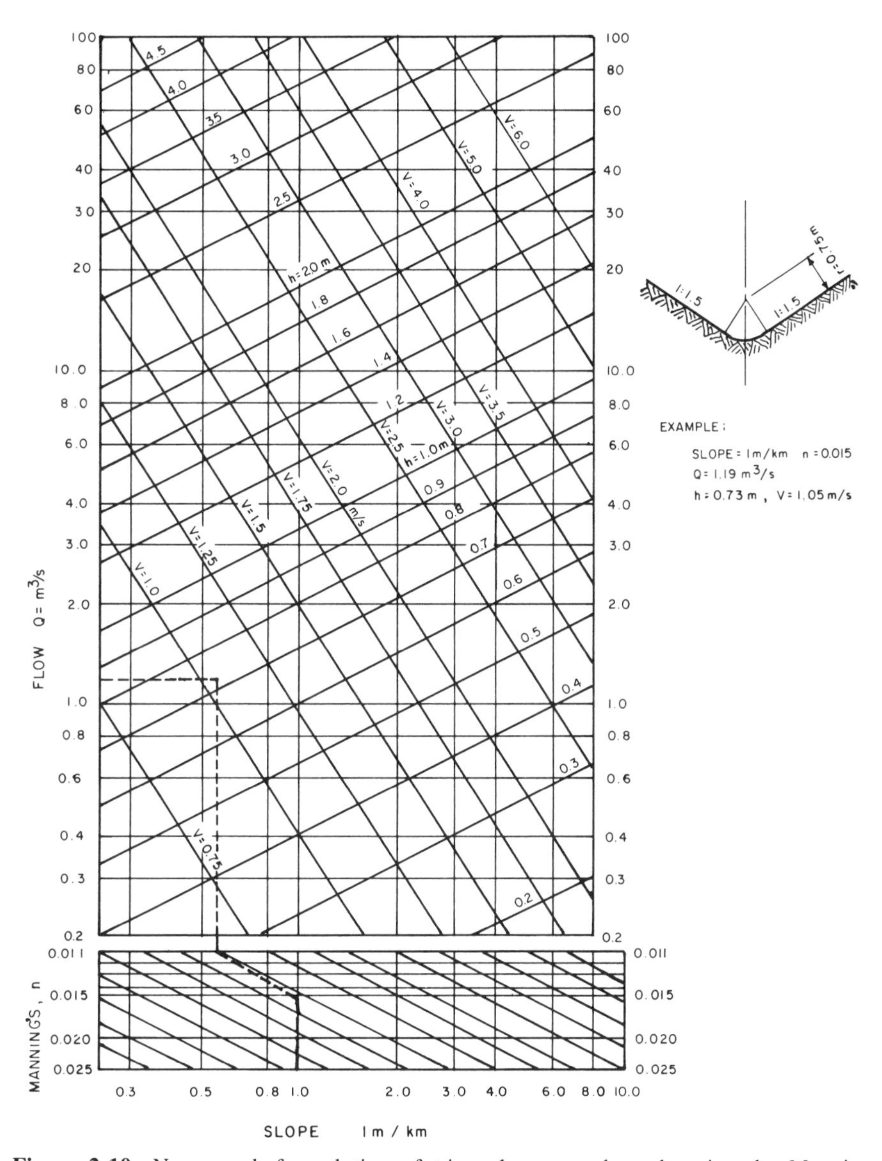

Figure 2.10 Nomograph for solution of triangular open channels using the Manning equation.

The flow area of a triangular channel with a bank slope of 1 to 3 is

$$A = h^2 Z = h^2 \times 3$$

The wetted perimeter becomes

$$P = 2h\sqrt{(1 + 3^2)}$$

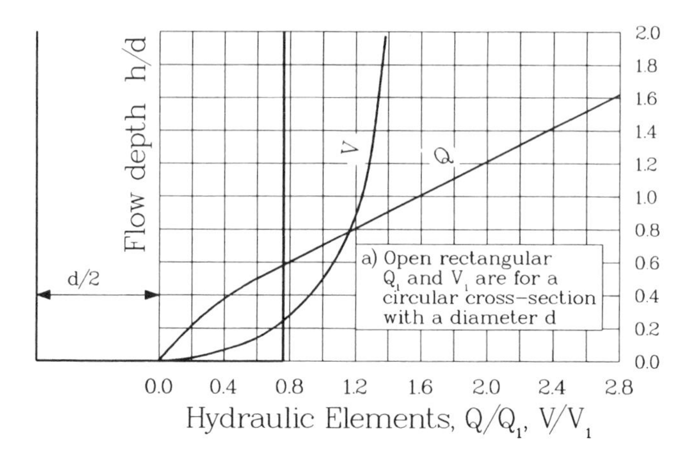

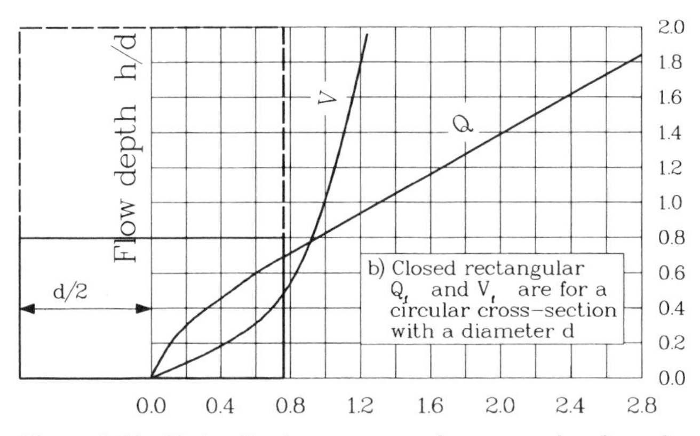

Figure 2.11 Hydraulic element curves for rectangular channels.

and the hydraulic radius is

$$R = 3h^2/\left[2h\sqrt{(1 + 3^2)}\right] = 0.474h$$

Substitute in the Manning equation

$$Q = Av = 3h^2 \frac{1}{0.04}(0.474h)^{2/3} \times 0.01^{1/2}$$

$$2 = 4.55h^{8/3} \qquad \text{or } h = 0.73 \text{ m}$$

Since this is a trial-and-error procedure check whether the roughness coefficient from Figure 2.13 is about the same as that initially estimated. Check also whether the velocity is nonerosive. Then

$$v = Q/A = 2/(3 \times 0.73^2) = 1.23 \text{ m/s}$$

$$R = 0.474 \times 0.73 = 0.35 \text{ m}$$

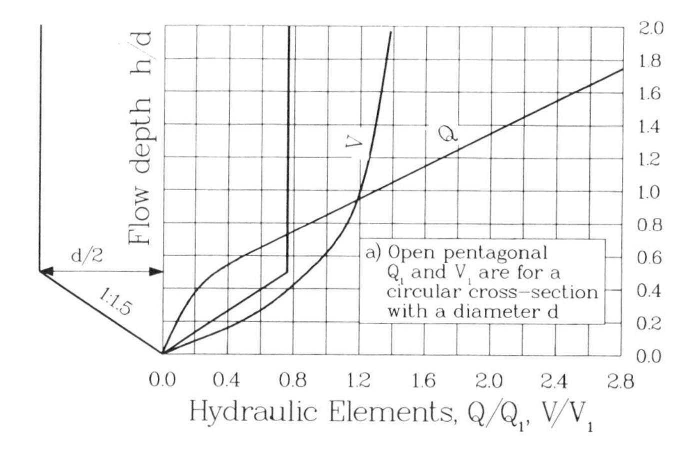

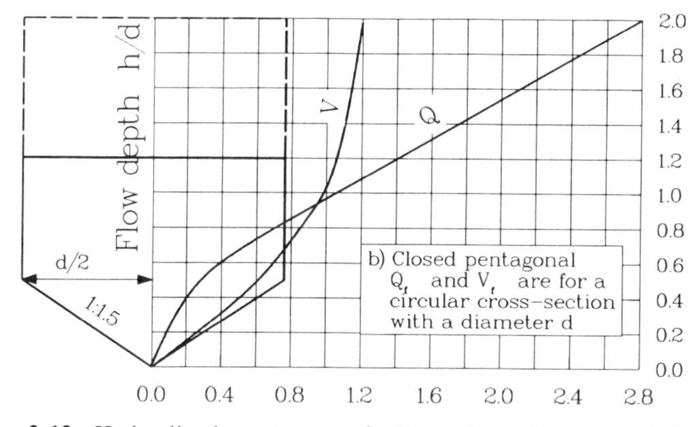

Figure 2.12 Hydraulic element curves for triangular and pentagonal channels.

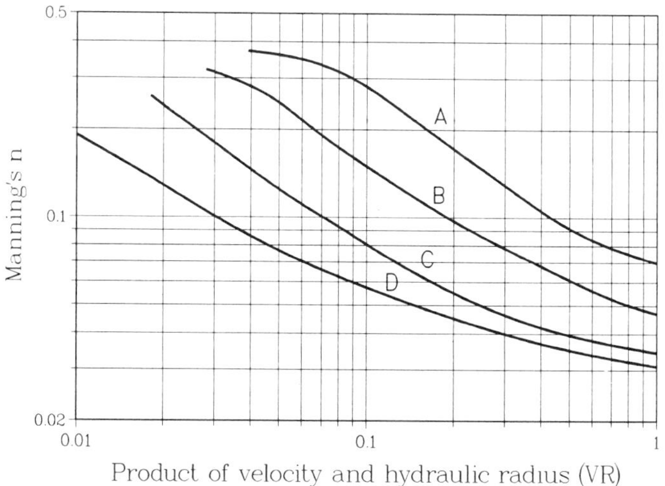

Figure 2.13 Relationship for determining Manning roughness factor, n, for grasses channels according to Ree and Palmer.[21]

For $vR = 1.23 \times 0.35 = 0.43$ the Manning roughness factor in Figure 2.13 is 0.036. Since this deviates from the initial assumption, recompute by substituting 0.036 instead of 0.04 for n in the Mannings equation

$$2 = 5.05h^{8/3} \qquad \text{or } h = 0.71 \text{ m}$$

$$v = 2/(3 \times 0.71^2) = 1.34 \text{ m/s} \qquad vR = 1.34 \times 0.474 \times 0.71 = 0.45$$

and from Figure 2.13 $n = 0.36$ or about the same as initially assumed after the correction. Therefore, the flow velocity is 1.34 m/s, which is still acceptable, and the flow depth is 0.71 m.

The Wallingford Design Procedure, ILLUDAS, SWMM, and SVK models introduced in the preceding section contain hydraulic components that can be conveniently used for the hydraulic design of sewer and channel systems.

2.3 COMPUTATION OF SAFETY OF SEWER PIPES AGAINST BREAK OR RUPTURE*

Installed buried pipes are subjected to three types of loads: (1) earth loads, (2) live loads from the impact of surface vehicles and railroads, and (3) surcharge loads from additional earth fill or building over an installed pipe.

Earth Loads. Marston[22] developed a general method for determining the vertical load on buried pipes. The theory assumes that the calculated load corresponds to the load after ultimate settling, that cohesion of the material is negligible, and the resulting load on the pipe equals the weight of the material above the top of the pipe minus the shearing forces on the sides of the trench.

The general form of the Marston equation is

$$W = C_{\text{d}} w B^2$$

in which W is the vertical load on the pipe per unit length (N/m or lb/ft), w is the specific weight of the earth (N/m^3 or lb/cuft), B is the trench width in meters or feet, and C_{d} is a dimensionless coefficient that measures the effect of the shearing forces between the fill in the trench and its interior walls and direction and amount of relative settlement. Figure 2.14 shows the types of installation of pipes.

In trench installations the pipes are buried in relatively narrow trenches covered with backfill that extends to the original ground surface. The load is affected by

*Detailed design procedures have been published in the ASCE-WPCF Manual No. 9 ("Design and Construction of Sanitary and Storm Sewers," ASCE-WPCF, 1982). Also review "Concrete Pipe Handbook" (Am. Concrete Pipe Assoc., Vienna, VA, 1980) and for design of corrugated steel pipes see "Modern Sewer Design" (Am. Iron & Steel Inst., Washington, DC, 1980). In Germany the safety of buried pipelines is estimated according to the "Guidelines for Static Computation of Sewers and Drainage Conduits (December 1984, Gesellschaft zur Förderung der Abwassertechnik, Markt 1, St. Augustin 1, FRG).

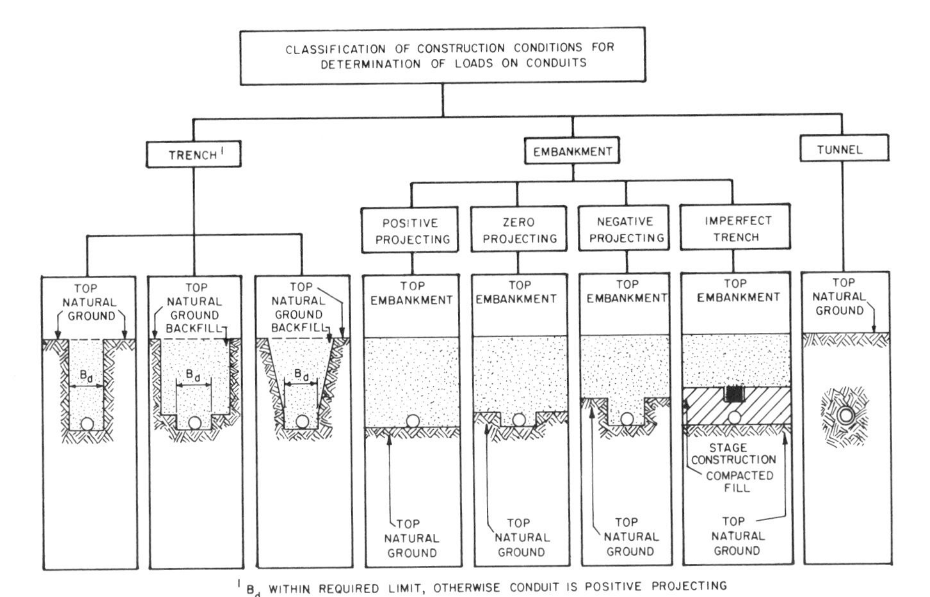

Figure 2.14 Classification of sewer trenches. (reploted with permission from reference 2).

the specific weight of the soil backfill. This value varies from a minimum of 15 700 N/m^3 (100 lb/cuft) to a maximum of 21 200 N/m^3 (135 lb/cuft). When the properties of the soil are unknown an assumed value of 19 000 N/m^3 (120 lb/cuft) can be used. The values of the load coefficient, C_d, for various types of soil backfill and various ratios of H/B_d, where H is the depth of the cover above the pipe, can be obtained from Figure 2.15.

The trench formula of Marston gives the total vertical load on a horizontal plane at the top of the pipe. If the pipe is rigid it will carry almost all of this load. If the pipe is flexible the trench formula can be modified to

$$W = C_d w B_c B$$

in which B_c is the outside width of the pipe.

Example: Determine the load on a 0.6-m-diameter rigid pipe under 4.2 m of cover in trench conditions filled with saturated soil.

Assume that the trench width leaves 30 cm on each side of the pipe for installation. The outside diameter of the pipe assuming a wall thickness of 5 cm is $B_c = 0.6 + 2 \times 0.05 = 0.7$ m and the trench width is $B = B_c + 2 \times 0.3 = 1.35$ m. Also select $w = 19\ 000\ N/m^3$ for saturated soil. Then $H/B_d = 4.2/1.35 = 3.11$ and from Figure 2.15 (Curve C) $C_d = 2.1$, from which $W = 2.1 \times 19\ 000 \times 1.35^2 = 72\ 713\ N/m^3$.

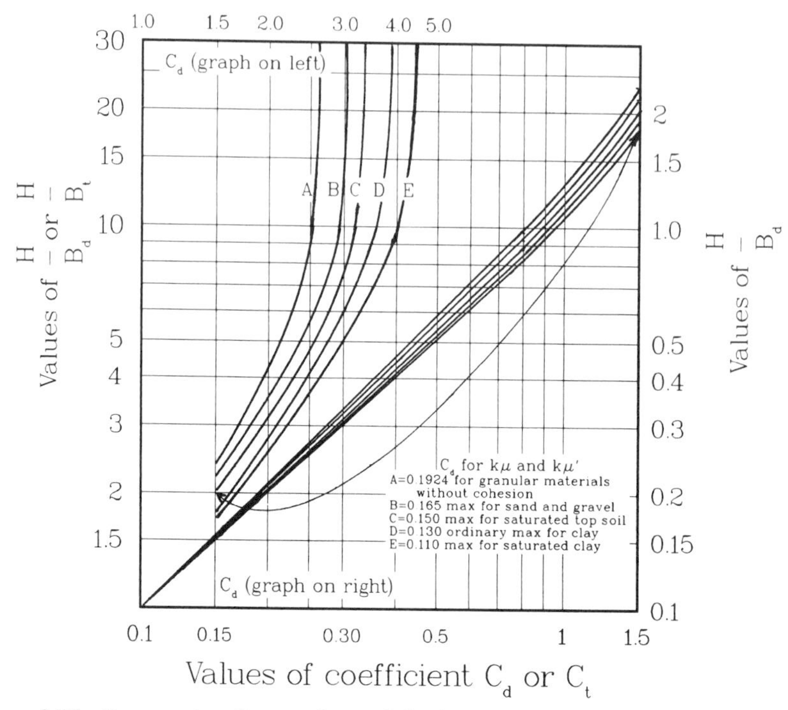

Figure 2.15 Computation diagram for earth loads on trench conduits (completely buried in trenches). (reploted with permission from reference 2).

The essential features of embankment installations are shown in Figure 2.14. The three categories of embankment and fill loading are (1) positive projection whereby the top of the pipeline is above the natural ground covered then with fill, (2) negative projection whereby the pipe is installed in a shallow trench, and (3) imperfect trench, which is usually installed as positive projection but a shallow trench is excavated over the pipe and backfilled to simulate a negative projection installation.

In addition to the factors discussed for trench installations, the magnitude of the C_d factor in the Marston formula for embankments is also affected by the direction and magnitudes of the relative settlement of the prism of soil directly above the conduit and of the soil adjacent thereto. The magnitudes of the C_d factor can be obtained from the ASCE-WPCF Manual No. 9.[2]

Live Loads can be either concentrated or distributed. Their effect is estimated usually by an application of an impact factor in a load formula such as Boussinesq's equation

$$W = C\frac{PF}{L}$$

in which C is the load coefficient, F is the impact factor, P is the concentrated load, and L is the effective length of the pipe.

REFERENCES

1. NASA, *Potential Climatic Impact of Increasing Atmospheric CO_2 with Emphasis on Water Availability and Hydrology in the United States.* U.S. Environ. Prot. Agency, Washington, DC, 1984.
2. *Design and Construction of Sanitary and Storm Sewers,* ASCE—WPCF Manual No. 9. Am. Soc. Civ. Eng., New York, 1982.
3. F. E. McJunkin, *J. San. Eng. Div. (Am. Soc. Civ. Eng.)* **96,** 1183–1210 (1964).
4. J. C. Kammerer, *Estimated Demand of Water for Different Purposes,* Water for Human Consumption, Proc. 4th World Cong. IWRA. Tycooly Int., Dublin, 1982.
5. M. J. Hall, *Urban Hydrology.* Elsevier, Amsterdam, 1984.
6. L. H. Watkins, *The Design of Urban Sewer Systems,* Road Res. Tech. Pap. No. 55. H M Stationary Office, London, 1962.
7. B. Anderl, *Mitt. Inst. Wasserbau* (Karlsruhe) **3** (7), (1975).
8. M. L. Terstriep and J. B. Stall, *J. Hydraulic. Div.* (Am. Soc. Civ. Eng.) **95** (HY6), 1909–1934 (1969).
9. *Urban Stormwater Management,* Rep. No. 49. Am. Public Works Assoc., Chicago, IL, 1981.
10. W. Viessman, Jr., J. W. Knapp, G. L. Lewis, and T. E. Harbaugh, *Introduction to Hydrology.* IEP-A Dun-Donnelley Publ., New York, 1977.
11. M. L. Terstriep and J. B. Stall, *Ill. State. Water Surv. Bull.,* No. 58, (1974).
12. *National Engineering Handbook,* Suppl. A. Sect. 4. USDA, Soil Conserv. Service, Washington, DC, 1968.
13. R. H. McCuen, *A Guide to Hydrologic Analysis Using SCS Methods.* Prentice-Hall, Englewood Cliffs, NJ, 1982.
14. R. K. Price in *Urban Stormwater: Hydraulics and Hydrology* (B. C. Yen, ed.), pp. 184–192. Water Res. Publ., Littleton, CO, 1981.
15. P. Jacobsen, P. Herremoës, and C. Jakobsen, *The Danish Storm Water Modeling Package: The SVK-System,* Proc. 3rd Int. Conf. Urban Drainage, pp. 453–462. IAHR-IAWPRC, Göteborg, Sweden, 1984.
16. Metcalf & Eddy, Inc., University of Florida, and Water Resources Eng., Inc., *Storm Water Management Model,* Rep. No. 11024DOC 07/71. U.S. Environ. Prot. Agency, Washington, DC, 1971.
17. W. C. Huber and J. P. Heaney in *Urban Storm Drainage* (B. C. Yen, ed.), p. 247, Water Res. Publ., Littleton, CO, 1981.
18. K. Maalel and W. C. Huber, *SWMM Calibration Using Continuous and Multiple Event Simulation,* Proc. 3rd Int. Conf. Urban Storm Drainage, pp. 595–604. IAHR-IAWPRC, Göteborg, Sweden, 1984.
19. P. Ackers and J. D. Pitt, *Segment Lined Tunnels: Field Scale Test to Determine Hydraulic Roughness.* Proc. 3rd Int. Conf. Urban Storm Drainage, pp. 147–156. IAHR-IAWPRC, Göteborg, Sweden, 1984.
20. K. Imhoff, *Wasser und Abwasser* **1,** 401 (1909).
21. W. O. Ree and V. J. Palmer, *U.S. Soil Conserv. Serv. Bull.* No. 967, (1949).
22. A. Marston, *Iowa State Univ. Eng. Exp. St. Bull.* No. 96, (1930).

CHAPTER 3

Urban Stormwater and Combined Sewer Overflow Management

3.1 POLLUTION BY RUNOFF AND COMBINED SEWER OVERFLOWS

Both combined sewer overflows and stormwater from separate sewers are significant sources of pollution. The average BOD_5 concentrations in the combined sewer overflows of U.S. cities are approximately one-half of that for raw sanitary sewage. Stormwater from separate sewers has average solids concentrations exceeding that of untreated sewage and BOD_5s similar to sewage receiving secondary treatment.[1,2] Urban stormwater also contains some toxic materials (lead, oil, and other hydrocarbons) in concentrations equaling or higher than those found in sanitary sewage. Bacterial contamination of combined sewer overflows is approximately one logarithmic unit less than that of raw sewage and even the bacterial content of urban stormwater runoff only is two to four orders of magnitude greater than the concentrations considered safe for swimming and other water contact recreation activities.

Furthermore, both combined sewer overflows and urban stormwater runoff must be considered in terms of their shock effects as a result of their intermittent nature. As shown in Section 2.1, the specific unit sanitary sewage flow has a magnitude of 1 liter/(s-ha). A storm with an intensity of 25 mm/hr (1 in./hr) from a 70% impervious urban watershed will result in a specific runoff of 0.7×25 mm/hr $\times$ 10 000 m^2/ha $\times$ 0.0001 m/mm $\times$ 1 000 liters/m^3 $\times$ 1/3 600 hr/s = 48.61 liters/(s-ha), which represents an almost 50 times larger volumetric and mass load to the receiving water bodies.

The frequency of combined sewer overflows is now regulated by the state governments of the United States and, generally, only 0.5–5 overflows in a year are allowed without treatment in most U.S. urban areas. Furthermore, in the United States overflows of stormwater from separate sewer systems will be regulated also. German practice considers the minimum specific unit flow as a criterion whereby runoff or overflow up to 15 liters/(s-ha) should be collected, stored, and conveyed for biological treatment without overflows. Recent German regulations require that 90% of all pollution load from urban areas, including overflows, should receive treatment.

Example: Compare volume and pollution potential generated by sewage and by a 50-mm (2-in.) 24-hr storm from an urban 70% impervious catchment. Assume the BOD_5 concentration in raw sewage is 200 mg/liter (mg/liter = g/m^3), 85% BOD_5 removal in the treatment plant (effluent BOD_5 = 30 mg/liter), and the BOD_5 concentration of stormwater is similar to that of the effluent. The population density is 100 persons/ha and the per capita sewage flow is 300 liters/(cap-day).

Sewage

$$
\begin{aligned}
\text{Flow} &= 300 \text{ liters/(cap-day)} \times 100 \text{ persons/ha} \\
&= 30\,000 \text{ liters/(ha-day)} \\
&= 30 \text{ m}^3\text{/(ha-day)} \\
\text{Raw sewage BOD}_5 &= 30 \text{ m}^3\text{/(ha-day)} \times 200 \text{ g of BOD}_5\text{/liter} \\
&= 6000 \text{ g/(ha-day)} \\
\text{Treated sewage BOD}_5 &= (1 - 0.85) \times 6\,000 = 900 \text{ g/(ha-day)}
\end{aligned}
$$

Stormwater

$$
\begin{aligned}
\text{Flow} &= 0.7 \times 50 \text{ mm/day} \times 10\,000 \text{ m}^2\text{/ha} \\
&\quad \times 0.001 \text{ m/mm} \times 1000 \text{ liters/m}^3 \\
&= 350\,000 \text{ liters/(ha-day)} = 350 \text{ m}^3\text{/(ha-day)} \\
\text{Stormwater BOD}_5 \text{ load} &= 350 \text{ m}^3\text{/(ha-day)} \times 30 \text{ g/m}^3 \\
&= 10\,500 \text{ g/(ha-day)}
\end{aligned}
$$

The stormwater BOD_5 load from this storm would be almost twice as large as the 24-hr load from raw sewage.

The pollution in urban stormwater and most of the pollution in combined sewer overflows originate from so-called nonpoint or diffuse sources. In contrast to point source pollution, such as a treatment plant outfall, these sources are difficult to locate and quantify. Furthermore, diffuse pollution is a hydrologic process that closely follows the statistic character of rainfalls and must be evaluated in a similar fashion.

The sources of urban diffuse pollution are numerous and can be classified as follows[2-4]:

- Wet atmospheric deposition is closely related to the levels of atmospheric pollution by traffic, industrial and domestic heating, and other sources. Urban rainfall is generally acidic with pH values below 5 pH units. The elevated acidity of urban precipitation damages pavements, sewers, and other building materials.

- Dry atmospheric deposition is a source of fine particles that may originate from distant (fugitive dust) or local (traffic on unpaved roads, construction, and industrial) sources.
- Litter deposition is a source of large-sized materials (greater that 3.2 mm) that contain items such as cans, broken glass, vegetation residues, and pet wastes. Pet fecal deposits can reach alarming proportions in urban centers in which a large number of people reside in highly impervious residential zones.
- Medium-sized deposits (street dirt) represent the bulk of the street surface accumulated pollution. The sources are numerous and very difficult to identify and control. They may include traffic, road deterioration, vegetation residues, pets and other animal wastes and residues, and decomposed litter.
- Traffic emissions are responsible for some potentially toxic pollutants found in urban runoff, including lead, chromium, asbestos, copper, hydrocarbons, phosphorus, zinc, and nickel.
- Road deicing salts that are applied in winter to maintain streets passable for traffic cause highly increased concentrations of salts in the urban runoff. Road salts are applied at rates 100–300 kg/km of highway and contain sodium and calcium chloride base. Some deicing salt mixtures used to contain several anticorrosion additives (cyanides, chromates, and phosphates). However, many locales have prohibited the use of such materials.
- Erosion of open urban lands. Erosion of urban lawns and park surfaces is usually very low. Exceptions are open, unused lands, and, above all, construction sites. Soil is a source of suspended solids, organics, and pesticide pollution. The soil loss by erosion depends on the energy of the rainfall, soil erodibility, slope of the surface, length of the eroding overland flow, and, to the greatest degree, on the soil surface cover and protection.

Erosion from construction areas represents the largest source of solids in the urban runoff. Reported unit loads of suspended solids from urban construction sites ranged from 12 to 500 tons/ha-year.[2]

3.2 PRACTICES FOR CONTROLLING URBAN RUNOFF

Urban runoff control measures can be divided into several categories. They can be either structural, nonstructural, or a combination thereof. Runoff (quantity and quality) control measures can also be categorized as to where the measure is implemented, namely into on-site source control and land management, hydrologic modifications, collection and drainage systems control, and end-of-pipe final control.[2,5–12]

Reduction of Peak Flow Rates and Volumes of Storm and Combined Sewer Overflows into Receiving Waters

Reduction of peak flow rates of stormwater and combined sewer overflows can be accomplished by a number of methods, including

- On-site infiltration by letting the runoff overflow on adjacent pervious lands and pervious pavements.
- On-site detention before runoff enters the sewer systems. These techniques rely on dry storage basins, inlet flow restriction, shallow storage on parking lots, rooftop storage on flat roofs, and retention ponds.
- In-line or off-line storage of peak runoff flows or combined sewer overflows in oversized sewers (tunnels) or groundlevel or underground storage basins. If the storage concept is used for controlling combined sewer overflows it must be followed by treatment or the capacity of the dry weather sewage treatment plant must be increased to accommodate the wet weather flows stored in the basin. The available storage capacity of sewers should be used as much as possible. Some sewer segments are oversized and could be used after minimal adaptation (e.g., by inflatable or collapsible weirs and gates). Computer-controlled in-line storage systems have been installed in Seattle, Washington and other major cities.

Stormwater Retention Basins

These basins are installed before stormwater enters the sewer or drainage systems and their primary objective is to prevent large peak flows from high-intensity rainfalls from entering the drainage sewers. This is accomplished by first storing a part of the stormwater runoff in the basin and subsequently discharging it when the runoff subsides in a connected sewer, drainage channel, or pumping station. The peak flow is thus reduced by prolonging the flow time and not by overflowing it into an adjacent receiving water body.

Stormwater retention basins should be distinguished from combined sewer in-line or off-line storage. Such basins should be located near receiving water bodies since they are often designed to store only the highly polluted ''first flush'' of the overflow while the larger tail portion is allowed to overflow into the receiving water body. In Milwaukee, Wisconsin, Chicago, Illinois, and other cities, the underground storage is designed to store and subsequently treat most of the combined sewer overflows.

Economic considerations and urban flood protection concerns are the major reasons for the use of retention ponds. Recently, their water quality benefits have come into the picture.[13] Only wet ponds (ponds that hold some minimum water volume permanently) are effective for pollution control. Dry ponds can be modified to include a filtration sand and gravel bed at the outlet to filter out solids during low and medium flows when no detention is provided (Fig. 3.1).

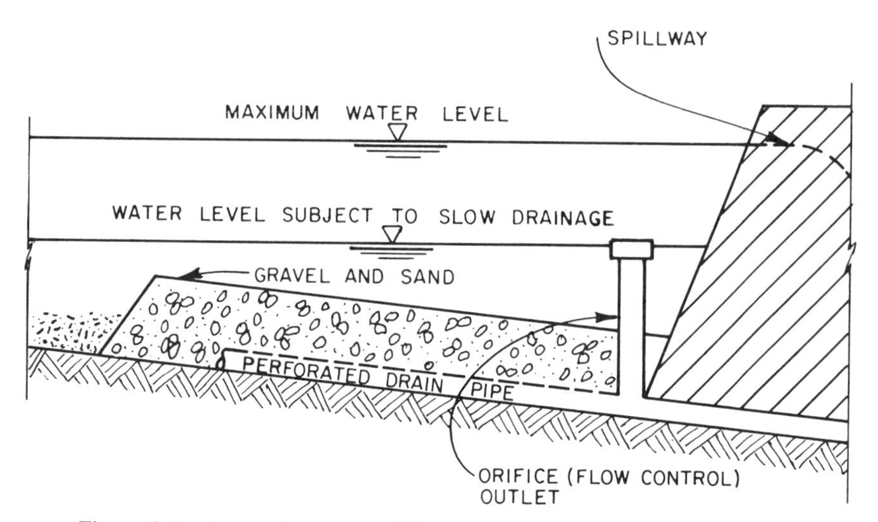

Figure 3.1 Dry pond outlet modification for stormwater pollution control.

The cost of construction and land for the pond should be less than the savings on the downstream drainage that can remain smaller and does not have to be expanded to accommodate the new flows. Furthermore, flood protection benefits must be considered as well. Situations favoring the use of detention basins would arise:

- in new sewer systems in which large storm water volumes are to be discharged into longer intercepting sewers;
- when a new development is added to an existing sewer system that is already loaded to its full capacity;
- when floods and sewer backups could occur as a result of increased imperviousness of new development or redevelopment of older urban sections;
- as a part of sewer rehabilitation;
- when large volumes of storm runoff are discharged into a smaller stream whereby the use of the retention pond would minimize or avoid channelization and/or expensive lining of the stream.

For design, the magnitudes and time course of the influent and effluent hydrographs of the basin are first determined and the storage requirement is generally estimated by a computer mass balance (continuity) model or graphically. However, in the case of retention pond design it is usually not known which design storm will require the largest retention storage volume. Also, the return (recurrence) period of the design storm is not the same as the recurrence period of the runoff resulting from these storms. Therefore, dimensioning must be carried out for several design storms with different intensities and durations and with different antecedent soil moisture conditions.

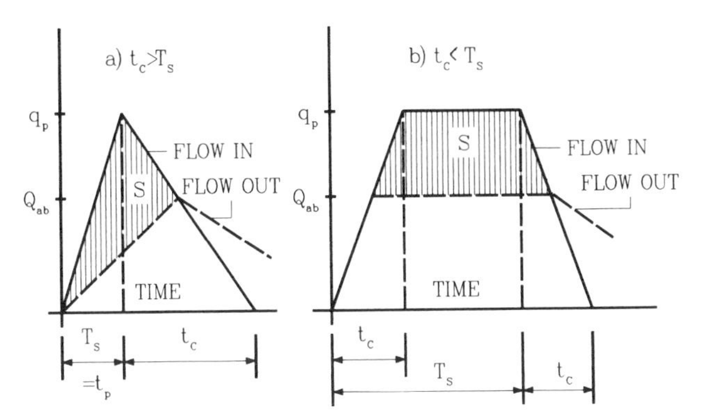

Figure 3.2 Determination of storage volume for stormwater detention (retention) basins.

The size of the ponds can be estimated using approximate methods with sufficient accuracy. A design storm runoff can be represented by a triangular (storm duration is less than the time of concentration) or trapezoidal (storm duration > time of concentration) inflow hydrograph (Fig. 3.2). For a simple triangular hydrograph the needed storage can be found from the relationship

$$S = 0.5(T_s + t_c)(q_p - Q_{ab})$$

where S = storage volume,
T_s = duration of the rainfall,
q_p = peak inflow rate, and
Q_{ab} = peak discharge (maximal outflow) rate from the pond.

This simple method was ranked by Boyd[14] as superior to several other approximate procedures. The trapezoidal inflow hydrograph assumption would lead to a formula

$$S = q_p T_s - Q_{ab}(T_s + t_c) + (Q_a b)^2 \, t_c / q_p$$

The procedure of Annen and Londong[15] is based on mathematically formulated relationships for the inflow, rainfall duration, flow time, and retardance coefficients for different contributing catchments, and on the computation of the maxima of the differential forms of these functions. The dimensioning is carried out with the aid of Figure 3.3 in the following steps[16]:

With the values

A_{red} = connected impervious area of the catchment in ha.
t_f = computed flow time of water in the sewer to the detention pond in minutes (time of concentration).
r_{15} = rainfall intensity of a design 15-min rainfall with the selected frequency (n) in liters/s-ha. If the U.S. weather service isopluvial map (Fig. 2.2)

is used then substitute $r_{15} = 6.5I$, where I is the rainfall intensity for a 1-hr rainfall in mm/hr.

Q_{ab} = outflow rate from the retention basin (for combined sewer systems without the dry weather flow).
Q_{ab} is entered as an arithmetic mean of the outflow rate at the beginning of the storage, min Q_{ab} (full capacity of the outlow at free surface flow), and the outflow rate at the highest water level in the basin, max Q_{ab}:

$$Q_{ab} = 0.5(\min Q_{ab} + \max Q_{ab})$$

Compute the normalized runoff rate

$$Q_{r15} = r_{15} A_{red}$$

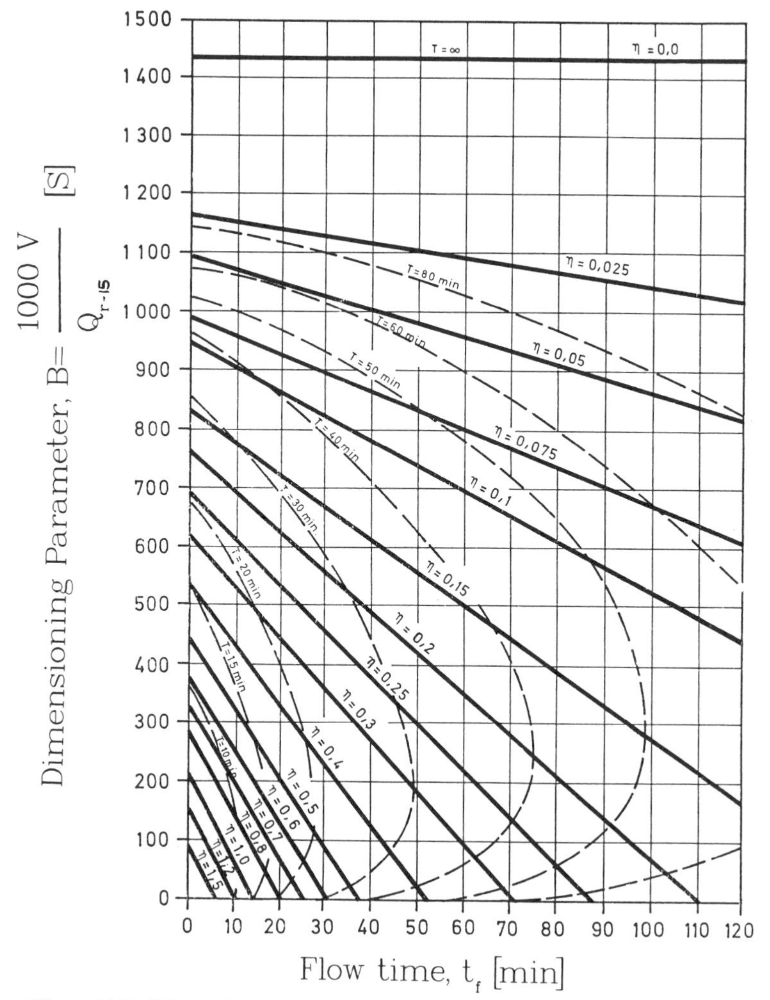

Figure 3.3 Dimensioning parameter, B, for stormwater retention basins.

and the detention coefficient

$$\eta = Q_{ab}/Q_{r15}$$

Using η and the critical flow time, t_f, the dimensioning parameter, B, can be read from Figure 3.3. Then the basin volume is

$$V = B \frac{Q_{r15}}{1\ 000} \qquad (m^3)$$

When selecting the overflow frequency, n, it should be remembered that the statistical probability of the rainfall intensity is not the same as the probability or frequency of the overflow in the sewer network or in the retention pond. Thus, for dimensioning of retention ponds, a somewhat smaller design rainfall frequency than that used for design of sewers should be selected. For enclosed or covered retention basins, use $n = 0.5$–0.2 (return period of the overflow 2–5 years), for open basins select $n = 0.1$ if local conditions do not require a higher safety factor.

Verworn[16] found that the ATV simplified procedure provides realistic results when the detention coefficient $\eta > 0.4$. Basins with a lower detention coefficient have a longer storage-emptying time and storms causing overflows may occur when part of the storage volume is still full from a preceding storm, which increases the overflow probability.

For multiple basins, the effectiveness varies greatly with the timing of the storms and the location of the basins.[7]

Example: A 25-ha urban area with 35% connected impervious area and flow time in the main interceptor of $t_f = 20$ min is drained by a combined sewer system. The dry weather flow is 10 liters/s. The area is to be connected to an existing sewer system that can accept only 340 liters/s as a maximum. The basin should be designed for an overflow frequency of $n = 0.5$ (one overflow in 2 years). For this frequency the design 15-min rainfall intensity is $r_{15} = 130$ liters/s-ha. Then $Q_{ab} = 340 - 10 = 330$ liters/s, $Q_{r15} = 130 \times 25 \times 0.35 = 1\ 138$ liters/s, and the detention coefficient $n = 330/1138 = 0.29$. From Figure 3.3, $B = 460$ s. The basin volume must be at least $V = 460 \times 1\ 138/1\ 000 = 523\ m^3$.

Example: Design a stormwater detention facility for a 4-ha portion of a shopping center parking lot such that the peak outflow will not exceed $0.7\ m^3/s$ in a 10-year storm. Water depth in the storage portion of the parking lot may not exceed 25 cm depth. The width of the basin should be 50 m. The inflow hydrograph had a peak time of 10 min which is less than the time of concentration of 20 minutes. The peak flow is $1.6\ m^3/s$.

The storage requirement can be computed using the triangular hydrograph concept (Fig. 3.2a) and assuming the $T_s = t_p$. Then

$$S = 0.5 \times (10\ \text{min} + 20\ \text{min}) \times (1.6\ m^3/s - 0.7\ m^3/s) \times 60\ s/\text{min}$$
$$= 810\ m^3.$$

The containing basin must be 64.8 m long to hold this quantity at a 0.25 m depth with a width of 50 m.

In-Line and Off-Line Storage of Combined Sewer Overflows

In-line and off-line storage facilities collect excess flow in combined sewer systems that would otherwise overflow untreated into receiving waters. An essential part of these facilities is a safety overflow that can be located either at the inflow or outlet section of the facility. The overflow is activated when the storage volume is full. In many cases, the storage is provided for collection of the "first flush" portion of the overflow. The sewage–runoff mixture collected in these facilities must be subsequently treated either in an expanded sewage treatment plant or in a satellite facility treating the overflows only.

For dimensioning determine first[17]

$$r_{ab} = \frac{Q_{ab}}{A_{red}} \qquad \text{(liters/s-ha)}$$

where Q_{ab} is the allowed flow from the storage facility downstream in the sewer network or toward the treatment plant (not including the dry weather sewage flow). The computation of Q_{ab} was presented in the preceding section. Using r_{ab} and the rainfall intensity at which the overflow is initiated, r_{crit} (see Section 2.1 for estimation of r_{crit}) the dimensioning volume, V_{sr}, can be then read from Figure 3.4. The required useful storage volume is then

$$V = V_{sr} a A_{red} \qquad (\text{m}^3)$$

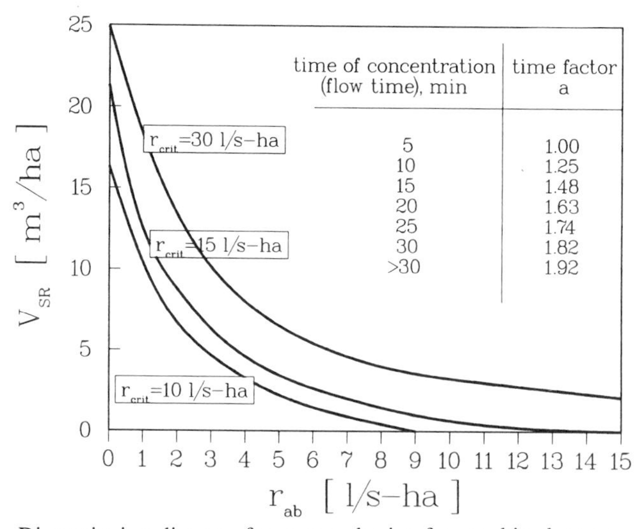

Figure 3.4 Dimensioning diagram for storage basins for combined sewer overflows for "first flush" control.

The time factor, a, depends on the time of concentration of the design rainfall to the storage facility. It can be estimated from the table in Figure 3.4. Depending on local conditions, the overflow storage basins can be installed as dry off-line or in-line flowthrough type (Fig. 3.5a,b). The off-line basins collect the "first flush" highly polluted part of the overflow during short rainfalls or during the first part of the rain. At longer rainfalls after the basin is filled, the safety overflow of the basin, BO, located on the inflow or upstream is turned on and the rest of the overflow is then discharged into the adjacent receiving water. Such basins are advantageous if the sewer system is not overloaded and the time of concentration is less than 15 min.

In-line flowthrough basins can have an additional clarifying overflow weir (OW), over which, after the basin is full, the mechanically clarified mixture or runoff and sewage is discharged in the receiving water. First, at flows greater than Q_{crit}, the overflow located upstream from the basin is activated. For German conditions, the detention times in the basin at the Q_{crit} should not be less than

$$\text{at } r_{crit} = 30 \text{ liters/s-ha} \qquad 10 \text{ min}$$

$$\text{at } r_{crit} = 15 \text{ liters/s-ha} \qquad 17 \text{ min}$$

$$\text{at } r_{crit} = 10 \text{ liters/s-ha} \qquad 20 \text{ min}$$

Dry weather flow usually flows through the in-line storage facilities. In contrast, in the off-line storage facilities the flow mixture during dry weather or low-flow

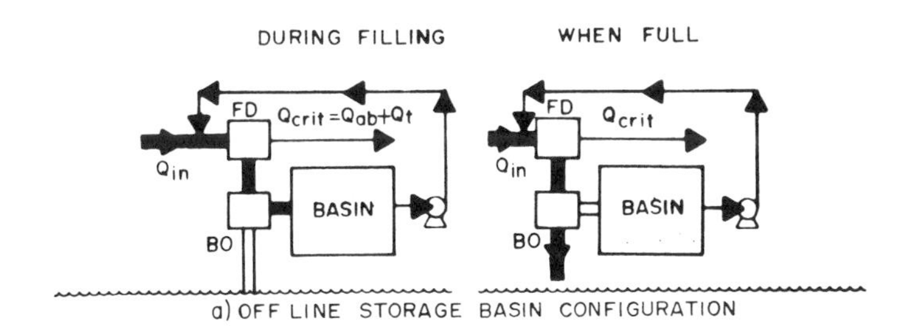

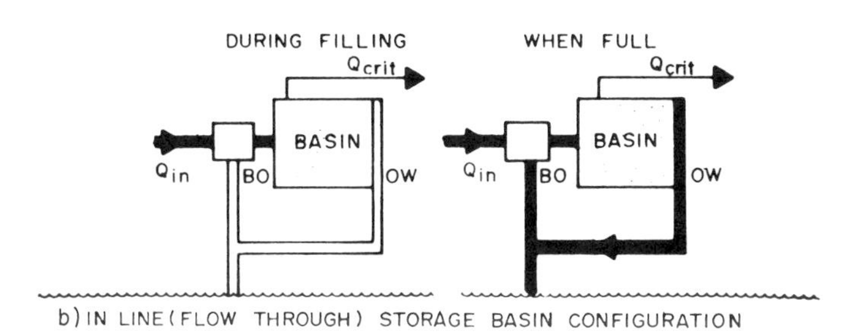

Figure 3.5 Schematics of in-line and off-line combined sewer overflow storage basins.

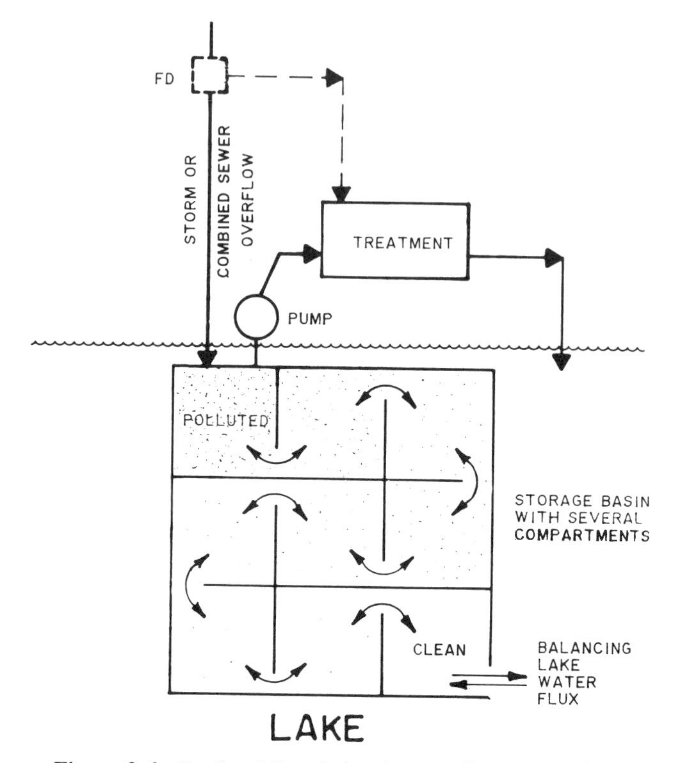

Figure 3.6 Dunkers' flow balancing overflow storage basin.

periods is diverted toward the treatment plant before the flow enters the basin. Therefore, a flow divider (FD) is necessary. According to slope conditions, the storage basins are emptied to the treatment plant either by gravity or by pumping.

The in-receiving water flow balancing method (Figure 3.6) has been developed in Sweden as an off-line storage alternative. In-receiving water storage facilities contain the combined sewer or stormwater overflow in the compartments made of plastic or structural curtains. After the overflow terminates, pumping starts automatically from the first compartment to convey polluted overflow to treatment and clean water from the surrounding water body will enter the last compartment and push the polluted water towards the pump. Thus the water body is used as a flow balance medium.[18]

Source Control Measures and Hydrologic Modifications

Source control measures include control of atmospheric deposition, particularly near the industrial zones and around construction sites, litter control programs, leaf and grass residue pickup, and street sweeping. Street sweeping has not been found to be an effective pollution control measure.[19]

Land management focuses on structural, semistructural, and nonstructural measures that reduce urban and construction site stormwater runoff volume and pollution before they enter the downstream drainage system. These measures include land use planning that permits the runoff to overflow on pervious grassed areas and controls development on tight soils, use of natural drainage, grassed buffer strips, and waterways, and use of natural and man-made retention.

Bare soils, particularly at construction sites, must be protected since otherwise, the soil loss could be considerable—more than 100 tons/(ha-year). Temporary or permanent seeding of grass, sodding, and mulching should be used on any unprotected soil surface within the watershed.

Hydrologic modification, which can be very effective for controlling both runoff quantity (volume and peak flow) and quality, implies a reduction of surface runoff. These measures can be divided into (1) practices that increase permeability and enhance infiltration such as the use of porous pavements and vegetation infiltration strips, (2) practices that will increase hydrologic storage, and (3) practices that will reduce directly connected impervious areas.

Porous pavements are made either from asphalt, in which the fine filler fractions are missing, or are modular or poured-in concrete pavements. These pavements exhibit very high permeability rates.[20] Brick, stone, or concrete tile pavements can be considered as semipermeable with a limited infiltration capacity ranging from 7 to 16 mm/hr.[21] Results from a study in Rochester, New York, indicated that peak runoff rates were reduced as much as 83% and the structural integrity of the porous pavements was not impaired by heavy load vehicles or by freezing. Periodic cleaning of porous pavement surface is necessary to prevent clogging.[22]

Available surface storage can be increased by storing rainwater on flat roofs, by restriction of stormwater inlets with a subsequent temporary ponding on the streets, parking lots, or in parks. As a water quality control measure in areas served by combined sewers, these practices are effective and also reduce requirements for in-line or off-line storage of the overflows. In areas served by storm sewers, they can control flooding and reduce requirements for detention basin volume, however, alone they do not have a significant water quality benefit.

Decreasing connected impervious areas can be accomplished by disconnecting roof drains from storm and combined sewers, by permitting surface runoff to overflow on adjacent pervious surfaces, and by using dry (french) wells and infiltration trenches into which the stormwater is directed. Since these measures rely on increased infiltration, the soils should have a sufficient permeability and the groundwater table should be low, basically well below the sewer line bottom elevation.[23]

Generally as a rule-of-thumb, the practices that reduce the runoff volume and pollution generation at the source are substantially cheaper than end-of-pipe final treatment and removal.

REFERENCES

1. R. Field and R. Turkeltaub, *J. Environ. Eng. Div. (Am. Soc. Civ. Eng.)* **107,** 83–100 (1981).
2. V. Novotny and G. Chesters, *Handbook of Nonpoint Pollution: Sources and Management.* Van Nostrand Reinhold, New York, 1981.

3. J. D. Sartor, G. B. Boyd, and F. J. Agardy, *J. Water Pollut. Control Fed.* **46,** 458–467 (1974).
4. P. A. Malmqvist, *Urban Stormwater Pollutant Sources,* Rep. Dept. Sanit. Eng. Chalmers University of Technology, Göteborg, Sweden, 1983.
5. R. Field, *The US EPA Office of Research and Development's View of Combined Sewer Overflow Control,* Proc. 3rd Int. Conf. Urban Storm Drainage, pp. 1333–1356. IAHR-IAWPRC, Göteborg, Sweden, 1984.
6. *Urban Stormwater Management,* Rep. No 49. Am. Public Works Assoc., Chicago, IL, 1981.
7. W. Whipple, N. C. Grigg, T. Grizzard, C. W. Randall, R. P. Shubinski, and L. C. Tucker, *Stormwater Management in Urbanizing Areas.* Prentice-Hall, Englewood Cliffs, NJ, 1983.
8. J. A. Lager, W. G. Smith, W. G. Lynard, R. M. Finn, and E. J. Finnemore, *Urban Stormwater Management and Technology: Update and User's Guide,* Rep. No 600/8-77-014. U.S. Environ. Prot. Agency, Cincinnati, OH, 1977.
9. M. P. Wanielista, *Stormwater Management: Quantity and Quality.* Ann Arbor Sci. Publ., Ann Arbor, MI, 1977.
10. M. J. Hall, *Urban Hydrology.* Elsevier, Amsterdam, 1984.
11. S. Fujita and R. Koyama, *Planning and Design of New Sewer Systems.* Kachima Publ. Assoc., Tokyo, 1984.
12. W. DeGroot (ed.), *Stormwater Detention Facilities.* Am. Soc. Civ. Eng., New York, 1982.
13. E. D. Driscoll, *Performance of Detention Basins for Control of Urban Runoff Quality,* Proc. Int. Symp. Urban Hydrol., Hydraul., Sediment Control. University of Kentucky, Lexington, 1983.
14. M. J. Boyd in *Urban Stormwater Quality, Management and Planning* (B. C. Yen, ed.), p. 370. Water. Res. Publ., Littleton, CO, 1982.
15. G. Annen and D. Londong, *Technische Wissenschaftliche Mitteilungen der Emschergenossenschaft und des Lippeverbandes* (*Technical and Scientific Communications of Emscher and Lippe River Agencies*), Essen, No. 3, 1960.
16. H. R. Verworn, *How Accurate are Simplified Detention Basin Procedures,* Proc. 3 Int. Conf. Urban Storm Drainage, pp. 1–9. IAHR-IAWPRC, Göteborg, Sweden, 1984.
17. *Guidelines for Dimensioning, Design and Operation of Stormwater Detention Basins,* ATV-(A117). Ges. Förderung Abwassertechnik, St. Augustin, FRG, 1977.
18. H. Söderlund, *Flow Balancing Method for Stormwater and Combined Sewer Overflows,* Proc. 3rd Int Conf. Urban Storm Drainage, pp. 57–66. IAHR-IAWPRC, Göteborg, Sweden, 1984.
19. U.S. Environmental Protection Agency, *Final Report of the National Urban Runoff Program.* USEPA, Washington, DC, 1983.
20. T. J. Jackson and R. M. Ragan, *J. Hydraul. Div.* (*Am. Soc. Civ. Eng.*) **12,** 1739–1752 (1974).
21. C. H. Van Dam and F. H. M. Van de Ven, *Infiltration in the Pavement,* Proc. 3rd Int. Conf. Urban Storm Drainage. IAHR-IAWPRC, Göteborg, Sweden, 1984.
22. R. Field, H. Masters, and M. Singer. *Wat. Res. Bull.* **18,** 264–270 (1982).
23. S. Fujita, *Experimental Sewer System for Reduction of Urban Storm Runoff,* Proc. 3rd, Int. Conf. Urban Storm Drainage. IAHR-IAWPRC, Göteborg, Sweden, 1984.

ADDITIONAL BIBLIOGRAPHY FOR PART I

J. Brix, K. Imhoff, and R. Weldert, *Die Stadentwässerung in Deutschland* (*Urban Drainage in Germany*). Fischer, Jena, 1934.

R. E. Featherstone and A. James (eds.), *Urban Drainage Systems.* Pitman, London, 1982.

W. Gujer and V. Krejci (eds.), *Topics in Urban Storm Water Quality, Planning and Management,* Proc. 4th Int. Conf. Urban Storm Drainage. IAWPRC & IAHR, Ecole Polytechnique Federale, Lausanne, Switzerland, 1987.

D. Stephenson, *Stormwater Hydrology and Drainage.* Elsevier, Amsterdam, 1981.

T. H. Y. Tebbutt (ed.), *Advances in Water Engineering.* Elsevier, London, 1985.

H. C. Torno, J. Marshalek, and M. Desbordes, *Urban Runoff Pollution.* Springer-Verlag, Berlin and New York, 1986.

B. C. Yen (ed.), *Topics in Urban Drainage Hydraulics and Hydrology,* Proc. 4th Interntl. Conf. Urban Storm Drainage. IAWPRC & IAHR, Ecole Polytechnique Federale, Lausanne, Switzerland, 1987.

PART II

SEWAGE AND WASTEWATER TREATMENT

CHAPTER 4

General Considerations

4.1 OVERVIEW OF THE PROBLEM

Planning. All municipal sewage and wastewater must eventually find their way into water courses or other bodies of water that constitute the natural drainage of the region. This can cause damages, including contamination and pollution of water supplies, swimming and bathing beaches, shellfish contamination, fishkills, creation of conditions offensive to sight or smell, and impairment of the use of natural waters for recreation, agriculture, commerce, or industry. The primary objective of sewage and wastewater treatment is to prevent such damages and injuries to receiving waters.

However, this problem is not solved by conveying the wastewater to the nearest discharge point nor it is solved by treating the waste to any arbitrary effluent quality. The treatment plants are an integral part of the entire sewerage system. Thus, planning, development, and operation of sewage treatment plants depend upon the character of the collecting system and upon the means for ultimate disposal of the wastewaters. In choosing the treatment process to be employed and in determining the degree of treatment, consideration must be given to the location of the plant, the nature of its present and anticipated future surroundings, the character and waste assimilative capacity of the available receiving water body, and the uses to which these waters are to be put.

It should also be remembered that sewage treatment is but one of many means for developing the full economy of regional water resources. Although this aspect of wastewater disposal is of special concern to state and regional water authorities and/or interstate commissions whose jurisdiction extends over the entire drainage basin, the engineer who designs a treatment plant for a single community or industry can best serve his client if he is familiar with the regional water development and wastewater disposal plans.

Engineers in general, and civil and environmental engineers in particular, should never forget that they must master not only the technical details but also the economic and social implications of their planning. To this end they must keep themselves free to seek the most economical and objective solution to all problems that are presented to them. They must also remember that protection of the environment and, hence, of water quality is a national and social objective. A book by Gunn and Vesilind[1] provides an insightful discussion of the ethical problems encountered in the practice of engineering.

Regulations. In the United States, the Water Pollution Control Act Amendments—passed by the U.S. Congress in 1972 with subsequent amendments—require that all public wastewater treatment and disposal plants must be in compliance with a regional (areawide) pollution abatement plan developed by a state-designated planning agency. In addition to the provisions dealing with regional wastewater disposal planning, this far-reaching act also requires that all municipal treatment plants discharging into navigable waters must have secondary (biological) treatment.

Also the effluent quality and, hence, the degree of treatment for industrial discharges is regulated by a permit system referred to as the National Pollutant Discharge Elimination System (NPDES). This system mandates that a wastewater discharger receives an NPDES permit. The permit is not necessarily related to, nor does it depend upon, the ability of the receiving water body to accept the waste discharge without harm to aquatic life or without an impairment of the beneficial uses of the water body. Instead, it is related to a specified treatment technology that would be expected to be installed in a well-functioning treatment plant. In addition, the permit specifies a compliance schedule that must be met and requires compliance with other relevant state and local pollution control laws. In situations in which the NPDES permit is based on treatment technology requirements and does not result in acceptable water quality, more stringent requirements may be issued by the regulatory agencies that would then be based on the limits imposed by the receiving water.

Institutions and Organizations. Both in the United States and in Germany, a number of regional water and wastewater disposal authorities have been established that are responsible for regional wastewater treatment and disposal and, to a lesser degree, for water quality management. An example of an integrated approach can be found in the Ruhr area of Germany in which several River Management Associations (Verbände and Vereine) have been established between 1910 and 1940.[2–4] The need for a drainage-areawide basin authority is particularly evident when the watershed is crossed by political boundaries, either national or international.

In the United States, many metropolitan municipalities have established sewerage districts. Such districts have been created by state legislatures and have, to some degree, a regional character.

4.2 METHODS OF SEWAGE TREATMENT

Treatment Processes

Sewage contains mineral and organic matter (1) in suspension (coarse and fine suspended matter), (2) in the colloidal state (very finely dispersed matter), and (3) in solution. Living organisms, notably bacteria and protozoa, find an abundant source of energy in the organic constituents of sewage. The foraging activities of these organisms result in the decomposition of the organic matter that can either be uncontrolled and offensive or controlled.

Three basic categories of treatment processes are employed: mechanical (physical), chemical, and biological.

Mechanical and Physical-Chemical Unit Processes. These include bar racks and screens, filters, sedimentation basins, and flotation. All mechanical unit processes can remove only particulates, including the suspended solids.

When the solids are dissolved in water they may aggregate (precipitate) to form larger flocs that can then be removed mechanically as suspended solids, or they may be broken down into submolecular or atomic particles (ions) that may then combine with other constituents to become harmless or escape as gas.

Between the dissolved and particulate solids are the colloids. Under the influence of physical, chemical, or electrochemical forces, the colloids can either flocculate to form larger suspended particles or can be broken down and become dissolved. A boiled egg white is an example of a flocculated colloid. It can be dissolved in the stomach by enzymes. Similarly, wastewater colloids can flocculate or be dissolved by bacterial enzymes.

Almost all sewage constituents are unstable and rapidly decay. The change of composition of the sewage constituents is sometimes purely chemical, such as the combination of acids and bases to form salts. In such cases chemical treatment units are important. In the chemical units, a coagulant, for example, ferric chloride, can be added that combines with other colloidal or dispersed wastewater constituents to form heavier flocs that subsequently settle to the bottom of the settling units. Preparation of sludges for dewatering can also be considered as a chemical treatment step. Here the colloidal matter flocculates, for example, with the aid of ferric chloride, and the water from the sludge is then more easily removed by filtration.

Biological Treatment

Biological Decomposition Processes. Separated from these purely chemical unit operations are a number of treatment steps in which treatment is accomplished by living microorganisms. Such unit operations are biological or biochemical.

The organic components are composed of carbon combined with other elements (hydrogen, oxygen, nitrogen, phosphorus, sulfur, and others). However, the organic constituents in sewage are not pure organic compounds in a chemical sense, they are complicated decomposable residues of living processes of humans, animals, and plants. From these constituents the most important are urea and proteins, which contain nitrogen in addition to carbon. Some proteins also contain sulfur (e.g., egg white) and, as a result of decomposition, produce offensively smelling hydrogen sulfide gas.

The type of process that breaks down the biodegradable wastewater constituents is related to the presence or absence of oxygen. Chemically, the waste components can handle oxygen in two different ways: they can gain oxygen and/or lose an electron (oxidation) or they can lose oxygen and/or gain an electron (reduction). As an example, aerobic biological treatment of sewage is primarily an oxidation process whereas anaerobic sludge decomposition is commonly a reduction process.

The organic components in sewage are mostly residues of decay reduction processes of plant and animal matter. The microorganisms and oxygen have the task of converting these rather unstable compounds into more stable, oxidized ones. The ultimate products of biochemical activities are gaseous carbon dioxide, mineral residues, and relatively stable (humus-like) organic materials.

Bacteria, particularly the aerobic (oxygen-requiring) species, are the means of executing these biochemical reactions, as long as free oxygen from the air or water is available. When the sources of oxygen are exhausted, the aerobic bacteria are replaced by anaerobic or facultative microorganisms that draw upon the oxygen contained in organic matter and in substances such as nitrates, nitrites, and sulfates. Aerobic decomposition is thereby supplanted by anaerobic decomposition; the sewage first becomes stale (dissolved oxygen just exhausted), and eventually septic (dissolved oxygen exhausted for some time and anaerobic decomposition well under way).

The decomposing bacteria are small cellular microorganisms that multiply by division of the cells. In a nutrient-rich environment, they can build colonies that are then visible to the eye. In sewage treatment units, these may be seen as suspended slimy flocs in the activated sludge units or slime on the media of trickling filters or on rocks and stones of land disposal–sewage irrigation fields. Higher organisms such as protozoa and (when oxygen has direct access) lower animal forms can live symbiotically with the bacteria.

The bacteria are sensitive to acids and bases in about the same degree. Therefore, the pH value of water should not substantially deviate from 7 and fluctuations and pH shocks should be avoided. The bacteria are also accustomed to a certain temperature range. In dry conditions the bacteria will perish. Disinfection additives such as chlorine incapacitate or kill them. The exchange of nutrients and waste materials during the living process occurs through the bacterial cell wall, therefore only water, gases, and dissolved materials can be exchanged. Solids and colloids must first be dissolved. This is accomplished by enzymes that are emitted through the walls of the cells. In a similar way, during the decomposition process, the used and decomposed residuals are released by the bacteria through their cellular walls as liquids or gases. When the surrounding liquid has less nutrients than the slime inside and around the bacteria, the cells become full. When the surrounding liquid layer thickens, for example, as a result of intake of salts or sugar, the cell volume shrinks and the life activity becomes inhibited. In a seawater environment the salt concentrations are still below the inhibiting levels.

The exchange of matter by bacteria is continuous. It works best when the nutrients are in motion or flowing, such as in trickling filters, digestion chambers, or experimental BOD or respirometer bottles.

Aerobic Treatment. What is known as "biological wastewater treatment" in almost all cases refers to aerobic treatment steps or to treatment steps in aerated water. This is the backbone of treatment; however, the removal of organic pollutants from wastewater cannot be attributed only to bacterial decomposition. Pollution removal is often accomplished by absorption or surface interaction of the pollutants with the slimy matter in which the bacteria live or are made of. This process, however, will quickly reach a standstill if the absorbed pollutants are not consumed by the bacteria. In order to oxidize and decompose these absorbed pollutants, bacteria require oxygen, which must continuously be supplied from the air and dissolved in the water.

The most important products of biochemical oxidation by bacteria are acid-forming oxides such as carbon dioxide (CO_2), nitric oxide (N_2O_5), and sulfur trioxide (SO_3). Since wastewaters usually contain enough alkaline compounds, these oxides are usually converted to dissolved salts (carbonates, nitrates, or sulfates) with an excess of CO_2 remaining as dissolved gas or escaping. Thereafter, the removal of pollutants in aerated water is completed.

Anaerobic Processes. The anaerobic and facultative bacteria that function without free oxygen are primarily used for pretreatment of high strength organic wastes and are also very important in the anaerobic digestion of sludges. They work best at higher temperatures (30° to 50°C) and decompose the organic matter at a substantially slower rate than that typical for aerobic treatment. However, the decomposition of the organics into gaseous byproducts (carbon dioxide and methane gases) is more complete, hence, less sludge is produced.

There are two types of anaerobic decomposition:[5] acid fermentation and methane fermentation. Both are bacterial reduction processes. Under normal conditions the acid fermentation sets in first followed by methane fermentation.

In acid fermentation, the microbes first attack the easily available food materials: sugars, starches, cellulose, and soluble nitrogen compounds. The products of decomposition are organic acids, acid carbonates, and gas consisting primarily of carbon dioxide and, in much smaller amounts, hydrogen, methane, and hydrogen sulfide. Since the oxygen is obtained by breaking up water molecules, the hydrogen ion and subsequent acid formation reduces the pH to 6 or lower. In addition, acids are formed by breaking grease molecules into organic fatty acids. As the decomposition in this stage depends only on carbon sources and not on nitrogen, the process resembles, to a certain degree, removal of organics in aerated water or a course of BOD degradation.

Acid fermentation is followed by a lengthy period of slow, acid regression, during which organic acids and nitrogen compounds are attacked and ammonia compounds and acid carbonates are formed. As the acidity recedes, alkaline or methane fermentation takes over, and the more resistant materials, including proteins and organic acids, are decomposed. Large volumes of gas, consisting mainly of methane and carbon dioxide, are released.

The ultimate products of sludge digestion (fermentation) are (1) more or less stable humus-like solid matter, (2) sludge liquor, including liquified and finely divided (more or less colloidal) solid matter, and (3) sludge gases. Further processing of dewatered sludge is by aerobic bacteria. Under moist conditions and good aeration sludge can be converted to organic compost.

4.3 WASTE TREATMENT PROCESSES AND THEIR EFFICIENCY

The pollution content of sewage consists of organic and inorganic compounds, which exist partially as suspended solids (settleable, suspended, or floating) that move with the water, and partially as dissolved solids. There are also small living organisms, primarily bacteria and protozoa that find nourishment in the sewage organic matter. The bacteria can cause uncontrolled decomposition of sewage, resulting in offensive odor problems and an unsightly appearance. There is also a potential for the presence of disease-producing organisms.

The removal or stabilization of sewage materials is accomplished in treatment works by a number of different operations:

1. Separation of coarse suspended and floating matter:
   - screening and bar racks
2. Separation of coarse settleable solids:
   - grit chambers
3. Separation of grease and oil and similar floating materials:
   - grease and fat traps
   - skimming (flotation) tanks
   - settling tanks with scum collectors
4. Separation of fine suspended solids:
   - settling tanks and clarifiers
   - skimming (flotation) tanks
   - chemical flocculation and coagulation
   - sand filtration
   - septic tanks
5. Separation of dissolved, colloidal, and very fine organic matter—biological treatment, namely:
   - land disposal by overland flow systems
   - slow rate irrigation with agricultural benefits
   - rapid soil infiltration
   - trickling filters
   - rotating biological contact units
   - oxidation ponds and lagoons
   - activated sludge units
   - septic tanks
6. Disinfection and odor control:
   - chlorination and other chemical means

The preceding unit operations can be categorized into the following groups of treatment processes:

1. Mechanical processes:
   a. Screening: Because of their size the particles will be trapped on screens or bar racks. Sand filters work predominantly in the same fashion.
   b. Flotation units: As a result of their buoyancy pollutants will rise to the surface where they can be skimmed off, for example, as in grease traps and skimming tanks.
   c. Sedimentation tanks and clarifiers: Particles sink to the bottom of the tanks because of gravity.
2. Chemical processes: chemical additives enhance the settleability of particles or kill microorganisms (chlorine).
3. Biological processes: The life processes of nature are used in sewage purification:
   a. Naturally in the soil or ponds and lagoons
   b. Artificially on trickling filters, activated sludge units, or septic tanks.

Individual treatment units can be compared according to their removal efficiency, which gives the ratio of the pollutant content in the effluent from the unit to that of raw waste or influent as a percentage. As a measure of the pollution content one can use the biochemical oxygen demand (BOD), the suspended solids content, or the bacteria count. The average efficiencies are given in Table 4.1.

TABLE 4.1 Typical Efficiencies of Waste Treatment Unit Operations

Treatment Operation or Process	Removal Efficiency (%)[a]		
	Biochemical Oxygen Demand[b]	Suspended Solids	Bacteria
1. Fine screens	5 to 10	2 to 20	10 to 20
2. Chlorination of raw or settled sewage	15 to 30	NA[c]	90 to 95
3. Plain sedimentation	25 to 40	40 to 70	25 to 75
4. Chemical precipitation	50 to 85	70 to 90	40 to 80
5. High-rate trickling filters[d]	65 to 90	65 to 92	40 to 80
6. Low-rate trickling filters[d]	80 to 95	70 to 92	90 to 95
7. High-rate activated sludge process[d]	50 to 75	65 to 95	70 to 90
8. Conventional (low-rate) activated sludge process[d]	85 to 95	85 to 95	90 to 98
9. Sand filter	90 to 95	85 to 95	95 to 98
10. Chlorination of biologically treated sewage	NA	NA	98 to 99

[a] Percentage removal or efficiency = 100 × (influent concentration − effluent concentration)/influent concentration.
[b] Five day, 20°C.
[c] NA, no or negligible effect.
[d] Followed by secondary clarification.

Discrepancies between these values and recorded results are to be expected. Efficiencies drop when treatment plants are overloaded and some of the sewage flow bypasses a treatment unit, or when the treatment plant has inadequate sludge disposal capacity where part of the sludge must be pumped or bypassed into the receiving waters. Also the sludge liquor discharged from sludge digestion tanks, mechanical sludge dewatering equipment, sludge thickening tanks, or inefficient drying beds generally has a high BOD content and it should not be discharged without treatment. Improper handling of sludge liquor may significantly lower the overall plant performance.

The role of the plant operator should not be overlooked inasmuch as in most instances, a well-operated plant demonstrates higher efficiencies than a poorly operated one.

Desired Treatment Plant Performance. Although sewage and wastewater can be purified to any desired degree, the economic considerations and the waste assimilative capacity of the receiving water body generally dictate the selection of the process. In the United States, the Water Pollution Control Act Amendments passed by the Congress in 1972 and The Clean Water Act Amendments of 1977 require that all publicly owned treatment works (POTW) provide treatment of sewage that will result in an average monthly effluent quality of 30 mg/liter each of BOD_5 and suspended solids, or better.

Effluent limitations imposed in the United States on private (industrial) sources are constructed around the previously mentioned NPDES. This system serves as the basic mechanisms for enforcing the effluent and water quality standards. The permit, among other things, establishes specific effluent limitations on a facility-by-facility basis that implement the technology-based standards. In the 1970s, the standards were based, industry by industry, on the "best practicable control technology currently available." Since the second part of the 1980s, the effluent standards and permits are based on "Best Available Technology Economically Achievable" (BATEA). The degree of effluent reduction required under these standards is imposed even though it may result in an effluent quality that is better than that required for protection of the receiving water body expressed by the stream water quality standards. In short, the point of reference is what technology should be applied to each source rather than what acceptable water quality requires. However, as a minimum, in cases in which the BATEA standards will not result in meeting the stream water quality standards, even more stringent controls may be required. These cases are called water quality limiting.

4.4 NATURAL OR ARTIFICIAL PROCESSES?

One should distinguish between natural treatment processes in which the natural processes prevail, and more technical and artificial treatment technology. In the following table the corresponding natural and artificial processes are grouped together:

Natural Processes	Artificial Processes
1. Anaerobic excavated basins Excavated settling ponds	1. Settling basins with sludge digestion chamber
2. Overland flow systems	2. Chemical coagulation
3. Soil infiltration (slow or rapid) Wetland systems	3. Biological trickling filters
4. Fish or algal ponds Reservoirs and lagoons	4. Activated sludge

In general, the natural processes require much larger land areas and the effluent is of poorer quality. For example, natural purification systems such as soil infiltration or reservoirs require almost 100 times more land area than their artificial counterparts, trickling filters or activated sludge units. Natural purification systems do not perform well in colder climatic conditions. However, when land is available they are much cheaper and can be feasible primarily in developing countries or for small and/or rural communities. Practically, they are as reliable as the artificial processes and do not require as much technical supervision.

The natural treatment processes are not odorless and must be located at a distance from population centers. Therefore, they may require longer sewer trunks.

4.5 UTILIZATION OF WASTEWATER

The sludge gas produced in the anaerobic digestion units of municipal treatment plants may be the most valuable commodity. In addition both sewage effluent and the decomposed sludge can be used in agriculture as fertilizer and/or soil conditioner. Finally, the treated effluent has value to the receiving water body, by providing needed flow during dry periods and nutrients.

At the plant, the methane gas, if scrubbed to remove odorous and toxic hydrogen sulfide, can be used for heating of the sludge digesters, for heating the buildings, or for generating power. The energy derived from the digester gas is sometimes sufficient to supply the entire treatment plant. The gas can also be sold as fuel.

The sludge can be used as a fertilizer–soil conditioner, however, the fertilizing value of digested sludge is less than that of mechanically dried sludge. The transport cost from the plant to the point of application must always be considered. Dried sludge from the Milwaukee, Wisconsin treatment plant has been commercially sold under the name "Milorganite" since the 1920s. In Europe, digested sludge has been mixed with municipal refuse and converted to compost.

The treated sewage effluent has an agricultural value due to its content of fertilizing compounds—nitrogen, phosphorus, and organic solids that have not been previously separated. In addition, the irrigation value of water in the effluent is considerable. For example, the City of Tucson, Arizona will be selling almost the entire effluent from the treatment plant for irrigation of golf courses. Effluents from several other cities in the arid parts of the United States are also being used or being considered for irrigation. Treated effluents may also be used for recharge of aquifers. In more humid regions of the world, the irrigation value of the effluent is not as high. Generally for public health reasons, only treated effluents can be used in these cases.

4.6 COST OF WASTE TREATMENT

The cost of a municipal treatment plant or of the entire wastewater disposal system can be divided into two basic categories—capital and operating cost.

The Capital or Installation Cost includes the cost of project construction; land, easement, right-of-ways, and water rights; capital outlays to relocate facilities or prevent damages; and all other expenditures for investigations and surveys, designing, planning, and constructing a project after its authorization.

Operation, Maintenance, and Replacement Costs include the value of goods and services needed to operate a constructed project and make repairs and replacement necessary to maintain the project in sound operating conditions during its economic life.

The investment for capital cost is spent during a relatively short time period during the initial phase of the project. The operation, maintenance, and replacement expenses (OMR) are usually spent in yearly installments throughout the lifetime of the project. The OMR cost includes outlays for maintenance, plant supply, labor cost, chemicals and energy, depreciation, real estate taxes, insurance, etc.

The capital costs include the cost of the construction elements such as settling tanks and clarifiers, activated sludge units, sludge handling, land, administration, maintenance and laboratory buildings, and other items. Construction costs depend primarily on the size of these units. However, unit costs (a cost per unit flow or served population) decrease with increased size of the plant. This is referred to as economy of scale, which implies that it is cheaper to build one or a smaller number of large treatment plants than a large number of smaller treatment plants. However, the cost of transporting sewage over large distances must also be included in an overall economic analysis.

Cost estimating in the planning and design of wastewater disposal facilities can be done with various levels of accuracy:

1. *Order of magnitude estimates* with an accuracy of plus or minus 50%, usually used in planning and development projects. To prepare this type of cost estimate, relatively little design information is required. Normally, the major unit operations and type of the system should be defined as well as its size and population served. The cost information can be derived from generalized cost figures such as those shown in Table 4.2
2. *Budget-type cost estimates* with an accuracy of plus or minus 25%. Such estimates require a substantial amount of preliminary engineering planning. The units should be sized and the layout of the system outlined.

3. *True cost estimate* with an accuracy of plus or minus 10% is usually done after the detailed plans of the system are prepared.

A report by Reed et al.[6] and references 7–9 contain detailed cost data and figures for environmental works and wastewater treatment.

Because of inflation and frequent price changes by equipment manufacturers and construction companies, it is necessary to reference the cost data to some fixed cost index. The Engineering News Record (ENR) Index published by the McGraw-Hill Co., New York, NY and Sewer and Sewage Treatment Plant Construction Index of the Environmental Protection Agency are the ones most commonly used in the United States.

As the costs in Table 4.2 are expressed in March 1980 dollars, conversion of this information to another year can be accomplished using the following relationship:

$$\text{New year cost} = \text{base (1980) cost} \times \frac{\text{new year index}}{\text{1980 index}}$$

The ENR Index for 1980 was 3150.

The cost of treatment works and other engineering systems is usually referenced in its present value, which means that the future annual expenses for operation, maintenance, and replacement cost are discounted by an inflation factor and converted thereby to an equivalent cost at the beginning of the project. Similar discounting is applied to any future income or other benefits of the project. Typical benefits that can be attributed to controlled sewage and wastewater disposal include improved water quality and availability of beaches for swimming, improved recreational opportunity, recreational and commercial fishing, lower treatment cost of

TABLE 4.2 Cost of Conventional Secondary Wastewater Treatment Systems[a] (March 1980 Dollars) according to U.S. Environmental Protection Agency[b]

Population Served	Capital Cost[c]		Annual Operating Cost[d]	
	$/cap	$/m³-day	$/cap-year	$/m³ treated
1 000	3 000	5 555	250	1.27
10 000	700	1 290	55	0.28
100 000	300	555	21	0.11
1 000 000	190	351	15	0.08

[a] The system includes the following process modules: preliminary treatment, influent pumping, primary clarification, activated sludge secondary treatment, secondary clarification, effluent disinfection by chlorination, and sludge treatment by thickening, digestion, dewatering, and final disposal by landfill. The effluent suspended solids and BOD concentrations should be below 30 mg/liter.

[b] Cost based on sludge dewatering by vacuum filtration.

[c] For small conventional secondary plants of 10 000 people served or less, sludge dewatering by drying beds would reduce capital costs by about 33%.

[d] The annual operating cost for the small plants defined above is about 23% less if sludge drying beds are used.

downstream water works, increased river front development, etc. Most important for engineering analyses are the benefits that can be expressed as monetary income or savings. In this way, the entire cost (capital and annual operation and maintenance costs) can be compared with the income and benefits of the project by engineering cost–benefit analysis.

Most of the treatment plants and sewage disposal systems are public utility organizations deriving their income from taxes and user charges. Funding of capital outlays is usually accomplished by borrowing the money with bonds and repaying the interest and principal from the annual income. Operation of the treatment plants by private companies (privatization) is feasible and sometimes a favorable alternative to public utilities.

4.7 SOLID WASTES, REFUSE

Effect on Wastewater Treatment Process. The municipal refuse that comprise solid wastes, including the household kitchen refuse and other garbage and street sweeping refuse, must be also disposed.[10,11] Only solid wastes from flushing toilets and the solids that pass through the kitchen and bathroom sink outlets and street stormwater inlets are flushed and carried toward the treatment plant. There is little distinction between the refuse carried by sewage and that disposed of on land or by incineration. Often, the deposited refuse will find its way into sewers as a result of flushing it into toilet bowls or through the street stormwater inlets and manholes. This explains large deviations and fluctuations in the volumes of shredded solids from the screens and comminuters and of the primary sludge in the treatment plants.

Quantity and Composition of Refuse. In a treatment plant, the trapped solids and refuse from screens and bar racks are shredded until they are small enough to pass easily through the pumps. Similarly, household garbage grinders installed in kitchen sinks have become very popular in the United States and in some communities they are mandated by local ordinances in order to keep food residues from collected solid wastes. In Germany and other European countries, the use of in-sink grinders is not as widespread as in the United States.

Application of kitchen in-sink grinders increases the per capita house sewage flow by about 2%, the BOD load by 30%, the total solids load by 50%, and the suspended solids load by 30%. Similar loading increases were also reported in Germany. U.S. data on municipal refuse production report an average per capita household refuse production of approximately 2 kg/cap-day, from which about 15% or 0.3 kg/cap-day is kitchen food residue with 70% moisture content. Thus, the potential solids addition to sewage from ground kitchen wastes is $0.3 \times 0.3 \times 100 = 90$ g/cap-day.[14] Approximately 80% or 70 g/cap-day will be settleable whereas 20% or 18 g/cap-day will be nonsettleable or dissolved with about 20 g/cap-day of BOD. These numbers, typical for U.S. conditions, are approximately 20% higher than German data. Because of the organic material added to sewage, the treatment plant could become overloaded and provisions must be made in the design to accommodate the increased loads. The cost of treatment of sewage that contains ground household garbage is therefore about 20–30% higher if compared to the cost of treatment of sewage without ground kitchen wastes.[15]

The total household and commercial refuse production in the United States (2 kg/cap-day) is about two times larger than that typical for German cities (0.84 kg/cap-day). The total refuse production in the United States, including industrial, street cleaning, demolition, and all other sources, is about 8 kg/cap-day. The density of compacted municipal refuse in a landfill is about 620 kg/m^3, giving the volume of compacted refuse of (2 kg/cap-day/620 kg/m^3) $\times$ 1000 ℓ/m^3 = 3.22 ℓ/cap-day. This is approximately 25 times more than the volume of a digested dewatered sludge (0.13 liter/cap-day) and about 10 times as much as the volume of moist settled sludge (0.26 liter/cap-day). The moisture content of refuse is about 30%, giving the weight of dry solids of $0.7 \times 2 \times 1000 = 1400$ g/cap-day. About 50% or 700 g/cap-day is combustible (volatile) matter. References 10 to 13 contain further and more detailed information on volumes and composition of refuse.

Disposal of Refuse—Concerns for Groundwater Safety. The disposal of refuse must be environmentally safe. Previous practices of dumping refuse on land or in uncontrolled landfills have resulted in severe contamination of groundwater resources and in a direct threat to health. In the United States, open dumping of solid wastes is prohibited. Present sanitary practices of solid waste disposal include controlled landfills, incineration, and composting. The solid refuse should also be considered as an important raw material and energy source. Therefore, thoughts should be given to recycling the valuable resources (paper, glass, aluminum and other metals, and fuel).

Sanitary landfills must be located and operated to prevent contamination of groundwater. As a result of surface runoff entering the landfill (in spite of daily coverage of the deposited refuse by clayey cover) highly contaminated liquid called leachate accumulates at the bottom of the landfill. This liquid must be pumped out and safely disposed of. Typical concentrations in the leachate are given in Table 4.3.

The amount of leachate depends on the hydrologic conditions of the landfill and on the amount of rainfall and surface water entering the landfill (these should be minimized as much as possible). Sometimes, liquid sewage is added to enhance methane production. If larger volumes of leachate are produced, pretreatment facilities may be necessary before the leachate is discharged into sewers and/or conveyed to a municipal treatment plant. The decomposition of refuse in landfills is anaerobic, hence, the methane production. The methane gas produced can be recovered and used for a beneficial purpose. Otherwise, safe release of the gas must be provided in the landfill to prevent gas explosions and fires.

The refuse can be composted jointly with treatment plant sludges. Pieces of metals, glass, and other nonbiodegradable solids must be removed from the refuse and the refuse should be shredded before the solids are mixed with digested sludge.

TABLE 4.3 Landfill Leachate Composition[a]

Constituent	Concentration (mg/liter) Range	Typical
BOD_5	2 000–30 000	10 000
COD (chemical oxygen demand)	3 000–45 000	16 000
Total suspended solids	200–1 000	500
Organic nitrogen	10–800	200
Ammonia nitrogen	10–800	200
Nitrate	5–40	25
Total phosphorus	1–70	30

[a] After Tchobanoglous et al.[11]

The industrial composting process is usually an aerobic thermophilic (higher temperature) degradation, the product of which is a humus-like matter that is a valuable soil conditioner and fertilizer.[10,16–18]

Composting is very popular in Europe, India, and China, however, in the United States, several pilot operations have failed because of an unfavorable economy when compared to cheaper landfill disposal. With a shortage of suitable landfill sites and stricter environmental regulations, composting may become a feasible alternative, considering that it represents an ultimate recycle of the sewage and refuse solids.

Refuse and dried sewage solids can also be jointly incinerated. With proper incinerating temperatures in the furnaces the air pollution hazard should be minimal, however, it is not negligible.

4.8 EMISSIONS

Odor Control

Treatment Plant Odors. Odor emissions from a well-planned and operated treatment plant are commonly not objectionable. However, rare operational failures combined with meteorological inversion conditions can result in insufficient dilutions of emissions of odor-causing components. For these reasons, the distance of a treatment plant from residential areas should be at least 300 m and a minimum 800 m for sludge disposal sites.

The task of the engineers is to keep odor at a minimum. Industrial spills and shock loadings should be eliminated before they reach the treatment plant. Strong organic and odorous wastes such as those from slaughter plants or fecal sludges should not be discharged in the sewers without pretreatment, or, they should be diverted directly to the aeration tanks or digestion units. It is important to keep sewage in the sewers fresh. If possible, sewage flow should be diverted into open lined channels with an adequate slope that will provide good aeration by the atmospheric oxygen, which prevents formation of hydrogen sulfide.

Hydrogen Sulfide is the most common cause of odor in the wastewater treatment systems and in sewers. The anaerobic conditions that lead to hydrogen sulfide production also favor the production of other, mainly organic odor-causing compounds.[19]

Sewer Gases. Control of the odor problem in closed conduits is more difficult. The buildup of hydrogen sulfide is caused by the bacterial slime layers and sludge deposits in the sewers.[19,20] The population of sulfate-reducing bacteria (the bacteria that reduce sulfate ions SO_4^{2-} to hydrogen sulfide H_2S) is low in fresh (aerobic) sewage but it is high in sludge. Therefore, accumulated sludge deposits in the sewers are particularly troublesome and the nuisance they cause increases with higher temperatures, higher content of suspended solids in the sewage, and with the decreasing slope of sewers and prolonged flow time. Hence, good aeration and ventilation of sanitary and combined sewers are needed. Special attention and care must be given to pressure pipes without aeration, particularly when they carry intermittent flows or when sludge deposits can accumulate. Hydrogen sulfide can also attack and corrode sewer concrete above the water level. This can be avoided or mitigated by periodic wetting of the concrete surface by increasing water level in the sewer.

The damages to sewers caused by hydrogen sulfide attack can also be mitigated by injection of pressurized air or by chlorination of sewage in the conduits.[21-23]

Mitigation of Odor. If sewage in the influent to the treatment plant is septic (anaerobic) the hydrogen sulfide must be removed by preaeration or by prechlorination.

If the influent to a treatment plant is anoxic two different remedial means are possible. In smaller treatment plants, the influent can be directly diverted in closed conduits to an underloaded activated sludge unit that was designed to permit recirculation (''carrousel''). Here, septic sewage will become aerobic in a relatively short time and odor-causing components will be absorbed by the sludge and subsequently degraded.[24] The second alternative involves aeration and oxidation of septic sewage by air and/or chemically. This must be done in specially covered tanks from which the exhaust gases must be scrubbed and deodorized. Mixing the septic influent with the aerated effluent from trickling filters or activated sludge units will have a similar effect. Chemically, 2.06 parts of chlorine or 0.47 parts of oxygen or 1.6 parts of anhydrid nitric acid (N_2O_5) or 1 part of sodium nitrate ($NaNO_3$) is needed to oxidize 1 part of the hydrogen sulfide (H_2S). The reaerated sewage does not differ from fresh sewage. However, it is more advantageous to abate the hydrogen sulfide formation at its origin. Heukelekian[25] has estimated that the amount of chlorine needed to remove hydrogen sulfide is 10 times as much as that for maintaining aerobic conditions that prevent its formation. This clearly shows the importance of keeping sewage fresh.

It is also necessary to keep sewage fresh (aerobic) during its residence time in the treatment plant. The conduits in the plant area should either be hydraulically designed to provide adequate velocities or be aerated by compressed air. Spaces with quiescent flow should be avoided. The residence time of sewage in settling basins and clarifiers ought to be limited to a few hours and the accumulated sludge must be removed in short time intervals before intensive gas bubble formation develops. Storage basins should also be without accumulated sludge deposits. High rate trickling filters with volumetric loadings over 900 g of $BOD_5/(m^3\text{-day})$ should be covered and the exhaust gases should be scrubbed and deodorized.

Sludge handling without odor emissions is not as simple. Raw sludge particles remain fresh and odorless as long as they are dispersed in an aerobic liquid environment. When removed from sewage, the sludge becomes relatively odor free only after digestion, aerobic stabilization, and dewatering on sludge beds, or after full mechanical dewatering and heat drying. Mechanically dewatered and dried sludge must be kept dry in order to remain without an objectionable smell. In all other stages the sludge emits an odor. After a few hours, raw sludge begins to undergo acidic fermentation, resulting in an odor resembling that of a pigsty. This odor appears any time when raw or partially stabilized sludge is exposed to the air, which can occur when cleaning screens and grit chambers. It can also emanate from floating sludge layers on the surface of sedimentation basins, from exposed unclean surfaces of the treatment plant when the water level drops, from irrigation ditches and reservoirs that accumulate sludge and septic sewage, and from dry places that accept fresh raw sludge.

In order to minimize sludge odor, the treatment plant must be kept clean. The solids from screening and bar racks must immediately be conveyed to a container and disposed of in a landfill or shredded and added to sewage for subsequent removal and handling as sludge solids. Only washed sand should be removed from grit chambers. Floating sludge mats and solids on the surface of the settling tanks should be periodically removed. All collected sludge particles should be immediately conveyed to further processing and storage of raw sludge should be avoided. Sewage for irrigation and land disposal should be relatively free of sludge particles and biologically treated. Sludge stored on dry places must be well stabilized. The drying of raw sludge (e.g., when sludge is converted to an organic fertilizer) must be rapid such as centrifuging. The vapor and gases from a heat drying process should be combusted. The temperature in the furnace for sludge incineration is about 800°C.

If these preventive measures are applied, the treatment plant should remain relatively odor free. In a tight, limited place, odor prevention is easier if the liquid raw sludge is pumped to a more convenient place for processing and disposal.

When an offensive smell has developed either as a result of malfunction of the operation or a poor design, it must be fully and rapidly abated. Dry places in which raw sludge has accumulated must be cleaned and sprayed with a chlorine solution or with chlorinated lime. Covering the plant or parts of it with a canvas or an enclosure has the benefit of reducing the warming effect of the sun that intensifies odor. However, this measure is not used too often because it reduces natural ventilation and artificial ventilation of the enclosed space must be installed. The outlet air can be deodorized by passing it through a carbon filter or by ozonization. The odor of the outlet air can also be removed by compressing it and passing it through

TABLE 4.4 Comparison of Three Biological Deodorization Processes[a]

	Compost Filter	Activated Sludge Tank (Diffused Air)	Washing with Activated Sludge
Dimensions	0.8 m deep	3 m depth	6 m pipe cascade
Filtration velocity, m/hr	50	3	Air: 2 400 Sludge: 100
Volumetric ratio, m^3 air/(m^3-hr)	62.5	1	24
Energy consumption, W-hr/m^3 air	2	(20)	2

[a] After Imhoff.[28]

a biological activated sludge unit or through a trickling filter. Sometimes, soil filters can be used.[26] Waltrip and Snyder[27] evaluated several methods of treatment plant odor control, including hydrogen peroxide addition, ferrous sulfate addition, chlorine addition, preaeration of the influent, and off-gas collection and scrubbing in packed beds. From these methods, preaeration followed by off-gas collection and scrubbing with caustic (sodium hydroxide) provided the most economical solution.

Design parameters of three biological deodorization processes are given in Table 4.4.

It can be noted that the diffused air units require the smallest filtration velocity and provide the highest contact opportunity of the sludge particles with the air. The high energy use requirement is for dissolution of oxygen.

Publications by Metcalf and Eddy[18] and by Imhoff[29] provide more detailed information on odor prevention, remedial measures, and odor measurements.

Noise Control

Measurements of Noise. Noise is commonly defined as unwanted sound. Modern treatment plants are equipped with a number of mechanical devices. These machines as well as the turbulence of treated sewage generate noise. Typical noise levels are given in Table 4.5.

The physical measure for noise is the sound pressure level, L. It is defined as a logarithmic ratio of the sound intensity I of the given noise and the sound intensity I_0 of the lowest audible noise. The unit expressing the noise level is decibel (to commemorate the inventor of the telephone, Alexander G. Bell). A decibel (dB) is generally used in practical measurements. As the noise intensity is proportional to the square of the noise pressure the following equation can be used for converting sound pressure to decibels:

$$L \text{ in dB} = 10 \log\left(\frac{I}{I_0}\right) = 10 \log\left(\frac{p^2}{p_0^2}\right) = 20 \log\left(\frac{p}{p_0}\right)$$

TABLE 4.5 Noise Levels from Mechanical Sources[a]

Source	Noise Level in dB(A) 1 m from the Machine
Electric motors	75–90
Gear boxes	75–85
Compressors	85–95
Blowers	100–105
Combustion engines	95–100
Conveyor belts	95–100
Pumps	80–85
Aerators	80–90

[a] After Porada and Schuller.[30]

where p_0 is the reference noise pressure at the lowest audible pressure of a sound with a frequency of 1 000 Hz. The value of p_0 is commonly chosen as 0.0002 μbars (1 μbar equals approximately 10^{-6} atm).

The difference between the noise scale in dB and actual noise levels must be noted. Doubling the intensity by two identical sources of noise will increase the noise level by approximately 3 dB. In terms of hearing, about a 10 dB increase is necessary to make a sound seem twice as loud to a listener. It should be also noted that the human ear is more sensitive to softer sounds and less sensitive to louder sounds. Also the sensitivity of the ear varies with the frequency of the noise. The ear is more sensitive to rapid than slow oscillations in air pressure. These oscillations are measured as number of pressure changes per second or, more scientifically, in hertz (Hz), which means the same thing. To adjust for these differences in the sensitivity of the ear, a scale known as dB(A) has been defined that reduces the noise level values at higher frequencies. Most noise ordinances express the standards in terms of dB(A).[33,34]

Noise Abatement and Control. In the United States, the noise levels are controlled by the Noise Pollution and Abatement Act of 1970. The standards for noise were issued by the Environmental Protection Agency.[34] An outdoor noise level criterion of 70 dB(A) has been recommended by the Federal Highway Administration for residential areas.[33] A corresponding interior noise level criterion is 55 dB(A). In Germany, a nighttime interior noise level of 40 dB(A) is recommended.

The noise levels of a treatment plant can be related to its size expressed in population equivalents as follows:

Population Equivalent Served by a Treatment Plant	Noise Level in dB(A)
10 000	95
100 000	100
1 000 000	105

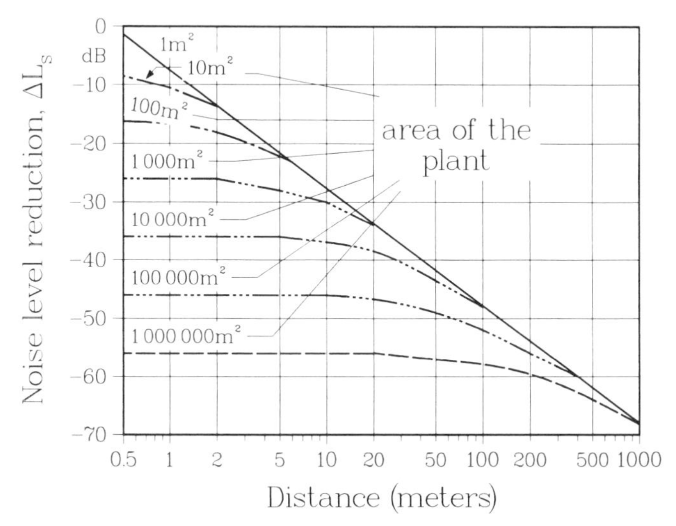

Figure 4.1 Noise level reduction related to the area of the plant and its distance from the receptor.

Noise levels are reduced with distance. Figure 4.1 can be used to determine the reduction of the noise level related to the area of the plant and its distance from the receptor.

Example: Assume a treatment plant serving a population of 100 000 with a 100 dB(a) noise level and an area of 5 ha (= 50 000 m^2). From Figure 4.1, the noise level difference at a distance of 250 m from the plant is -57 dB and the noise level of the plant is the $100 - 57 = 43$ dB.

If the noise level is too high when compared with a standard or criterion, noise abatement measures must be implemented. Such measures work best if they are aimed at the source of the noise. This can be accomplished by good maintenance of the machinery or, in some cases, by replacement. If the reduction of the source noise emissions is not enough, interception or diversion of the noise may be feasible. Simple walls and sound barriers and/or thick hedges can reduce the noise levels by approximately 7 dB. References 31, 32 and 35 contain comprehensive reviews of noise pollution abatement.

Aerosol Emissions

Aerosols can emanate from the surface of the basins filled with sewage, particularly when the turbulence in the basin is very high, such as that in aeration basins. They evolve in a form of mist or fog either as a result of higher turbulence in the basin or wind effects. Based on scientific knowledge the risk of adverse health effects of aerosol emissions can be minimal. However, it is recommended that in order to minimize the exposure, surface aerators causing high turbulence should be turned off when work is done in the vicinity of the aerated basins. Use of protective sprays is recommended. Emissions from trickling filters can be minimized by installing a protective surrounding wall.

REFERENCES

1. A. Gunn and P. A. Vesilind, *Environmental Ethics for Engineers.* Lewis Publ., Chelsea, MI, 1986.
2. G. M. Fair, *J. Water Pollut. Control Fed.* **34,** 749–766 (1962).
3. A. V. Kneese and B. T. Bower, *Managing Water Quality: Economics, Technology, Institutions,* 2nd ed. Johns Hopkins Press, Baltimore, MD, 1971.
4. K. R. Imhoff, *J. Water Pollut. Control Fed.* **46,** 1663 (1974).
5. P. L. McCarthy, *Public Works* **95,** No. 9, 107–112, No. 10, 123–127, No. 11, 91–95, No. 12, 95–99 (1964).
6. S. C. Reed et al., *Cost of Land Treatment Systems,* EPA 430.9-75-003. U.S. Environ. Prot. Agency, Washington, DC, revised September 1979.
7. *Construction Cost for Municipal Wastewater Conveyance Systems,* EPA 430/9-81-003. U.S. Environ. Prot. Agency, Washington, DC, 1981.
8. *Estimating Water Treatment Cost,* EPA 600/2-79-162 a, b, c. U.S. Environ. Prot. Agency, Washington, DC, 1979.
9. *Dodge Guide for Estimating Public Works Construction Costs,* Dodge Building Services. McGraw-Hill, New York (published annually).
10. *Municipal Refuse Disposal.* Am. Public Works Assoc., Chicago, IL, 1970.
11. G. Tchobanoglous, H. Theisen, and R. Eliasen, *Solid Wastes.* McGraw-Hill, New York, 1977.
12. T. R. Haseltine, *Water Sewage Works,* **97,** 467 (1950).
13. F. Pöpel, *Einflusse auf Menge und Zusammensetzung von Hausmull, Sperrmull, und Industrieabfallen* (*Factors affecting Quantity and Composition of Household Refuse, Litter and Industrial Solid Wastes*). Kommissionsverlag R. Oldenbourgh, Munich, FRG, 1969.
14. A. E. Zanoni and R. J. Rutkowski, *J. Water Pollut. Control Fed.* **44,** 1756–1762 (1972).
15. W. Bucksteeg and K. R. Imhoff, *GWF, das Gas-und Wasserfach* **105,** 1226 (1964).
16. A. Seifert, *Der Kompost,* H. G. Müller Verlag, Munich, 1957.
17. E. Spohn, *Compost Sci.,* **18** (3), 25 (1977).
18. Metcalf & Eddy, Inc., *Wastewater Engineering: Treatment, Disposal, Reuse,* 2nd ed. McGraw-Hill, New York, 1979.
19. R. R. Dague, *J. Water Pollut. Control Fed.* **44,** 583–594 (1972).
20. R. Pomeroy and F. D. Bowlus, *Sewage Works J.* **18,** 597 (1946).
21. F. S. Taylor, *Sewage Works J.* **20,** 917 (1948).
22. D. R. Kennedy, *Sewage Works J.* **20,** 104 (1948).
23. E. M. Lemcke, *Sewage Works J.* **20,** 111 (1948).
24. M. K. H. Gast, *Prog. Water Technol.* **11** (3), 223 (1976).

25. H. Heukelekian, *Water and Sewage Works* **95,** 17 (1948).
26. D. A. Carlson and C. D. Leiser, *J. Water Pollut. Control Fed.* **38,** 829 (1966).
27. C. D. Waltrip and E. G. Snyder, *J. Water Pollut. Control Fed.* **57,** 1027 (1985).
28. K. R. Imhoff, *Report of Institute WAR* (*Wasserversorgung, Abwasserbeseitigung und Raumplannung*), No. 9, p. 7, Technical University, Darmstat, FRG, 1982.
29. K. Imhoff, *Ges. Ing.* 64, 610 (1941).
30. Porada W. and W. Schuller, *GWF, das Gas-und Wasserfach: Wasser/Abwasser,* **123** (12), 594–599 (1982).
31. W. Tempest, *The Noise Handbook.* Academic Press, Orlando, FL, 1985.
32. A. R. Thumann and R. K. Miller, *Fundamentals of Noise Control Engineering.* Fairmont Press, Atlanta, GA and Prentice-Hall, Englewood Clifs, NJ, 1986.
33. National Research Council, *Noise Abatement and Control,* Highway Res. Rec. No. 448. Highway Res. Board, Washington, DC, 1973.
34. W. S. Gatley and E. E. Frye, *Regulation of Noise in Urban Areas.* U. S. Environ. Prot. Agency, Office of Noise Abatement and Control, Washington, DC, 1971.
35. D. N. May, *Handbook of Noise Assessment.* Van Nostrand Reinhold, New York, 1978.

CHAPTER 5

Wastewater Characterization

5.1 QUANTITY AND QUALITY

Municipal Sewage. Fresh domestic sewage has little odor, it is gray in color, the sewage solids are only slightly disintegrated, decomposition is not evident, and dissolved oxygen is present. As sewage becomes stale and finally septic, its odor intensifies, the color turns black, the solids disintegrate, decomposition becomes active, dissolved oxygen eventually disappears, and hydrogen sulfide is then evolved.

The actual composition of municipal sewage and its strength depends on a number of factors such as sewer separation (presence or absence of urban stormwater runoff), industrial inputs and whether they are pretreated or not, water use that depends on geographic location, economy and standard of living, time factors (time of the day and time of the year), and use of in-sink garbage grinders. The average composition of untreated domestic sewage for U.S. conditions is given in Table 5.1.

These numbers are based on a typical U.S. domestic water consumption of 380 liters/cap-day. As mentioned in Part I, the average water consumption in the United States varies greatly, resulting in wide ranges in wastewater strength. It should be noted that typical European water consumption (200 liters/cap-day) is about one-half of the United States, hence the concentrations in raw sewage—for example, from a typical German municipality—is about twice that for a typical U.S. city. Concentration ranges for typical U.S. municipalities are 110–400 mg/liter for BOD_5 and 100–350 mg/liter for suspended solids, respectively.

Rather than using concentrations that can vary greatly, it is more convenient to express the raw sewage contributions in so called unit loads, using grams per capita per day as a unit. The typical raw wastewater loading factors are presented in Table 5.2.

Again, some variations in these numbers can be expected. The loading values in the table correspond to domestic sewage without in-sink grinders and for separate sewers. If wastewater also contains urban stormwater laden with street solids, the average BOD_5 is then about 75 g/cap-day. Similar numbers are also used in Germany, Switzerland, Sweden, and Great Britain. If garbage disposals are installed, the per capita loadings should be increased by the following factors.[1]

Wastewater Characteristic	Percentage Increase for 100% In-Sink Grinders Use
Flow	2
BOD_5	30
Total solids	50
Suspended solids	21

In determining the loadings to surface waters, it is necessary to subtract the pollution removed by treatment from the raw sewage load. A properly designed and operated secondary sewage treatment plant should remove 85–90% of BOD_5 and about the same fraction of suspended solids from the raw sewage influent, resulting in respective average effluent BOD_5 and suspended solids concentrations between 20 and 30 mg/liter. In addition, phosphorus removal and disinfection are required in some parts of the United States.

As the water demand for urban areas varies with the time of day and the season, so does the sewage flow. The highest water demand and pollution load is near noon and in the afternoon hours. Figure 5.1 shows a typical loading pattern for a medium-size midwestern city.

Industrial Effluents. In most industries, wastewater effluents result from the following water uses: sanitary wastewater (from washing, drinking, and personal hygiene), cooling (from disposing of excess heat to the environment), process wastewater (includes water used for making goods, washing the products, waste and byproduct removal, and transportation), and cleaning (includes wastewater from cleaning and maintenance of industrial areas).

Excluding the large volumes of cooling water discharged by the electric power industry, the wastewater production from urban areas is about evenly divided between municipal and industrial sources.

Commonly, it is beneficial to jointly treat municipal sewage and industrial effluents in one plant. Attention must be given to the presence of toxic and other

TABLE 5.1 Typical Composition of Sewage

	Concentration in mg/liter (g/m^3)			
	Mineral	Organic	Total	BOD_5
Settleable solids	40	100	140	55
Nonsettleable	25	70	95	65
Dissolved	210	210	420	40
Total solids	275	380	655	160
Nitrogen	15[a]	20	35	
Phosphorus	5	3	8	

[a]As ammonia.

TABLE 5.2 Unit Loads by Sewage

	Unit Loadings in g/cap-day			
	Mineral	Organic	Total	BOD_5
Settleable solids	15	39	54	20
Nonsettleable	10	26	36	25
Dissolved solids	80	80	160	15
Total solids	105	145	250	60
Nitrogen	9.5[a]	5.7	15.5	
Phosphorus	1.9	1.1	3.0	

[a] As ammonia.

harmful compounds in the industrial wastewater that could damage the treatment plant operation, biota in the tanks, or cause a violation of effluent and/or stream standards. In such cases, pretreatment may be necessary. In the United States, such pretreatment and pollutant elimination will be specified in the discharge permit issued by the regulatory agency or by local ordinances.

Industrial wastewater effluents are usually highly variable, with quantity and quality variations caused by batch discharges, operation start-ups and shut-downs, working hours distribution, etc. There are many possible in-plant changes, process modifications, and water-saving measures by which industrial wastewater loads can be significantly reduced. Up to 90% reductions have been achieved recently by industries employing methods such as recirculation, operation modifications, effluent reuse, or changing to more efficient operation. As a rule, treatment of an

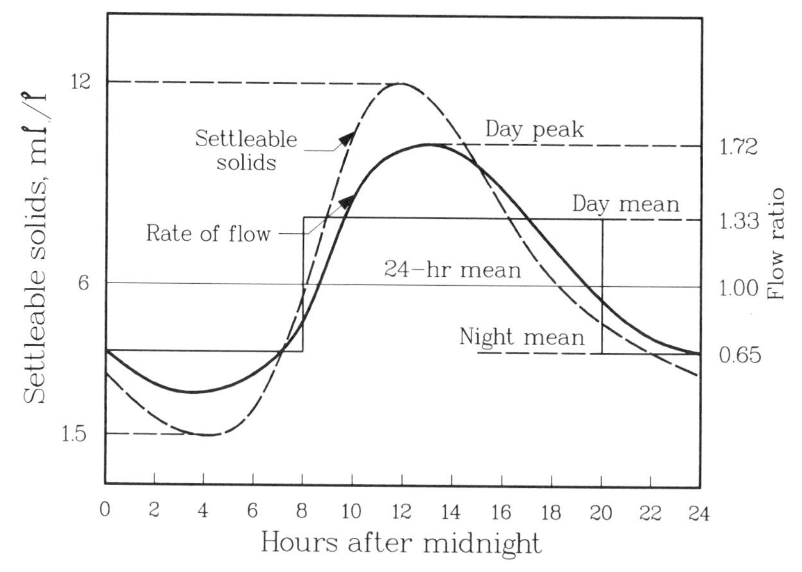

Figure 5.1 Hourly variations in municipal wastewater flow and strength.

industrial effluent is much more expensive without water-saving measures ‚ diluted wastes) than the total cost of in-plant modifications and residual effluent treatment. A long-term detailed survey is usually necessary before a conclusion on the pollution impact from an industry and the possibility of joint treatment is reached.

5.2 WASTEWATER CHARACTERIZATION

The objectives of wastewater characterization are to (1) provide pertinent information on the wastewater for the design of the treatment process; (2) serve as a basis for estimating effluent strength with respect to the waste assimilative capacity of the receiving water body and/or with respect to regulatory effluent and stream standards, (3) provide the operator with information on the treatment process performance, and (4) provide a data base needed for calibration and verification of mathematical models used for design and control of the treatment process or for waste assimilative capacity studies.

For details of individual procedures see the Standard Methods for Examination of Water and Wastewater[2] or its counterparts in other countries. These detailed procedures will not be repeated here. Instead, the methodology that should be utilized in wastewater and surface water characterization will be discussed so that an engineer or a treatment plant operator has an understanding of the reliability and limitations of the measured values.

Several other handbooks describing the analytical methods for water and wastewater characterization can also be consulted.[3–6] For a textbook explanation of the theory and techniques of chemical water and wastewater analyses see Sawyer and McCarthy.[7]

Sampling and Monitoring

As wastewater flow and quality continuously change a single grab sample collected at a particular time and place can represent the composition of the wastewater flow only at that time and place. When the wastewater flow is variable, grab samples collected at time intervals according to a predetermined plan and analyzed separately can document the extent, frequency, and duration of these variations. The sampling interval should be selected according to the frequency at which these changes occur and can range from 5 min (for flush intensive stormwater contributions) to as long as 1 hr or more. When the source composition varies in space (as it does in a typical treatment plant) samples should be collected at appropriate locations.

Great care should be given in sampling to obtain a representative sample that will conform to the objectives of the sampling program.

A composite sample refers to a mixture of grab samples collected at the same point at different times. The intervals at which the individual samples are collected are based either on a predetermined time (e.g., every 1 hr for 24 hr) or proportional to flows. Time-composite samples are most common and their measured values represent an average over the sampling period.

Automatic sampling devices for both time composite and flow proportional measurements are available. Such samplers are reliable and effective and can significantly increase the frequency of sampling. They may also provide refrigeration for sample preservation.

Integrated samples are obtained by simultaneously collecting grab samples at predetermined locations and mixing them to obtain an average sample. An example of the need for such sampling occurs in a river receiving a wastewater discharge that is not uniformly mixed with the river flow. The samples are then taken at specified cross-sectional locations along with the flow velocity measurement at that point. Then the mixture is made by proportioning the samples according to the corresponding flow (velocity times the cross-sectional area).

Chemical Analyses

Suspended solids determination and BOD are the most common and most important analyses for evaluation of the performance of a sewage treatment works or in assessing the waste assimilative capacity of a receiving water. It should be noted that these two parameters are commonly mandated and regulated by the discharge permit. The values on suspended solids concentration also give a measure of the sludge production in the clarifiers.

Suspended Solids are characterized as filterable, settleable, and nonsettleable. To find the concentration of suspended solids in a wastewater sample, a measured volume of the sample is filtered through a preweighed standard glass-fiber filter and the residue retained on the filter then constitutes the suspended (or filterable) solids. Total suspended solids are obtained by weighing the filter with the retained solids and then subtracting the weight of the filter.

The Settleable Solids are measured in a conical cylinder 40 cm (15 in.) high that has a graduated tip in ml and holds 1 liter of sample (Imhoff cone—Fig. 5.2). The volume of solids settling into the tip of the inverted cone is normally read after 1 hr. The sides of the cone should be gently stirred with a rod after a lapse of 45 min in order to effect the deposition of the solids that would otherwise stick to the sides. A Zone Settling Test is similar except it can be performed in a standard 1-liter graduated cylinder and the settling of the top of the solids layer is recorded periodically in shorter time intervals. This test provides information on the design and performance of settling units. Between 2 and 5 ml of sludge is ordinarily deposited by 1 liter of raw domestic sewage. The performance of settling tanks is generally considered to be satisfactory when the effluent contains no more than 0.5 ml of settleable solids. The percentage removal of the settleable solids in the treatment plant is computed as $100 \times (a - b)/a$, where a is the volume of settleable solids of the raw influent, and b is that of the effluent.

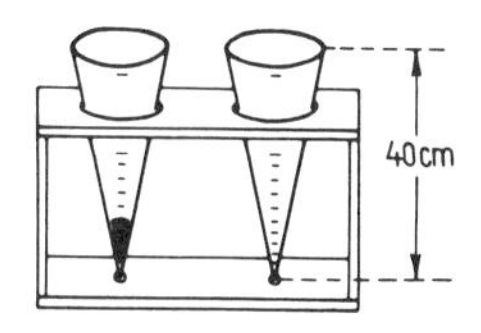

Figure 5.2 Imhoff's laboratory settling cones for the volumetric determination of settleable solids.

Turbidity caused by suspended solids and other components (some soluble organic compounds) can be related to transparency. For this purpose, a Secchi disc is used. This white rounded plate is submerged during the test in the water or wastewater and the depth at which visibility of the disc is lost is then the measure of turbidity. These depths are relatively small for wastewater (a fraction of 1 m). For surface waters, they range from about 1 m for typical slightly polluted waters to over 10 m for clean stream and lakes. The Secchi disc procedure is primarily a field method for turbidity estimation. In the laboratory, turbidity is measured visually by a Jackson candle turbidimeter or by special turbidimeters or nepthelometers measuring optical light-scattering properties of turbidity causing particles in water.

Volatile Suspended Solids are determined by igniting the filter residue at 550°C. The remaining residue (ash) is then weighed and represents the fixed (mineral) solids. The difference between the total suspended solids and mineral solids concentrations is the volatile solids. This value is often related to the organic content of the suspended solids although this may not be correct since the loss by ignition also contains some volatile mineral components. The organic suspended solids are of importance as they have a lower specific gravity and do not settle readily.

When all the suspended particles are removed from the sewage, the remaining solids are either colloids or dissolved ions and salts. There are no specific analyses available for determining the content of colloidal particles. In general, the difference between the total solids and the suspended solids is considered to be the dissolved solids content. The total dissolved solids content is determined by evaporating water from the filtered sample of a known volume and weighing the residue (concentration = weight of the residue/volume of the sample). The volatile dissolved solids content is then the difference between the weights of the residues before and after ignition (at 550°C) divided by the volume of the sample. With the same reservations as stated in the previous paragraph, the volatile dissolved solids values can again be related to the dissolved organic content, whereas the residue concentration after the ignition represents the mineral content.

Routine observations of *pH (hydrogen ion content)* provide information on whether the sewage is acidic ($pH < 7$), neutral ($pH = 7$), or alkaline-basic ($pH > 7$). The pH value also shows stages of various biochemical reactions taking place in the treatment plant, in the sewage itself, or in the receiving water. In addition, pH affects the performance of many treatment processes since practically every phase of wastewater treatment is pH dependent. On-site pH estimation can be easily accomplished by pH indicator papers (litmus paper). Electrochemical pH estimations are more accurate.

The Dissolved Oxygen content of water and wastewater determines whether the water or treatment process is aerobic [dissolved oxygen (DO) concentration >> 0] or anaerobic (DO = 0). The DO analysis is one of the key tests in water pollution and wastewater treatment control. The wet DO content determination is by the Winkler (iodometric) titration method. Today, the DO test is performed mostly by membrane electrode probes. The oxygen-sensitive membrane electrodes are either of the polarographic or galvanic type. These probes are submersible and can be used for both field (portable battery operated) and laboratory analyses. In investigations of treatment plant operations, membrane electrodes have also been used for continuous DO analyses such as BOD and oxygen uptake tests. Care should be taken in calibration of the probes, however.

The dissolved oxygen content in waters cannot usually exceed its saturation value, which depends on temperature, salinity, and, to a lesser degree, barometric pressure. The DO saturation values are given in Table 5.3.

The seawater values are reported for a chloride concentration of 20 g/liter, which corresponds to a salinity content of about 3% (1% by weight = 10 g/liter). As the average DO saturation value of 10 mg/liter represents only 0.001% content by weight, dissolved oxygen has a low solubility in water.

The Oxygen Deficit, D, is the difference in mg/liter between the measured oxygen concentration and the saturation concentration at the sample temperature. The reaeration rate, that is, the rate at which new oxygen is supplied to the water from the atmosphere, is proportional to the oxygen deficit.

The degree of organic pollution of water or wastewater is best expressed as the amount of oxygen needed to oxidize the organics. Several analytical measurements are available from which the chemical oxygen demand (COD) and biochemical oxygen demand (BOD) are the most common.

Total oxygen demand (TOD) and total organic carbon (TOC) are analyses that rely on laboratory instruments and are becoming popular as a result of their ease of measurement and short time requirements.

The Chemical Oxygen Demand (COD) is a measure of the oxygen equivalent in mg of O_2/liter of the organic matter of the sample that can be oxidized by a strong chemical oxidant. As an oxidant, Standard Methods (APHA, AWWA, WPCF) recommend potassium dichromate ($K_2Cr_2O_7$), whereas German and British methods recommend potassium permanganate ($KMnO_4$). The COD analysis is a "wet" laboratory procedure. Oxidation of most organic compounds is 95–100% of the theoretical value. Some organic components are not completely oxidized by

TABLE 5.3 Oxygen Saturation

Temperature	0°C	5°C	10°C	15°C	20°C	25°C	30°C
Freshwater	14.6	12.8	11.3	10.1	9.1	8.3	7.6 mg/l
Seawater	11.3	10.0	9.0	8.1	7.4	6.7	6.2 mg/l

this method but few are not oxidized at all. These organics include benzene, pyridine, and toluene. On the other hand, the chemical oxidant employed in the analysis will oxidize some inorganic compounds such as ferrous iron, sulfides, and secondary manganese. Because of these problems, the COD values of wastewaters with different compositions cannot be compared; however, the COD values do represent the relative strength of organic pollution of samples of similar composition.

The Biochemical Oxygen Demand

Nature of Test. The BOD test measures the oxygen required for the biochemical degradation of organic materials (carbonaceous demand). This test can also reflect an immediate oxygen demand by inorganic materials such as sulfides and ferrous iron. This inorganic demand should be accounted for by subtracting the oxygen used during the first 15 min of the test, however, the present U.S. Standard Methods includes the 15 min demand in the BOD test.

Measurement. The methodology utilizes a laboratory technique quite similar to the natural biooxidation processes taking place in surface waters and in aerobic biological treatment units. The BOD value does not depend only on the kinds and concentrations of the organic pollutants but it is also affected by the number and activity of microorganisms participating in the process, on temperature, mixing, presence of toxic materials, pH of the sample, and several other factors. Therefore, to avoid large discrepancies in the BOD values, the test is standardized.

The conventionally used bottle measurement consists of placing a known quantity of the waste or river water, some microbiological "seed," and nutrients in one bottle and all but the wastewater in a second, identical bottle (blank). The bottles, with a volume of approximately 300 ml, are placed in an incubator, with a water seal at 20°C. Both bottles are then analyzed for DO, and the difference between the initial DO reading and the reading after 5 days incubation, accounting for dilution and seed, comprises the standard 5-day 20°C BOD. For long-term tests designed to determine the ultimate BOD or the reaction rates at which the microorganisms consume the oxygen, sufficient samples and blanks are prepared in order to plot the removal of oxygen each day for a designated time period.

The biochemical oxygen demand can also be measured manometrically using Warburg or Hach instruments. Both work on the same principle. The process involves placing the sample and seed into a flask that is agitated at a constant temperature. A vial in the center of the flask contains potassium hydroxide (KOH) which absorbs the CO_2 produced by the bacteria. The prime advantage of the test is that the manometer, which is connected directly to the sample flask, can be calibrated to read the oxygen uptake. Furthermore, several dilutions can be prepared and analyzed at the same time. These analyses are used mostly for special studies such as toxicity investigations and sludge activity and treatability. It should be noted that the manometric method progresses at a higher rate and yields a higher ultimate first stage BOD than a conventional BOD bottle method.

BOD Reaction and Its Rates. The biooxidation reaction occurs in two distinctive stages: the first stage during which readily available carbonaceous organic matter is broken down by the microorganisms, and the second stage during which unoxidized nitrogen (organic nitrogen and ammonia) is oxidized to nitrates. The two BOD stages are plotted in Figure 5.3, which shows results of long-term BOD experiments performed at 9°, 20°, and 30°C.

The first (carbonaceous) stage of the BOD reaction at 20°C is completed in approximately 20 days and the second stage (nitrogenous) usually begins at about 7–10 days after the beginning of the BOD experiment. Therefore, the standard 5-day test is not normally affected by nitrification in the sample, however, nitrification may occur and should be accounted for.

The lag time between the beginning of the BOD reaction and the beginning of the second nitrogenous stage depends on the temperature and on the degree of treatment. For some treated effluents or river samples containing treated effluents, the nitrogenous stage may start in less than 5 days after the beginning of the test. As most of the effluent and stream standards refer only to the first stage, carbonaceous BOD, occurrence of the nitrogenous oxygen demand may distort the data and interpretation of the results. To inhibit nitrification in the BOD test, the U.S. Standard Methods recommends the addition of 10 mg/liter of 2-chloro-6-(trichloromethyl)pyridine to the dilution water or to the sample.

If all the above prerequisites are satisfied the carbonaceous BOD reaction stage can be described as a first-order chemical reaction that follows the formula

$$y = L(1 - 10^{-kt}) = L(1 - e^{-Kt}) \tag{1}$$

where y = oxygen demand satisfied at time t,
L = ultimate first-stage carbonaceous BOD.

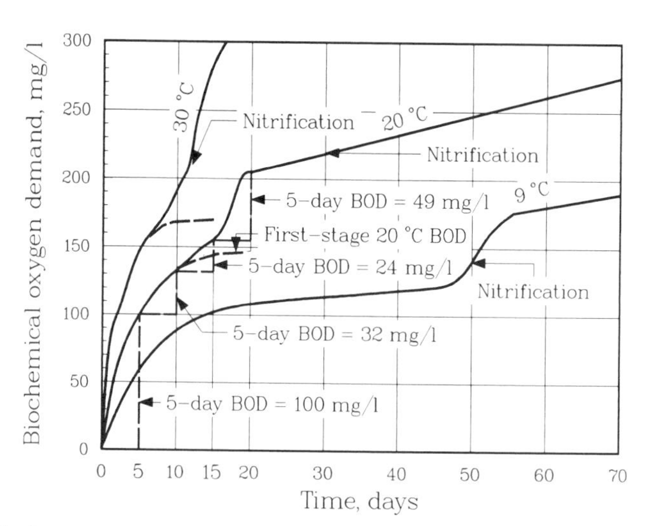

Figure 5.3 Progress of biochemical oxygen demand at 9°, 20°, and 30°C (after Therinault[11]).

K and k represent the average reaction rate coefficients, k being to log base 10 and K being to log base e. Note that $K = 2.31k$.

A similar equation can be used for the description of the second (nitrogenous) stage, provided that the lag time is subtracted from t.

One-half of the oxygen demand is satisfied after the half-time $t = 0.301/k$.

Inasmuch as the reaction-rate coefficient represents an overall reaction rate that varies throughout the reaction, the mean k may vary markedly depending on the quantity and nature of the organics present in the sample. Typical values of the mean-rate coefficient were given by Eckenfelder[8] as

	$k_{\text{base 10}}$ at 20°C
Untreated sewage	0.15–0.28 day^{-1}
High-rate filters and biological contact units	0.12–0.22 day^{-1}
High-degree biological treatment effluents	0.06–0.10 day^{-1}

The BOD curve actually yields the oxygen consumed during the reaction. Therefore, both ultimate first-stage carbonaceous BOD and the reaction coefficient are unknown and must be computed or determined graphically. A comprehensive description of many methods available for determining these parameters was presented by Gaudy et al.[9] A rule-of-thumb approximation of the BOD reaction rate coefficient was suggested 50 years ago as $k_{(\text{log base 10 at 20°C})} = 0.1\ \text{day}^{-1}$, which was established by extensive studies on polluted river waters and domestic wastes in the United States and England. The average reaction rate for domestic sewage is about 0.15 day^{-1}.

As the magnitude of the reaction rate differs from sample to sample, the proportion between the $\text{BOD}_{\text{ultimate}}$ and BOD_5 changes as well. As a result, the proportion of the first-stage demand reached in 5 days at 20°C has been observed to vary from 55 to over 90%. The BOD exerted in receiving waters is still more variable because it depends upon the nature of these waters (whether the stream is fast or sluggish, shallow or deep, smooth or filled with rocks) as well as upon the nature of waste matters discharged into them.

The importance of the reaction rate k with respect to the BOD developed at any time is presented in Table 5.4.

Effect of Temperature. Temperature affects the BOD reaction rate as well as the magnitude of the ultimate first-stage carbonaceous BOD. The conventional relationship for the BOD reaction rate is

$$k_{\text{T}} = k_{20°\text{C}} \times 1.047^{(T-20)} \qquad (2)$$

where T is the sample temperature in °C.

From this relationship it follows that the reaction rate changes 4.7% for each 1°C change of temperature. The thermal coefficient 1.047 can be used for the

TABLE 5.4 BOD Exerted[a]

Time (Days)	Percentage of BOD Exerted Related to $BOD_{ultimate}$ day^{-1} $k = 0.05$	$k = 0.10$	$k = 0.15$	$k = 0.20$	$k = 0.25$	$k = 0.30$
1 (daily)	10.9	20.6	29.2	36.9	43.8	50
2	20.6	37	50	60	68	75
3	29	50	64	75	82	87
5	44	68	82	90	94	97
10	68	90	97	99	99+	99+
20	90	99	99+	99+	99+	99+

	Ratio of BOD_t to BOD_5 day^{-1} $k = 0.05$	$k = 0.10$	$k = 0.15$	$k = 0.20$	$k = 0.25$	$k = 0.30$
BOD_1/BOD_5	0.24	0.30	0.36	0.41	0.47	0.52
BOD_2/BOD_5	0.46	0.54	0.61	0.67	0.72	0.77
BOD_3/BOD_5	0.66	0.73	0.78	0.83	0.87	0.90
BOD_u/BOD_5	1.76	1.46	1.22	1.11	1.06	1.03

[a] k log base 10.

temperature range between 5° and 30°C. For the nitrogenous (second-stage) BOD reaction, the thermal coefficient is about 1.097 between the temperature ranges of 5° and 22°C and 0.877 if the temperature is greater than 22°C.[10]

The relation between the reaction rates at 20°C and other temperatures is shown in Figure 5.4 and tabulated in Table 5.5.

Theriault[11] found the BOD_u to increase with temperature, whereas Gotaas[12] demonstrated that BOD_u did not vary, but the process took longer to reach the ultimate at lower temperatures.

There are other influences that can affect the progress of the BOD. In samples that contain an inadequate flora and fauna, the BOD reaction has a slower start shown as a lag period on the BOD curve. On the other hand, if the sample is anaerobic or contains reducing chemical substances (e.g., sulfides, secondary iron, or manganese) an immediate oxygen demand can occur as previously mentioned. Also, pH outside the optimum range or the presence of toxic compounds can affect both the reaction rate and the magnitude of the BOD.

Example 1: The BOD_5 of a typical municipal sewage was measured as 200 mg/liter at 20°C. Since the average BOD reaction rate coefficient is about 0.15 day^{-1} the ultimate first-stage (carbonaceous) $BOD_u = 1.22 \times 200 = 244$ mg/liter, using the multipliers from Table 5.4, the first day oxygen demand is $0.36 \times 200 = 72$ mg/liter, the second day demand is $(244 - 72) \times 29.2/100 = 50.2$ mg/liter [also $= 200 \times (0.61 - 0.36)$], and in 10 days the reaction is essentially completed.

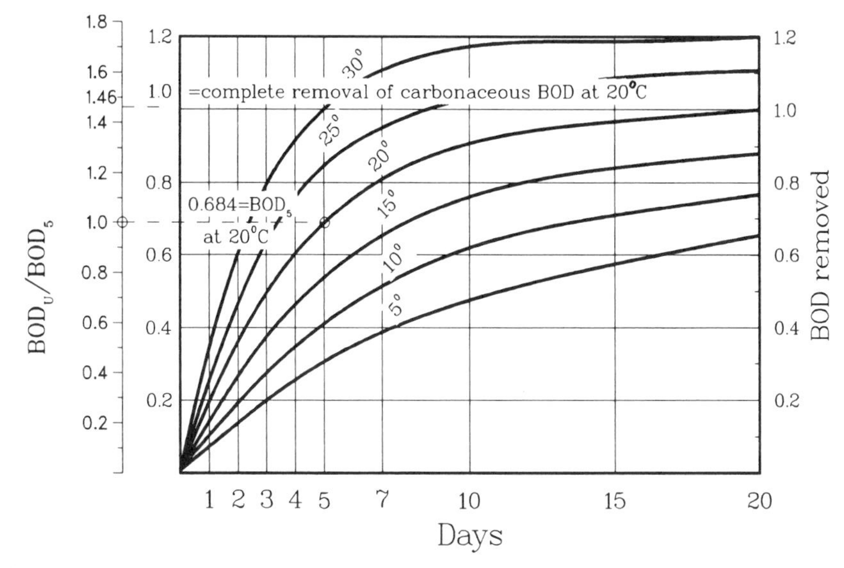

Figure 5.4 Removal of carbonaceous (first-stage) BOD at different temperatures ($k_{\text{base 10}} = 0.1\ \text{day}^{-1}$).

Example 2: BOD_5 determination of a stream sample was not performed and is missing. However, 1-day oxygen uptake was measured as 2.5 mg/liter at 20°C. Estimate the approximate BOD_5 assuming a typical BOD reaction rate of $0.1\ \text{day}^{-1}$.

From Table 5.4 the BOD_1/BOD_5 ratio for $k = 0.1$ is 0.30. Therefore the $BOD_5 = 2.5/0.3 = 8.33$ mg/liter.

Example 3: Using the standard laboratory test, the BOD_5 of sewage was measured as 150 mg/liter. Estimate first day oxygen uptake of the sewage at 20° and at 25°C. Assume that the average BOD reaction rate of $0.15\ \text{day}^{-1}$.

Again, from Table 5.4, at 20°C, the ratio $BOD_1/BOD_5 = 0.36$, hence $BOD_1 = 0.36 \times 150 = 53.4$ mg/liter. At 25°C, the reaction coefficient increases to [according to Eq. (2) or Table 5.5] $k_{25} = 1.25k_{20} = 1.25 \times 0.15 = 0.19\ \text{day}^{-1}$. This is very close to $0.2\ \text{day}^{-1}$. Then the ratio BOD_1/BOD_5 for $k = 0.2^{-1}$ is 0.41, or the $BOD_1 = 0.41 \times 150 = 61.5$ mg/liter. Better results can be obtained by interpolation.

TABLE 5.5 Temperature Conversion of the BOD Reaction Rate

Temperature	5°C	10°C	15°C	20°C	25°C	30°C
$k_T/k_{20^\circ C}$	0.50	0.63	0.79	1.0	1.25	1.58

Example 4: The 2-day oxygen demand in a standard BOD test was measured as BOD_2 = 110 mg/liter and the 5-day BOD_5 as 170 mg/liter. Estimate an approximate magnitude of the reaction constant, *k*, and the ultimate oxygen demand of the sample.

$BOD_2/BOD_5 = 110/170 = 0.65$. By interpolating in Table 5.4, this ratio corresponds to the reaction rate $k = 0.19\ day^{-1}$. From the same table, $BOD_u/BOD_5 = 1.12$, hence $BOD_u = 1.12 \times 170 = 190.4$ mg/liter.

For the purpose of this analysis, it is better to fit a curve through the points on the BOD versus time graph to smooth out the relationship rather than to use direct measurements.

Other Measurements of Organic Content of Wastewater

The Stability and Relative Stability of sewage is best measured by the BOD test. Stability was defined by Imhoff and Fair[13] as the converse of putrescibility. It can be also measured by a simpler methylene blue test, in which decolorization of this dye (0.4 ml of a 0.05% aqueous solution in a 150-ml glass stoppered bottle) records the appearance of hydrogen sulfide, or disappearance of available oxygen. The bottle is incubated at room temperature (20°C). The appearance of hydrogen sulfide can also be detected by its odor.

Stability tests are generally performed to determine whether or not plant effluents have been adequately treated. The longer the time required for the appearance of a hydrogen sulfide odor or for decolorization of methylene blue, the greater the stability.

Example 5: A water sample from a large polluted stream had a dissolved oxygen content of 15 mg/liter (the water was oversaturated because of the photosynthesis of plants and algae). The methylene blue test indicated that the sample had turned anaerobic after 3 days at the room temperature of 20°C. The 3-day oxygen demand was therefore about 15 mg/liter. Then, assuming that the sample reaction rate typical for larger polluted streams is $k = 0.1\ day^{-1}$, and using the multiplier from Table 5.3, the 5-day BOD is approximately $15/0.73 = 20.6$ mg/liter.

So far three tests that measure the quantity of biodegradable organics have been introduced: the test for volatile suspended solids (VSS), for biochemical oxygen demand (BOD), and for chemical oxygen demand (COD). All three tests estimate organics by oxidation, which in the first test is accomplished by ignition, in the second by bacterial oxidation, and in the third by chemical oxidation. However, there is commonly no close relation between the results of the three tests and, at best, only statistical correlation may be detected. It appears that the compounds volatile by ignition are not identical to those biodegradable by bacteria or oxidizable by a strong chemical oxidant.

The fourth analysis for organic matter is the *Total Organic Carbon* (TOC) test. In the TOC analysis, the carbonaceous material is oxidized at 950°C and the CO_2 gas is then measured. The inorganic carbon present in the sample is either removed by pretreatment of the sample or analyzed separately and then subtracted. The method is rapid and relatively simple and the instrument is not excessively expensive, however, the test requires an experienced technician. Since the test measures the amount of organic carbon and not the oxygen utilization, the results are not directly comparable to the BOD. One would expect the stochiometric COD/TOC ratio of a wastewater to approximate the molecular ratio of oxygen to carbon ($32/12 = 2.66$). Practically, the ratio may range from zero, for components that are resistant to chemical oxidation, to about 5.[14] Although it has been difficult to correlate BOD with TOC for industrial wastes, relatively good correlations have been obtained for domestic wastewater. A 5-day BOD/TOC ratio of 1.8 is typical for raw domestic sewage whereas a ratio of 1.0 would correspond to treated effluent.

The fifth technique is the *Total Oxygen Demand* (TOD) test. In this test, the oxygen loss during ignition at 900°C with a platinum catalyst is measured in a laboratory analyzer. The reduction of oxygen is measured by a silver–lead fuel cell detector. In this process, the total oxygen demand is measured without the shortcomings typical of the wet COD analysis and the results can be obtained in minutes. The TOD test also measures nitrogenous demand.

The theoretical oxygen demand is the amount of oxygen needed for full oxidation of the organic carbon. It can be estimated only if the chemical composition of the compound is known. In this case it is assumed that the oxidation is complete, carbon is oxidized to carbon dioxide, and organic nitrogen is converted to ammonia and then to nitrates. Table 5.6 shows some relationships between the BOD, COD, TOC, and the theoretical oxygen demand for some organic compounds of known composition.

The *chlorine demand* of a water is the amount of chlorine that must be applied to yield 0.3 mg/liter of the total available chlorine after 10 min of contact time. The chlorine demand of fresh raw domestic sewage is about 2–3 g/cap-day. It is higher for anaerobic sewage.

TABLE 5.6 Oxygen Demand of Some Organic Compounds

	Theoretical Oxygen Demand (mg/g)	Measured Oxygen Demand			Total Organic Carbon (mg/g)
		With $KMnO_4$ (mg/g)	With $K_2Cr_2O_7$ (mg/g)	As BOD_5 (mg/g)	
Lactic acid	1067	260	970	540	400
Glucose	1067	600	990	580	400
Lactose	1122	390	920	580	421
Dextrin	1185	220	950	520	444
Starch	1185	120	990	680	444
Phenol	2383	2360	2340	1700	766
Casein	1410[a]	150	1150	580	560

[a]Without nitrification.

The Redox Potential (oxidation–reduction potential—ORP) provides information on the effective reduction and oxidation strength of a wastewater or sludge. It is measured either by a chemical wet titration or by an electrode. The common units for ORP are millivolts. A negative ORP value means that most of the components present in the waste are reduced, whereas a positive ORP implies mostly oxidized components. ORP can be also expressed according to a scale—r-H—introduced by Clark[15] that represents the negative logarithm of the hydrogen pressure. The range of the scale is 0–42, whereby the values below 15 indicate reduced components, whereas values above 25 indicate oxidized components. The methylene blue indicator loses color when the r-H value is between 13.5 and 15.[16]

Biological and Microbiological Investigations

Bacteriological Analyses. These tests are needed to reveal the presence or the potential for presence of pathogenic (disease-causing) microorganisms. Water may be a carrier of pathogens including viruses, bacteria, protozoa, and the helminths. A number of diseases including typhoid, infectious hepatitis, giardiasis, and gastroenteritis can be transmitted by sewage and sewage-contaminated water.

For practical purposes, it is not realistic to monitor for all potential pathogens that may be present. Therefore, for routine wastewater analyses and water quality evaluations, indicator organisms are used, including total coliform bacteria, fecal coliform bacteria, and fecal streptococci.

The Coliform Group comprises all of the aerobic and facultative anaerobic, gram-negative, nonspore-forming, rod-shaped bacteria that ferment lactose with gas formation within 48 hr at 35°C. These bacteria can be found both in the guts and feces of warm-blooded animals (including humans), cold-blooded animals, and in soils. Thus, a positive coliform bacteria test indicates only a potential for contamination. The coliform test is performed using a multiple-tube fermentation procedure or by membrane filter techniques and the results are expressed in a number of organisms per 100 ml of sample. An inverse value—colititer—in ml per one coliform bacteria is used in Germany: 0.1 colititer corresponds to 1,000 coliforms/100 ml.

The fecal coliform test is performed at a higher temperature (44.5°C) as it has been found that coliform bacteria from sources other than gut and feces of warm-blooded animals are not capable of producing gas from lactose at this elevated temperature.

Fecal Streptococci live and multiply in the intestines of humans and other warm-blooded animals but, unlike the coliforms, they do not multiple in surface waters. Of interest is the fecal coliform–fecal streptococci ratio. An FC/FS ratio of less than one may indicate contamination by animal wastes typical for urban and agricultural runoff, whereas an FC/FS ratio greater than four could indicate the presence of human sewage.[17]

For surface water, biological analyses and evaluations are necessary to determine whether the aquatic life of a water body is healthy or if damage has been done by sewage and/or discharges of toxic materials. It should also be noted that the biological analyses reflect long-term impact of adverse water quality whereas chemical grab samples, as stated before, describe the state of the water quality only at the time the sample was taken.

Biological Investigations. These analyses and surveys commonly focus on the types of the organisms living in water (plankton), attached organisms (periphyton), and organisms living on the bottom (benthos). Data are gathered from the analyses on the organism types, numbers, and their tolerance to pollution. For this purpose, several classification systems have been introduced of which the Kolkwitz–Marson saprobity system is widely used throughout Europe. It has also been recommended in the latest edition of Standard Methods.[2] A Mean Saprobien Index can give a numerical value by which waters, including sewage, can be classified. Other numerical indices of biological quality include the Species Diversity Index introduced by Shannon and Weaver[18] and the Sequential Comparison Index recommended by Cairns et al.[19]

In addition to the estimation of organism population and density, limnologists can also measure metabolic rates of the aquatic biota, including nitrogen fixation and organic matter production by algae and oxygen uptake. The biological state of the water body can be also related to its chemistry and vice versa. Thus, both chemical and biological analyses should be performed and reported at the same time.

Toxicity Bioassays. These tests are needed to assess the potential harmful effects of wastewater discharges on the aquatic biota and subsequently on humans. In addition, toxic materials can affect the biological processes occurring in the aquatic environment and reduce the water body capability to safely accept wastewater discharges.

The toxicity test is a procedure in which the responses of aquatic test organisms are used to detect the presence or effects of one or more substances, or environmental factors, alone or in combination. The test organisms recommended by Standard Methods include algae, zooplankton species, aquatic insects, and a number of fish. The prime considerations in selecting test organisms are their sensitivity to the factors under consideration, their geographic distribution, their abundance, their economic and recreational value and ecological importance, and freedom from diseases and parasites. In fish tests the organisms should not be more than 5–8 cm long and should have a relatively short life cycle.

The test organisms are placed in containers with various dilutions of the tested (toxic) compound or wastewater plus one container with test dilution water only. The numbers of organisms surviving after the specified time period (preferable 48–96 hr) are then plotted versus the logarithm of the concentration of the toxicant in the containers. Biological toxicity of the wastewater is then expressed by the following parameters:

LC_0—maximal concentration or dilution of the wastewater at which all test organisms will survive,

LC_{50} or TLm^t—concentration or dilution at which 50% of the test organisms will survive (the exponent t denotes the duration of the test in hours),

LC_{100}—minimal known concentration or dilution at which all test organisms will be killed.

The graphic plotting technique (shown in Fig. 5.5) is recommended in order to eliminate possible errors by one sick organism, different tolerances, and/or different conditions in one container.

In addition to the Standard Methods,[2] the U.S. EPA has developed its own procedures.[20,21] Numerical information on the toxicity of various compounds is contained in Quality Criteria for Water.[22] The results obtained by these observations represent so called acute toxicity. In the United States, acute toxicity is commonly determined as 0.1 TLm^{96}. Chronic toxicity involves measurements of long-term effects and often is caused by bioaccumulation of the toxic substances in the tissue or in the vital parts of the organisms. Chronic toxicity evaluations are needed to establish maximum allowable concentrations of various components in drinking water supplies and in fish designated for human consumption. Chronic toxicity is sometimes related to the acute toxicity concentration, LC_{50}, multiplied by an application factor ranging from 0.001 to 0.1. If not specified otherwise,

$$\text{Chronic toxicity} = 0.01\ TLm^{96}.$$

Chemical, biological, and toxicological analyses of the wastewater are indispensable tools in the design of modern treatment plants, particularly in situations

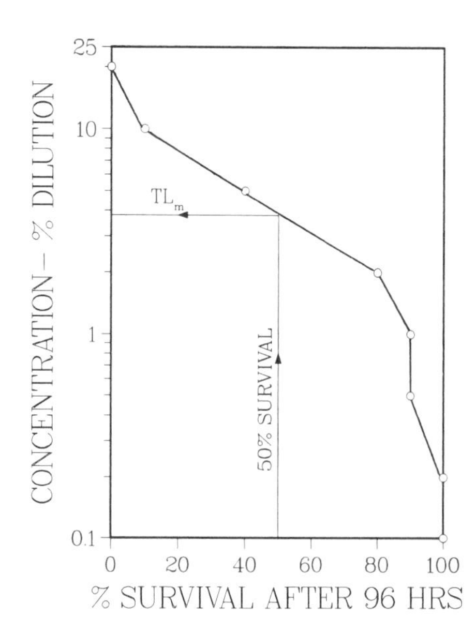

Figure 5.5 Wastewater toxicity plot.

in which significant industrial contributions are expected or in which the treatment plant is solely designed to treat industrial wastes. Unfortunately, instances in which a new treatment system can be directly designed based on previous knowledge of the wastewater composition and condition are rare. Wastewater can change when the treatment plant is constructed. The designer must commonly rely on other characteristics and parameters such as wastewater flow, number of population served, and the character of the connected industrial discharges.

Also, in cases in which a wastewater survey is feasible, its results are useful only for estimating the averages and ranges of the parameters measured. It is quite difficult to establish the characteristics of the wet weather flow contribution and of the often highly variable industrial discharges. In larger systems, dynamic storm-water models calibrated and verified by special surveys may provide suitable means for determining the variability ranges and possible scenarios for the design of treatment and disposal systems handling wet weather flows.

REFERENCES

1. A. E. Zanoni and R. J. Rutkowski, *J. Water Pollut. Control Fed.* **54,** 1756–1762 (1972).
2. *Standard Methods for Examination of Water and Wastewater,* 16th ed. APHA, AWWA, WPCF, Washington, DC, 1985.
3. *Methods for Chemical Analysis of Water and Wastes,* EPA Rep. No. 600-4-79-020. U.S. Environ. Prot. Agency, Cincinnati, OH, 1979.
4. C. I. Weber, *Biological Field and Laboratory Methods for Measuring the Quality of Surface Water and Effluents,* EPA Rep. No. 670/4-73-001. U.S. Environ. Prot. Agency, Cincinnati, OH, 1973.
5. R. H. Bordner, J. A. Winter, and P. Scarpino, *Microbiological Methods for Monitoring the Environment-Water and Wastes,* EPA Rep. No. 600/8-78-017. U.S. Environ. Prot. Agency, Cincinnati, OH, 1978.
6. *Annual Book of ASTM Standards—Water and Environmental Technology.* Am. Soc. Test. Mater., Philadelphia, PA, (published annually).
7. C. N. Sawyer and P. L. McCarty, *Chemistry for Environmental Engineering,* 3rd ed. McGraw-Hill, New York, 1978.
8. W. W. Eckenfelder, Jr., *Water Quality Engineering for Practicing Engineers.* Barnes & Noble, New York, 1970.
9. A. I. Gaudy et al., *Methods for Evaluating the First Order Constants, K_l and L for BOD Reaction,* Bioenvironmental Engineering, Oklahoma State University, Stillwater, 1967.
10. A. E. Zanoni, *J. Water Pollut. Control Fed.* **41,** 649–659 (1969).
11. E. J. Theriault, *U.S. Public Health Bull.,* 173 (1927).
12. H. B. Gotaas, *Sewage Works J.,* **20,** 441 (1948).
13. K. Imhoff and G. M. Fair, *Sewage Treatment,* 2nd ed. Wiley, New York, 1956.
14. D. L. Ford, *Application of the Total Carbon Analyzer for Industrial Wastewater Evaluation,* Proc. 23th Purdue Ind. Waste Conf. **23** p. 989–999, Lafayette, Ind. 1968.

15. W. M. Clark, *Oxidation-Reduction Potential of Organic Systems.* Williams & Wilkins, Baltimore, MD, 1960.
16. J. W. Hood, *Sewage Works J.*, **20,** 640 (1948).
17. E. E. Geldreich, L. C. Best, B. A. Kenner, and D. J. Van Donsel, *J. Water Pollut. Control Fed.* **40,** 1861 (1968).
18. C. E. Shannon and W. Weaver, *The Mathematical Theory of Communication.* Univ. of Illinois Press, Urbana, 1963.
19. J. Cairns, Jr., D. W. Albaugh, F. Busey, and M. D. Chanay, *J. Water Pollut. Control Fed.* **40,** 1607–1613 (1968).
20. *Methods for Accute Toxicity Tests with Fish, Macroinvertebrates, and Amphibians,* EPA Rep. No. 660/3-75-009. U.S. Environ. Prot. Agency, Cincinnati, OH, 1975.
21. *Methods for Measuring the Accute Toxicity of Effluents to Aquatic Organisms,* EPA Rep. No. 600/4-78-012. U.S. Environ. Prot. Agency, Cincinnati, OH, 1978.
22. *Quality Criteria for Water.* U.S. Environ. Prot. Agency, Washington, DC, 1976.

CHAPTER 6

Mechanical and Chemical Treatment Units

6.1 SCREENING AND PRETREATMENT

Screening is the first unit operation in a wastewater treatment process. Removal of larger objects in the wastewater stream such as sticks, wood, and rags protects the mechanical equipment of the plant (particularly the pumps) and prevents clogging of the pipes and weir overflows. There are two basic types of screens: the bar or rack screens and the mesh screen or sieves. The term screen more commonly refers only to the mesh cloth or wire plates (sieves) rather than to bar racks.

Mesh screens or sieves with openings of less than 10 mm are primarily used for pretreatment of industrial effluents. It is expected that floating objects greater than 3 mm in size can be collected on the mesh screens. Very fine screens with openings of less than 0.1 mm are used for final effluent polishing.

Coarse racks are made of steel bars with openings in excess of 50 mm (2 in.). Their primary purpose is to protect the sewage pumps against trash. Finer screens with bar spacing between 10 and 30 mm ($\frac{3}{8}$ to $1\frac{1}{4}$ in.) are used in advance of grit chambers and settling tanks. These openings are narrow enough to hold sticks, rags, and other larger trash but wide enough to allow objectionable excreta and toilet paper to pass through and be removed with the sludge.

Hand cleaned racks (Fig. 6.1) have a slope of 1 vertical to about 3 horizontal. This increases the effective screening surface, facilitates cleaning, and prevents excessive loss of head by clogging. Mechanically cleaned racks (Figs. 6.2–6.4) could be mounted vertically but a more common mounting is at 60–80°. The channel in which the racks are located should be designed to maintain a self-cleansing velocity of approximately 0.6 m/s (2 fps) between the racks, which should keep the screens free of sand deposits. The maximum velocity should not exceed 1 m/s to minimize the hydraulic head loss on the racks and loss of screenings into the waste stream.

To compensate for the head loss through the racks, the floor of the rack chamber is generally lowered 8–15 cm (3–6 in.) below the invert of the entering sewer. Stop gate slots should be provided ahead and behind the screen chamber so that the unit can be dewatered for maintenance. To avoid overtopping of the approach

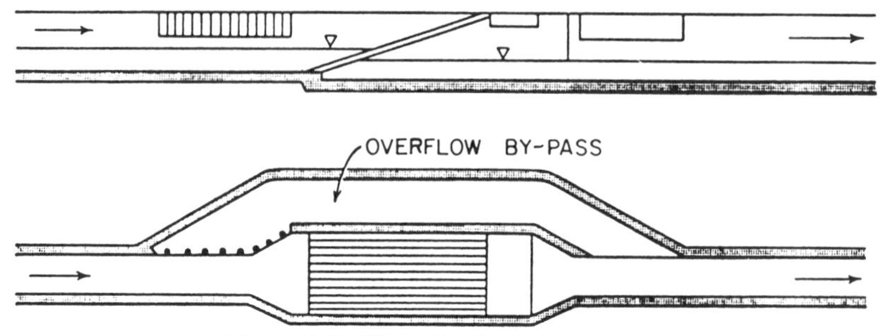

Figure 6.1 Manually cleaned bar rack.

channel or placing the entrance sewer under pressure, a bypass channel equipped with vertical bars 100 mm (4 in.) apart may be provided (Fig. 6.1).

The amount of collected materials (screenings) depends on the width of the openings between the bars. Approximate amounts of screenings are given in Table 6.1.

Normally, screenings contain about 85–90% water and upon drying they are about 85% organic.

Fine screens can collect up to 20% of the total solids load to the plant and remove up to 5% of the BOD.

In small plants, manually or mechanically collected screenings are generally wheeled away in barrows or movable containers. Segment screens with a mechanical rotating cleaning rake are popular in Europe (Fig. 6.2). In larger plants, the platform for the container can be elevated (Fig. 6.3) to allow gravity transport of the screenings in containers on rails. Oil-hydraulic cleaning rakes (Fig. 6.4) that empty the screenings in the upstream direction are also popular. This arrangement saves space. Heating of the screen compartment may be necessary in areas with low winter (subfreezing) temperatures.

Disposal of Screenings. Acceptable disposal methods are by burial, incineration at a temperature of at least 820°C, mixing with other organic sludge in the digesters, hauling to a landfill, and composting. Problems with direct disposal can

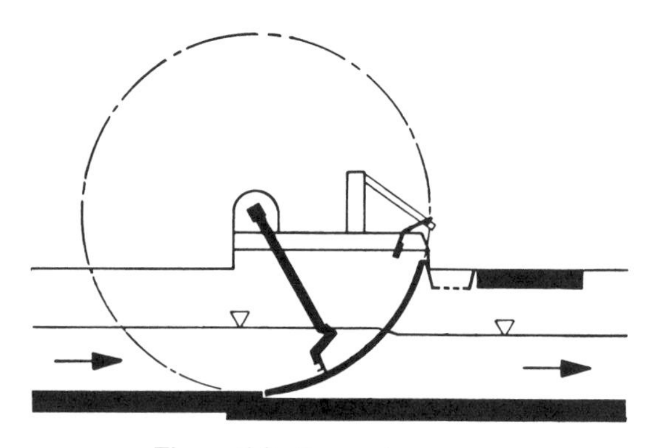

Figure 6.2 Segment screens.

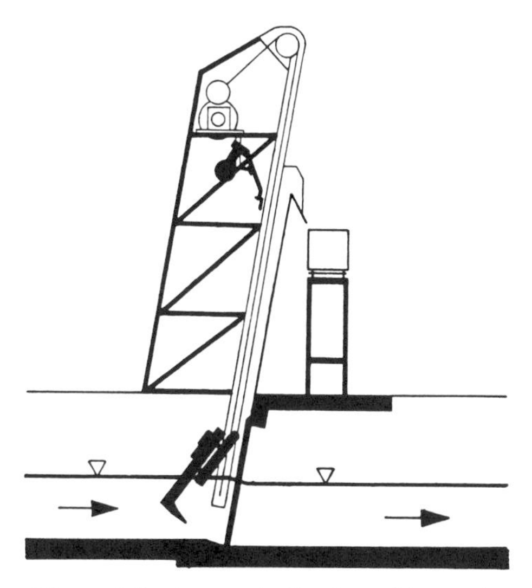

Figure 6.3 Automatically cleaned bar rack.

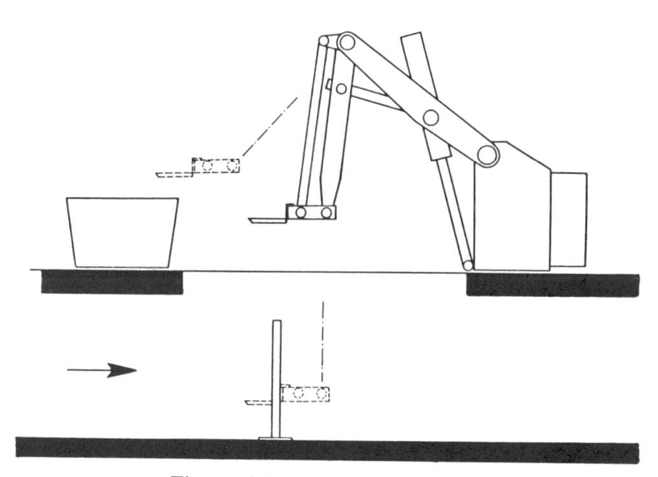

Figure 6.4 Counterflow screen.

TABLE 6.1 Quantity of Rakings and Screenings

Quantity of Screenings	Opening (mm)		
	50–75	25–50	10–25
1/1 000 m^3 of sewage	2–5	5–22	22–60
1/capita-year			
Typical United States	<0.6	0.6–3.0	3.0–8.0
Typical German	2–5	5–15	>15

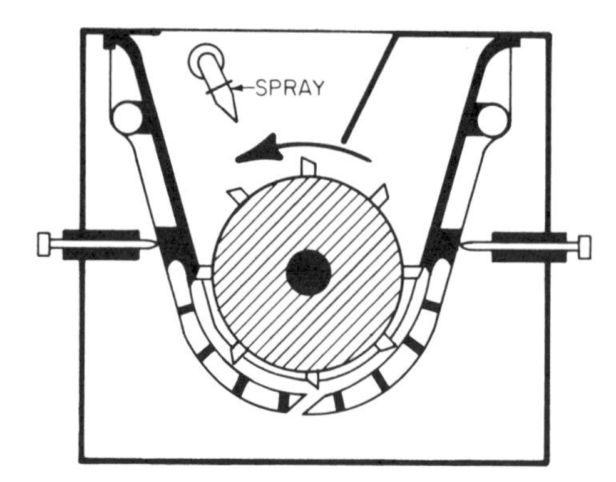

Figure 6.5 Shredding mill.

be minimized if the collected screenings are shredded and/or ground. One way of disposing of rakings or screenings is to return them, after shredding, back to the wastewater stream, where they will then behave as settleable sewage solids and will be incorporated into the primary sludge.

Examples of disintegrators (comminutors) are shown in Figures 6.5–6.7. These units can be classified as follows: (1) shredding mill (Fig. 6.5), that consists of a high-speed cylinder with rotating cutting knives, (2) a disintegrator pump (Fig. 6.6), which breaks the screenings and returns them into the wastewater stream ahead of the screens, and (3) a comminutor (Fig. 6.7), consisting of a vertical trommel or cylindrical screen rotating around a vertical axis.

In smaller plants, screenings are collected in a container and shredded once a day and in larger plants, shredding is continuous.

Grit chambers that collect larger mineral particles should be located ahead of the comminutors.[1]

Larger pieces of rags and textiles can cause problems in the shredding process. Therefore, it is advisable to remove them prior to shredding.

Preaeration

Preaeration of wastewater influents is sometimes used to ''freshen'' the sewage and control odor created in long sewer lines by formation of hydrogen sulfide and

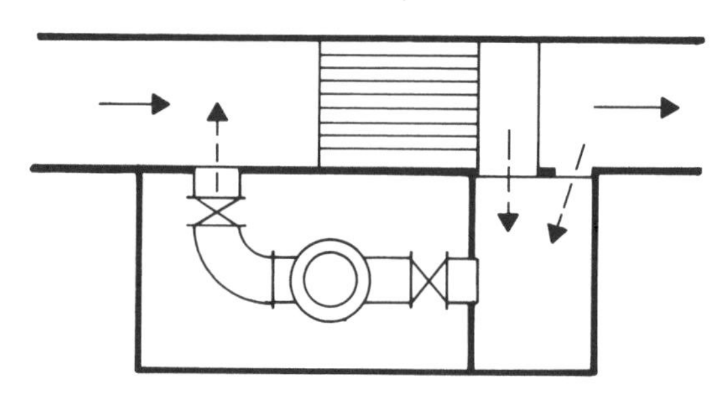

Figure 6.6 Disintegrator pump.

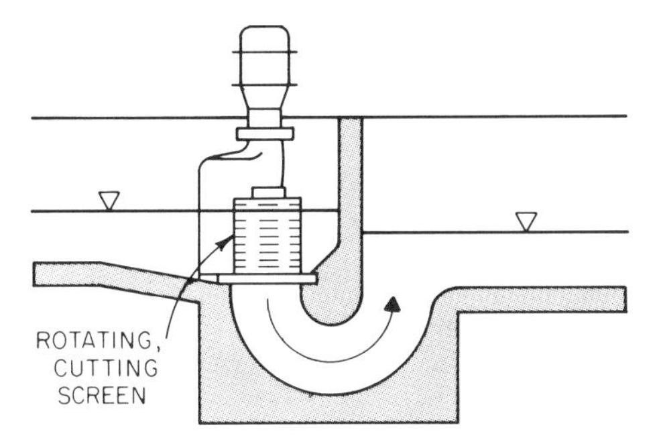

Figure 6.7 Comminuter.

other odor-causing components. Preaeration also enhances flocculation of suspended solids.

Preaeration is usually accomplished in a separate aeration chamber. The amount of air flow may range from 0.7 to 2.8 liters of air per liter of wastewater flow. If the preaeration is designed for enhancing flocculation then the detention time in the chamber should be about 45 min and the air flow/wastewater flow ratio should be 0.6–1.2. Too much turbulence is not advisable since it breaks up the solids, which may result into lower removal efficiencies of the clarifier.

Addition of chlorine or hydrogen peroxide to the influent is also effective to control odor. A more detailed discussion on odor control is contained in Section 4.8.

6.2 SKIMMING AND FLOTATION

Oil, grease, and other matter lighter than water can rise to the surface of quiescent sewage where they can form a floating surface layer and subsequently be skimmed. Every unit in which the wastewater flow is slowed down or stopped can act as a grease trap. This is especially true for primary settling tanks that should include means for skimming off the floating solids. Therefore, specially designed grease traps are commonly not needed in treatment plants dealing with municipal sewage. Grease traps should be installed in restaurants and automobile repair shops where most of the grease is discharged, however.

The efficiency of the grease-removing units depends on their surface area rather than on the detention time. This is similar to granular settling that will be discussed in the following section. The minimum surface area of a grease trap or skimming tank should equal the rate of wastewater flow divided by the smallest rising velocity, or

$$\text{Surface area } (\text{m}^2) = \frac{\text{flow rate of wastewater } (\text{m}^3/\text{hr})}{\text{smallest rising velocity } (\text{m/hr})}$$

The smallest rising velocity is the rate of ascent of the smallest grease particle that must reach the surface if a given efficiency of removal is to be obtained. This velocity can be found by recording the time required for a given portion of the total grease, oil, or other light material in the wastewater to rise to the surface of a tall glass cylinder. Then

$$\text{Smallest rising velocity (m/hr)} = \frac{\text{cylinder height (m)}}{\text{time (hr)}}$$

Example: In an experiment conducted by Zunker,[2] 95% of the linsed oil dispersed by him in water at 15°C was removed by a grease trap when the smallest rising velocity of the oil globules was 4 mm/s = 14.4 m/hr. Then at the wastewater flow of 1 liter/s = 3.6 m^3/hr the necessary surface area of the trap is 3.6/14.4 = 0.25 m^2.

Separation of solids can be enhanced by flotation. This is accomplished by blowing finely dispersed compressed air into the wastewater or by applying a vacuum above the wastewater surface. The air bubbles that are released attach themselves to the suspended matter and accelerate the rise rate of the particles. The principles of flotation are known from the processing of coal and ore. Addition of skimming flotation aids can improve the efficiency. These chemicals enhance bonding of fine particles with air bubbles and increase foam formation. If the oil in wastewater is emulsified, chemicals such as ferric chloride or alum may be added to break the emulsion. The efficiency of the flotation units depends on the air to solids ratio. This must be determined by laboratory experiments.[3] Generally, flotation treatment processes are relatively rare in municipal wastewater treatment plants but they are frequently found as a part of industrial wastewater treatment processes. Examples are mining, textile, meat packing, rendering, and pulp and paper production.

In rare instances, a treatment plant could be required to install a separate skimming tank after the comminutor and ahead of the primary settling tanks. These independent units are usually in the form of long, trough-shaped structures (Fig. 6.8) with a detention time of about 3 min. To keep heavy solids from settling to the bottom, air is blown into the wastewater from diffusers placed in the tank, however, it has been found that air addition with such a short detention time in a skimming tank does not improve the BOD or grease removal.[4] The scum that collects on the surface is retained by baffles, drawn into a separating tank in which the excess water is removed and pumped to disposal by burial with screenings or in anaerobic digester units.

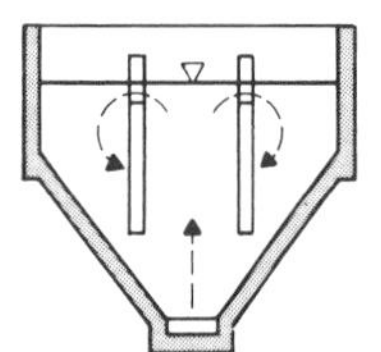

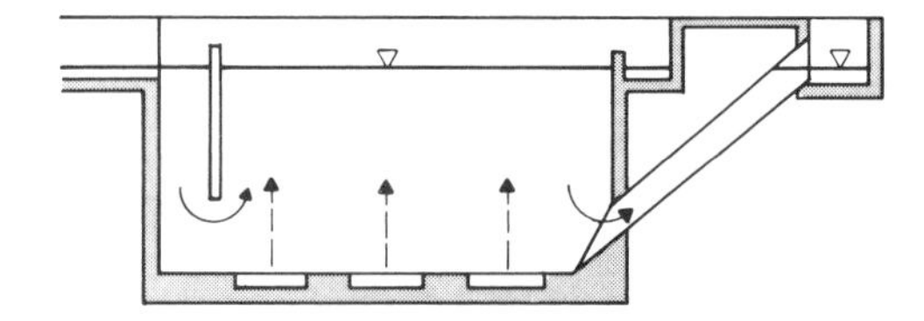

Figure 6.8 Aerated skimmer tank.

The addition of chlorine to the compressed air tends to improve the skimming efficiency, however, this practice is not commonly used.

Special grease traps are commonly installed in some industries (rendering plants, meat packing, and machine operations), restaurants, and hospitals. They could be constructed in the form of small skimming tanks with submerged inlet and bottom outlets. Presently, most of such oil- and grease-producing operations install flotation units. Gasoline and other lighter oils should not be introduced into sewers because they create fire and explosion hazards.

6.3 SEDIMENTATION

Particle Settleability and Type of Settling

Most of the solids suspended in a wastewater flow are too fine to be removed by screens and too heavy to be removed by flotation and/or skimming. However, a significant portion of the suspended solids can be removed by sedimentation (settleable solids). There are basically two types of sedimentation basins: (1) horizontal flowthrough basins (Fig. 6.9) and (2) upflow clarifiers (Fig. 6.10).

Settling or sedimentation units in a typical treatment plant are used for several purposes and at several locations throughout the treatment process: (1) in grit chambers located at the beginning of the treatment process after screening, the heavy mineral or otherwise inert solids are separated, (2) in primary sedimentation, settleable organics and finer mineral solids are removed in advance of biological treatment, and (3) in secondary or final clarification, flocculent biological solids are removed from the effluent before its discharge into the receiving water.

Settleability characteristics of wastewater solids are usually determined in laboratory or pilot experiments. Most common are the experiments in the standard 1-liter glass cylinders [Imhoff cones (Fig. 5.2)] in which the settleable portion of the solids and the time of settling (settling velocity) can be determined. However, settling in the laboratory in relatively small cylinders may not be identical to a real prototype situation. Therefore, scaled-up settling tests employing plexiglass settling columns or chambers with the same height and flow velocities as the prototype clarifier are commonly used in pilot treatability studies.[3] The results are more representative.

The design of a clarifier or settling tank also depends on the type of solids to be removed. The particles in wastewater can be characterized as (1) granular or (2) flocculent. Granular suspensions are composed of discrete particles that settle independently at a constant rate that depends on the particle mass and size (discrete

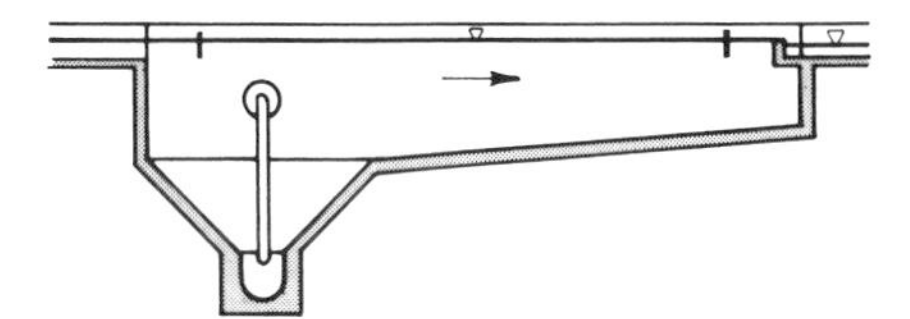

Figure 6.9 Simple horizontal flow settling tank with manual cleaning.

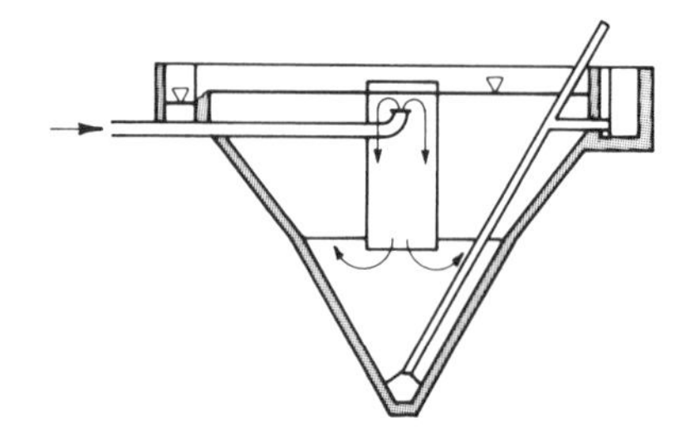

Figure 6.10 Simple upflow clarifier, hopper bottom, circular or square tank with hydrostatic sludge removal.

settling). Examples include sand, fly ash, coal, or heavier soil particles such as result from the washing of beets in sugar beet processing plants. Flocculent suspensions are composed of particles that aggregate and increase in mass and size during the settling process. Examples include iron, alum, and biological activated sludge flocs at concentrations below 500 mg/liter. The difference between discrete and flocculent settling is shown in Figure 6.11.

Flocculation can be enhanced by adding coagulating chemicals such as ferric chloride or alum (aluminum sulfate) or organic polymers.

The sedimentation of granular particles follows Stokes law. Hazen[5] was the first to recognize that the efficiency of a granular settling process depends only on the surface area of the tank and that the depth or detention time is irrelevant or secondary. The surface area can be found from

$$\text{Surface area } (\text{m}^2) = \frac{\text{wastewater flow } (\text{m}^3/\text{hr})}{\text{smallest settling velocity } (\text{m}/\text{hr})}$$

Here the smallest settling velocity is the velocity maintained by the last particle of suspended matter to reach the bottom of a settling cone or cylinder within the period of time allowed for clarification, or

$$\text{Minimum settling velocity } (\text{m}/\text{hr}) = \frac{\text{cylinder height } (\text{m})}{\text{settling time } (\text{hr})}$$

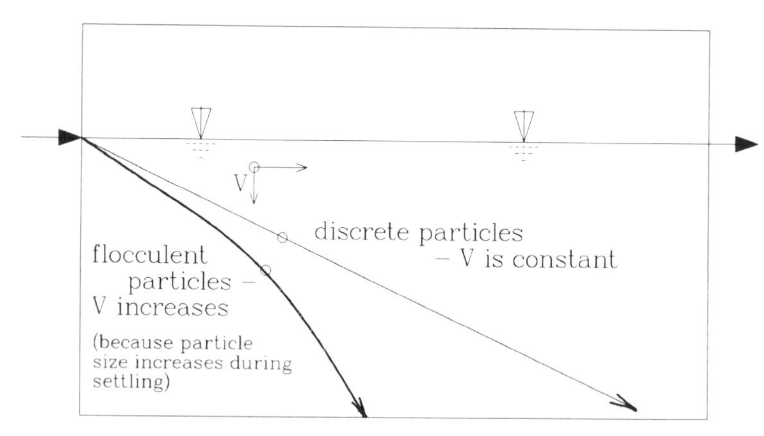

Figure 6.11 Discrete versus flocculent settling.

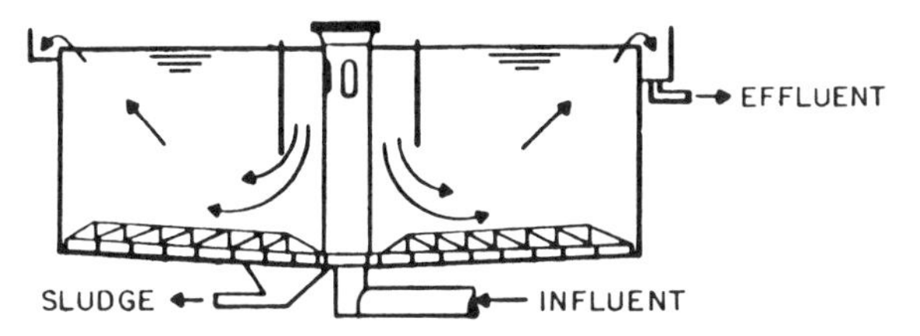

Figure 6.12 Circular center feed clarifier with a scraper sludge removal system.

In the design, the loading of the clarifier or its *overflow rate* in $m^3/(m^2\text{-hr})$ is made equal to the settling velocity of the smallest particle to be removed, or

$$\text{Surface overflow load } [m^3/(m^2\text{-hr})] = \text{smallest settling velocity } (m/hr)$$

Example: In a laboratory experiment with a coal washing wastewater performed in an Imhoff cone, the desired clarification of granular suspended solids was achieved after 4 hr of settling. This was ascertained by measuring the turbidity of the supernatant liquid in the cylinder and by recording the volume of the deposited sludge collected on the bottom. It can be expected that the solids removed in the clarification process will have settling velocities greater or equal to 0.4 m/4 hr = 0.1 m/hr. Hence, the design surface loading of the clarifier is then 0.1 m/hr = 0.1 $m^3/(m^2\text{-hr})$ and if the wastewater flow is, for example, 10 liters/s = 36 m^3/hr, then the corresponding surface area of the clarifier should be (36 m^3/hr)/[0.1 $m^3/(m^2\text{-hr})$] = 360 m^2.

Upflow clarifiers (Figs. 6.12 and 6.13) are designed by determining the surface area only. Their upflow velocity should not be greater than the smallest design settling velocity in order for the particles to be retained effectively in the clarifier. For discrete settling, depth has no effect on the design and, therefore, the most efficient clarifiers could be rather shallow with a large surface area. Larger depths may be necessary for reasons other than clarification efficiency, such as the requirements for the bottom slope for an effective transfer of solids toward the sludge hopper or providing volume for the accumulated sludge. For flocculent settling, depth and detention time must be considered.

Tube Settlers and Lamella Separators. The fact that the most important parameter in the design of settling tanks is their surface area led earlier designers to suggesting placement of horizontal or inclined trays in the clarifier. This would effectively increase the surface area and, hence, the removal efficiency. However,

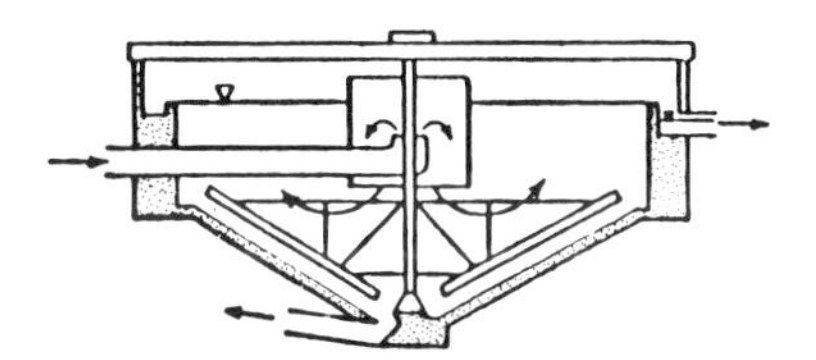

Figure 6.13 Upflow clarifier with a conical bottom.

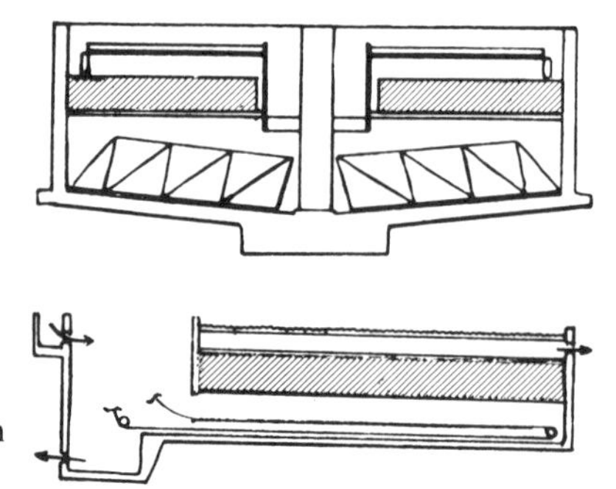

Figure 6.14 Tube settlers installed in an existing clarifier.

problems were encountered with the removal of solids accumulated on the trays. Tube settlers, which are now becoming very popular devices for improving clarifier efficiency, work on the same principle (Fig. 6.14). The tube settlers are made of plastic modules that, when combined, form a large number of tubes with 2.5–3.5 cm (1–1.5 in.) squares. The effective surface area is then increased several times as each tube plate vertically projected area could theoretically be considered. Practically, the effective area of tube settlers is up to four times the projected area.[4,6]

The tubes are inclined at an angle of 45 to 60°, which is steep enough to cause the settled sludge to slide down the tubes. Biological growths in the tubes can cause operational problems, such as odors and a reduction of efficiency.

Settling Velocity

Approximate settling velocities of particles of different sizes and specific gravity are shown in Table 6.2.

The specific gravity of quartz sand is about 2.65, that of coal is 1.5, and that of settleable sewage solids is 1.2 or lower. Specific gravity is a measure of relative density of the compound relative to that of water. In the previous example, the settling velocity of 0.1 m/hr would correspond to coal particle diameters below 0.01 mm.

The settling velocity of smaller particles is affected by viscosity and, hence, by

TABLE 6.2 Settling Velocities (m/hr) of Particles of Different Size (mm) and Specific Gravity at 10°C

	Diameter (mm)						
	1.0	0.5	0.2	0.1	0.05	0.01	0.005
Quartz sand	502	258	82	24	6.1	0.3	0.06
Coal	152	76	26	7.6	1.5	0.08	0.015
Typical settleable sewage solids, less than	122	61	18	3	0.76	0.03	0.008

temperature. If the settling velocity at 10°C is 1.0 then at 0°C the settling velocity would be 0.73 and at 20°C it would be 1.3.

Besides the settling velocity, performance of clarifiers with horizontal flow is affected by other factors. The horizontal flowthrough velocity may affect the turbulence level in the clarifier and cause resuspension of the sludge accumulated on the bottom. However, the limiting forward velocities are quite large and usually are not exceeded. Typically, the limiting horizontal velocity should not be more than 1.2 m/min = 72 m/hr. The horizontal velocity in m/hr can be simply computed from

$$\text{Velocity (m/hr)} = \frac{\text{wastewater flow (m}^3\text{/hr)}}{\text{cross-sectional area (m}^2\text{)}}$$

Grit Chambers

Design Considerations. The purpose of grit chambers is to remove sand and other heavy, primarily inert mineral and organic matter. Examples of organic matter removed herein include seeds and coffee grounds. The grit chambers are installed if large quantities of grit are expected from combined sewers during stormwater runoff events. Removal of these granular substances prevents wear of machinery, such as pumps and comminuters, limits grit accumulation in sludge digesters, and facilitates handling of primary sludge. Coarse racks should be placed in advance of grit chambers.

At normal flow velocities, granular grit particulates are settled onto the bottom of the conduits leading to the treatment plant. With this assumption, it is possible to size the grit chamber, considering the flow in the lower, grit-containing portions of the cross section of the conduits.

The amount of grit collected in a grit chamber depends on factors such as the contributing area, climate, street cleaning, garbage grinders, and discharge of industrial wastes. Typically, the amount of grit is 2–10 liters/capita-year or 2.3–180 liters per 1 000 m^3 combined sewage treated (0.3–24 ft^3/mg). The lower numbers are typical for established population centers with a higher population density whereas higher values are for lower density urban areas.

The collected grit should be relatively clean and without significant quantities of organic matter that would make it offensive. By experiments it has been established that at velocities of about 0.3 m/s (1 fps), most of the lighter organic particulates will remain in the flow whereas most of the grit will settle out. Therefore, the cross-sectional area of a grit chamber (width × depth) should be

$$\text{Cross-sectional area (m}^2\text{)} = \frac{\text{flow (m}^3\text{/s)}}{\text{velocity (0.3 m/s)}}$$

The surface area of the grit chamber should then be computed from the smallest settling velocity of the grit particles to be removed in the grit chamber. Typical overflow rate is 20–40 m^3/m^2-hr. From the information given in Table 6.2, this

TABLE 6.3 Experimental Efficiencies of Grit Chambers[a]

	Removal at Overflow Rates of		
Diameter	100%	90%	80%
0.16 mm	12	16	20 m/hr
0.20 mm	17	28	36 m/hr
0.25 mm	27	45	58 m/hr

[a] After Kalbskopf.[7]

overflow rate would correspond to settling velocities of sand particles 0.1–0.15 mm in diameter. This overflow rate is about 30 times larger than is typical for primary clarifiers.

Kalbskopf[7] found that the removal efficiency of a given sand particle will fall into a certain probabilistic range, which means that, for example, not all sand particles with a diameter of 0.2 mm will be removed at an overflow velocity of 82 m/hr, as indicated in Table 6.2. The removal efficiencies measured by Kalbskopf are given in Table 6.3.

Types of Grit Chambers. There are basically two types of grit chambers generally in use: (1) Narrow rectangular channels and (2) aerated chambers. Typical examples of grit chambers are shown in Figures 6.15 and 6.16. Narrow basins (Essen type) should have a length long enough to provide storage for several days of the accumulated grit. These narrow tanks provide the best performance at typical detention times and overflow rates.

A problem in the design of grit chambers is to keep the velocities reasonable uniform during highly variable storm flow events. If the flow does not fluctuate significantly, one grit channel is sufficient. To maintain reasonably uniform velocities at variable flows, such channels are equipped with a flow restricting outlet or a flow proportional weir (Fig. 6.17). When the flow increases, the depth and flow area of the grit chamber rise proportionally. With these arrangements, another channel can be added for safety and to handle extreme flows. Otherwise, single grit chambers should be equipped with a bypass or dry weather channel. With more variable flow conditions, multiple channel grit chambers should be designed. In

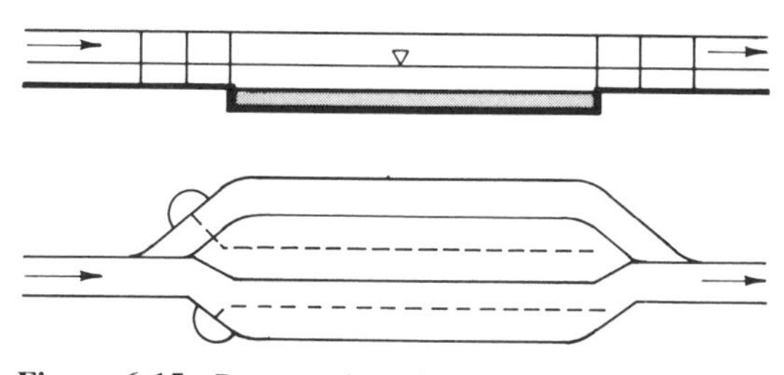

Figure 6.15 Rectangular grit chamber (Essen type).

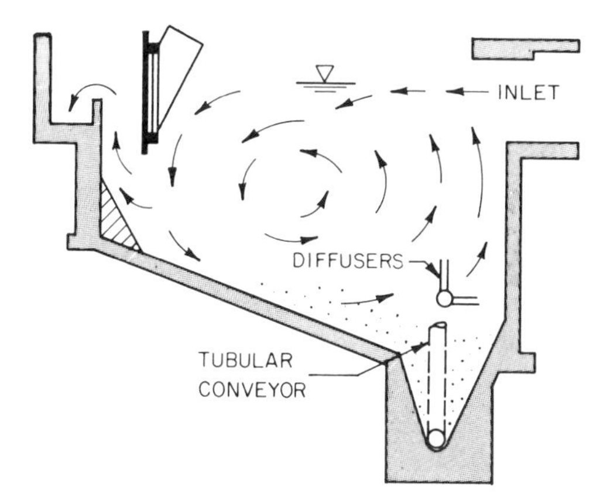

Figure 6.16 Aerated grit chamber.

such systems, the channels are switched on and off depending on the rate of the incoming flow.

The bottoms of grit channels should be located 15–45 cm (6–28 in.) lower than the invert of the inlet and outlet channels, but it should be horizontal and should not contain deep storage hoppers. The grit from the bottom can be removed manually or mechanically. Mechanical conveyors similar to those employed in settling tanks (Fig. 6.20) have been commonly used to convey the grit to a sump or hopper from which it can be removed by an air lift pressure pump or jet pump. The volume of the hopper should not be too large or the organic matter will accumulate and decay.

Example: A grit chamber serving a community of 50 000 people is 20 m long and 1.2 m wide. Typical dry weather flow during daily hours is 556 m^3/hr. What is the minimum size of the sand particles removed in the chamber.

The surface area is $20 \times 1.2 = 24\ m^2$. The overflow rate that corresponds to the settling velocity of the particles is $556/24 = 23$ m/hr. From Table 6.3 it can be estimated that about 95% of the 0.2-mm-diameter sand particles will be removed in the chamber. During wet weather conditions the flow in the plant will be about five times greater (assuming that the input is from combined sewer systems), therefore a second chamber should be included. The overflow rate for both chambers

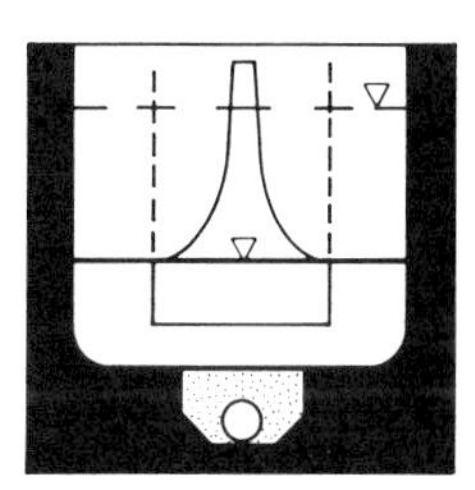

Figure 6.17 Flow control weir.

is then $(556 \times 5)/(24 \times 2) = 58$ m/hr. Table 6.3 then indicates 80% removal of sand particles with a 0.25-mm diameter.

In aerated flow chambers, spiral circulation induced by the air blown into the chamber controls the quantity and organic content of the collected grit. The air is introduced along the side of the chamber as shown in Figure 6.16. Because the grit removal is controlled by the roll velocity and not by the forward velocity as in rectangular grit chambers, higher removal efficiencies can be maintained over a larger range of flows. The roll velocity and hence the removal efficiency can be adjusted by the rate of air input. The surface area of an aerated grit chamber is computed similarly to that of rectangular chambers as shown previously. The cross-sectional area of the chamber should provide a flowthrough velocity of at least 0.2 m/s at the highest design flow. The air flow is about 0.5–1 m^3 per m^3 of the grit chamber volume per hour. The detention time in the chamber should be 2–5 min, typically 3 min. If the grit chamber is also used for preaeration of sewage the detention time should be longer.

Example: Design an aerated grit chamber for a dry weather flow of $Q_d = 200$ liters/s and wet weather flow $Q_r = 1\ 000$ liters/s $= 1\ m^3/s = 3\ 600\ m^3/hr$.

Select two chambers, each 2 m wide, 3 m deep, and 20 m long. Assuming that for accumulation of the collected sand, the bottom of the chamber is lowered 0.5 m below the invert of the inlet and outlet conduits, the effective cross-sectional area of the chambers is $2 \times 2 \times (3 - 0.5) = 10\ m^2$. The surface area is $2 \times 2 \times 20 = 80\ m^2$ and the wet weather surface overflow rate is $(3\ 600\ m^3/hr)/(80\ m^2) = 45$ m/hr. According to Table 6.3, sand particles with a diameter of 0.25 mm will be removed with approximately 90% efficiency. The horizontal velocity is $v = 1\ m^3/s/10\ m^2 = 0.1$ m/s, which is less than 0.2 m/s. The detention time is $20\ m \times 10\ m^2/(1\ m^3/s) = 200$ s, which corresponds approximately to 3–4 min. The recommended air flow volume is $200\ m^3 \times 1\ m^3/(m^3\text{-hr}) = 200\ m^3/hr$. Select two compressors each with $100\ m^3/hr$ air flow capacity.

Morales and Reinhart[8] conducted a full-scale evaluation of five different types of aerated grit chambers. The study indicated that narrow configuration channels provided the best removal efficiency with a good quality of collected grit. It appears that the longer narrow channels provide an opportunity for washing the collected sand and minimizes short-circuiting. The study also showed that the position of the air headers in relation to the direction of flow was crucial in establishing the proper flow patterns. The quality of grit also depends on the grit removal equipment. In addition, aerated grit chambers provide preaeration of the incoming sewage.

In many U.S. installations, flow measuring devices, such as Parshal flume or Palmer–Bowles flume, are incorporated into the grit chamber.

The grit is typically disposed of by burial and/or on landfills.

Flocculent Settling

Thus far the discussion has focused on granular discrete settling during which the particles settle independently from each other and with a constant settling velocity. However, settling of organic solids present in wastewater may not conform to this model because such solids are more flocculent than granular ones. Settling of chemical sludges and solids from biological treatment units in lower concentrations is strictly flocculent and has different settling properties than granular solids.

With flocculent settling, the depth and detention time are very important design parameters. As the flocs settle to the bottom, they coagulate and thus become larger and heavier. Therefore, their settling velocity increases with depth (Fig. 6.11). Under these circumstances, the surface overflow rate is not as important as for granular settling. Therefore, the design for flocculent settling must also consider the depth and volume (detention time) of the clarifier.[9,10]

In upflow clarifiers, the upward movement of water is important as the incoming fine flocs come in contact with the heavier ones that settle from the surface to the bottom (Figs. 6.12 and 6.13). It is then advantageous to introduce the inflow in the lower portion of the clarifier in order to receive the benefit of enhanced flocculation by contact of new finer flocs with older heavier flocs in the tank. Also, a sludge blanket can form in the mid-section of the tank through which the incoming water and solids are filtered as they move vertically upward.[11] The sludge blanket forms a distinct interface between the clarified surface water and the entire mass of the blanket more-or-less settles as one.

Clarifier Operation, Malfunctions, and Remedial Measures

A massive spill of the sludge blanket solids over the weir into the effluent is a primary cause of clarifier failure.[12,13] To prevent spillage of solids, German practice proposes occasional interruption of the flow into the clarifier,[14] which puts responsibility on the operator to observe the levels of the sludge blanket in the clarifier and adjust the flow accordingly. Keinath[15] developed a computerized operational technique for control of secondary clarifiers.

Density currents can reduce the efficiency of clarifiers treating flows with higher concentrations of incoming solids. Since the influent has a higher density compared to the density of the clarified liquid in the tank, the denser water sinks quickly to the bottom and moves along the bottom toward the effluent weir. The detention time is then shorter and the particles do not have enough opportunity to flocculate. Furthermore, the clarified lower density water is forced to circulate back toward the inflow. Density currents can also be caused by differences in the temperature and/or salinity. To minimize the effect of density currents, the settling tanks should be relatively shallow and elongated. For rectangular clarifiers, the recommended depth-to-length ratio should be about 1/20.

A dye test can detect short-circuiting of flows in a clarifier. In the test, Rhodamine-WT dye is injected in the influent to the clarifier and the distribution of the effluent concentration of the dye is then measured and plotted versus time. A well-functioning clarifier would exhibit a short duration, spike-resembling concentration curve. Elongated and tailing curves indicate short-circuiting or density currents.

Earlier design practices of clarifiers handling flocculent settling (primarily secondary clarifiers removing biological solids and clarifiers dealing with chemical sludges) relied on the surface loading (velocity) parameter. Typical "settling" velocities for activated sludge solids are in the range of 0.5–3 m/hr, depending on the size of the flocs. U.S. design practices give preference to circular clarifiers with radial flow from the center (Fig. 6.18). Upflow clarifiers with a larger depth (e.g., 5 m) require less surface area if the inflow is located in the lower portion of the tank and the direction of the flow is vertical.[16]

The solids flux diagram that can be developed from pilot studies with a laboratory scale clarifier provides a better basis for clarifier design in which flocculent settling and sludge thickening take place.

Design and Selection of Settling Tanks

Design Parameters. Typical municipal sewage contains both granular and flocculent solids. For settling of these materials, the required capacity of primary settling tanks can be determined either

1. from the necessary detention time (e.g., 2 hr) for flocculent settling. Then, increasing the surface area will improve the settling efficiency, or
2. from the hydraulic surface loading (e.g., 1 m/hr = 1 $m^3/(m^2\text{-hr})$. It should be noted that a deeper tank will perform better than a more shallow tank with the same surface area.

The detention time is computed according to the following simple formula:

$$\text{Detention time (hr)} = \frac{\text{settling volume } (m^3)}{\text{wastewater flow } (m^3/\text{hr})}$$

The actual detention time is usually shorter because of short-circuiting or density currents. It can be measured by a dye test.[17]

Figure 6.19 shows approximate removal efficiencies that were measured by Sierp in laboratory settling cylinders.

Laboratory experiments using 1-liter cylinders (Fig. 5.2) may not be the best

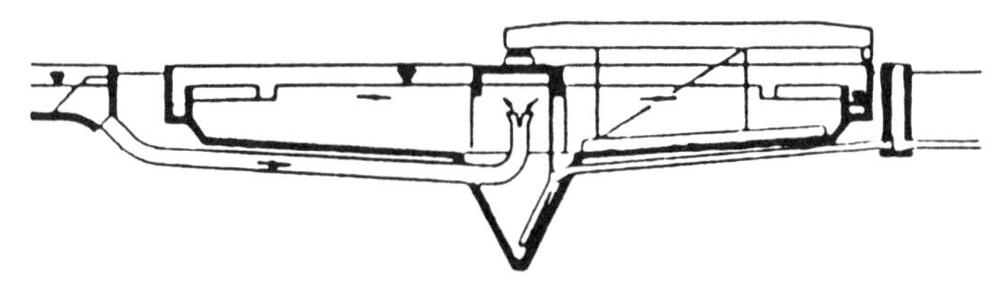

Figure 6.18 Radial flow clarifier.

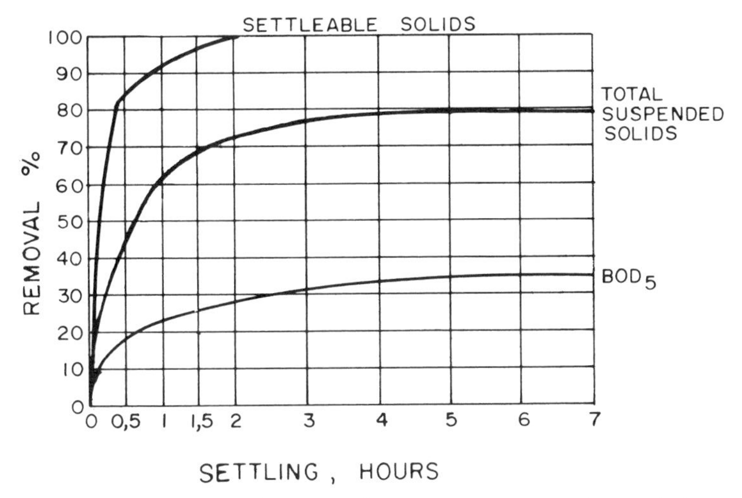

Figure 6.19 Effect of retention time on settling efficiency for typical municipal wastewater (after Sierp).

means of determining the smallest design settling velocity. Such experiments do not account for the increase of the settling velocity resulting from flocculation. Also, the settling time in a 0.4-m-high cylinder is not the same as the detention time, for example, in a 2-m-deep tank. For the design, pilot studies using scaled down pilot clarifiers (usually consisting of cylinders with the same depth and hydraulic loading as the prototype) will provide better design parameters.

The additional increase in efficiency of upflow clarifiers can be determined only by larger scale pilot studies. Commonly, the increase in efficiency may be negligible unless flocculated solids, for example, from secondary clarification of biological flocs or chemical sludge, are present. Van der Zee investigated the settling efficiency of primary sludges in 1.2-m-high laboratory cylinders and found that if some biological solids from secondary clarifiers (or from sludge return pipe) are introduced into the settlers with primary sludge, the best improvement in settling efficiency was achieved when the hydraulic surface loading (upflow velocity of water) was kept at or below 2.2 m/hr.[18]

For the detention time of primary settling tanks treating a typical municipal sewage, European practice recommends an useful average value of about 1.5 hr, provided the active depth of the tank does not differ substantially from 2 m (6.5 ft). The 1.5-hr detention time is based on the 12 hr daily average flow value (Fig. 5.1). This would correspond to an approximately 2-hr detention time if the 24-hr average daily flow value is used as the basis for the design. The minimum depth parameter of a settling tank should also consider an allowance for sludge accumulation. U.S. practice recommends depth of a settling tank to be between 2.1 m (7 ft) and 3.6 m (12 ft).[19]

In cases in which it is anticipated that the sludge is mostly of granular character rather than flocculant, it is better to base the clarifier design on a hydraulic surface loading value. The recommended hydraulic surface loading value for primary clarification in the United States is 1 m^3/m^2-hr (600 gpd/ft^2). Assuming 2 m effective depth of the settling tank, the surface loading rate can be also related to the previously specified detention time as

$$1.5 \text{ hr detention} = 2 \text{ m}/1.5 \text{ hr} = 1.33 \text{ m/hr } (800 \text{ gpd/ft}^2) \text{ surface loading}$$

$$2 \text{ hr detention} = 2 \text{ m}/2 \text{ hr} = 1 \text{ m/hr } (600 \text{ gpd/ft}^2) \text{ surface loading}$$

Using these established design values for settling tanks—1.5 hr detention and 1.33 m/hr hydraulic loading based on the daily 12-hr average flow rate (or 2 hr detention and 1 m/hr surface loading based on a 24-hr average flow)—settling tanks in most cases will be economically sized. It is not advisable to deviate substantially from these values. If the tanks are smaller, a sizable portion of the suspended matter and BOD will escape over the weir into the effluent and impose an unnecessary burden on the subsequent biological units. If the units are made larger, their efficiency will not be appreciably better, and there is a danger that the effluent and water in the tank can turn anaerobic and offensive. Also, the construction cost is unnecessarily increased.

These rules of design are based on presumptions that the treatment plant handles a typical aerobic (fresh) municipal wastewater that is not excessively diluted by stormwater.

There are special cases in which larger tanks may be required. These include cases in which industries may discharge wastes that include toxic particulates that inhibit the decomposition of sludges (e.g., copper- or cadmium-containing wastes), or when the industrial contributions to the municipal wastewater are toxic. In these cases, larger detention times may provide a buffering effect and a time for consolidation of the accumulated sludge. However, it should be noted that present U.S. environmental regulations prohibit discharges into municipal sewers of toxic substance or substances that could adversely interfere with the treatment process in a municipal treatment plant. Such components must be removed by pretreatment at the site of their origin.

Also, when the influent contains abnormally large contributions of stormwater (more than twice the daily dry weather flow), a larger than normal settling capacity may be required.

In such special cases, the detention time in the settling tanks is longer than the recommended 2 hr. For example, in England, it is possible to find treatment plants with primary settling tanks that have detention times ranging from 6 to 12 hr.

The inlet portion of a typical settling tank is usually equipped with a baffle to slow down the incoming velocity and distribute the flow uniformly over the width and depth of the basin. The baffle should extend above the water surface to trap oil and grease. Perforation of the baffles is often used to promote even distribution of flow. When several tanks in parallel are used, a flow divider channel with baffles is needed. Such channels work best at slow velocities. To prevent sludge accumulation in these channels, aeration may be needed.

The effluent from the tank is withdrawn across a horizontal weir that stretches across the entire width of the outlet. A scum-board or shallow baffle, placed in front of the weir and submerged about 0.25 m (10 in.), will keep the floating solids, scum, and grease away from the effluent. When the length of the weir in relation to flow is very large, such as in circular tanks where the weir encompasses the entire perimeter of the tank (Figs. 6.10 and 6.18), it is very difficult to maintain uniform weir loadings and, as a result, the performance of the tank can be adversely affected. To minimize short-circuiting it is common to break up the weir crest into triangular notches spaced about 0.3 m apart that will release the effluent in jets that are not affected by small rates of flow, wind, and other effects, and provide more even distribution of the effluent flow. The weir plates should also be vertically adjustable.

The length of the weir is determined from the recommended values of the weir loading, which for a primary clarifier should be between 5 and 15 m^3/hr per 1 m of the weir length (10 000–30 000 gpd/ft). In Germany, higher weir loading rates are common.

With a fixed overflow weir, the water surface level remains constant. However, when the upper portion of the settling tank serves for retention or equalization of the flow, two overflow weirs and effluent channels located at different elevations should be installed, or the tank should be equipped with a floating overflow channel with stops at two endpoints. The sewage surface level in the tank can then, for example, fluctuate between the dry and wet weather levels.

Types of Settling Tanks. Settling tanks differ in shape, inlet and outlet arrangement, and method of sludge removal. As the accumulated sludge undergoes decomposition—particularly during warmer periods—it should be allowed to remain in the tank only as long as the production of sludge gas is minimal and sludge gas bubbles are not formed. Methods of sludge removal and types of sludge—removing devices lead to the following categories of settling tanks:

1. Tanks that are emptied for sludge removal (Fig. 6.9). These more-or-less archaic tanks are usually rectangular with a bottom slope of 1–2% toward a sump. Sludge is allowed to accumulate until gas evolution into the overlying sewage is noticeable. In colder climates this usually happens in 1–2 weeks and in warmer climates in several days. The basin is then put out of service, the overlying sewage is emptied, and the sludge is flushed into a sump from which it is withdrawn by gravity or by pumping, or by hydrostatic pressure after the tank has been refilled. Allowance in the tank volume (detention time) must be provided for 1–2 weeks storage of the accumulated sludge.
2. Funnel-shaped upflow clarifiers—Dortmund type (Fig. 6.10). These tanks have a funnel-shaped sludge hopper at the bottom from which the sludge is removed daily without interrupting the operation. The effective volume of the clarifier is between the top of the hopper and the surface. A typical hydraulic loading is 1–3 m/hr (600–1 800 gpd/ft^2).
3. Tanks with sludge scrapers. The scrapers push the sludge along the tank bottom to the sludge outlet. The scrapers are attached either to rotating arms (Fig. 6.18), endless chains (Fig. 6.20), or a traveling bridge (Fig. 6.21).

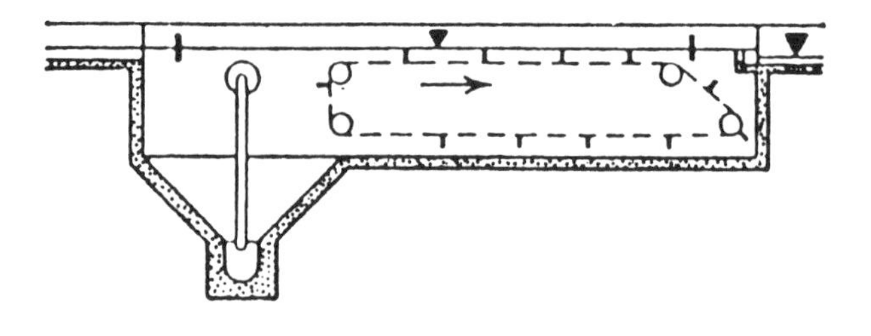

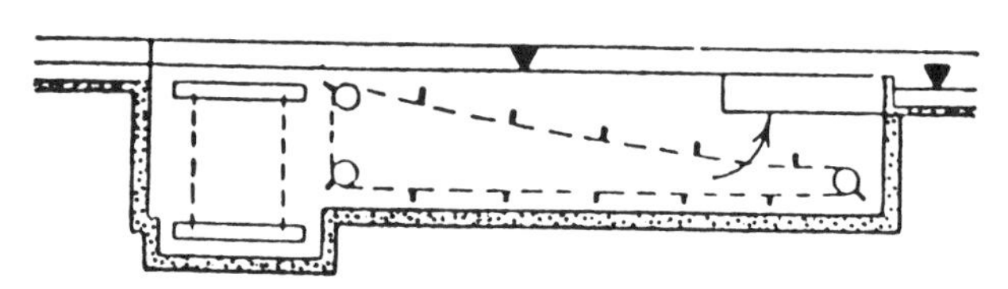

Figure 6.20 Rectangular clarifiers with sludge scrapers and scum collectors.

The velocity of the scraper under water is 10–60 mm/s (2–12 ft/min), depending on the specific weight of the sludge. For rotary scrapers the tank must be either circular or square with rounded corners. The sewage is directed through the tank either transversely from one side to the opposite side (Fig. 6.20) or from the center outward (Fig. 6.18). Vertical, horizontal, or inclined flows are possible. Radial flow inward from the tank periphery is also possible.[20] The tanks may be equipped with a sludge sump that will hold about 1 day's storage of sludge. The sump acts as a thickener and reduces the sludge volume. Light flocculant sludges such as activated sludge and mixtures of activated and primary sludges can be removed continuously. Heavier primary sludges in larger rectangular tanks may need cross-conveyors or scrapers attached to mechanical pulling vehicles. The scrapers can also become scum collectors.

According to investigations by Groche[21] rectangular or funnel-type settling tanks perform best if the effective tank volume is less than 1 000 m^3

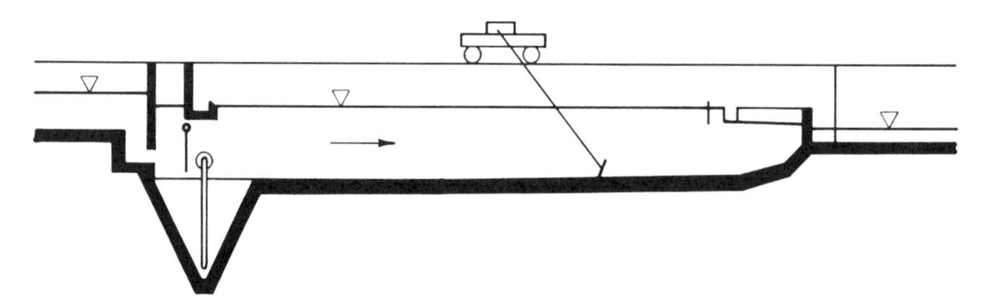

Figure 6.21 Rectangular clarifier with traveling bridge for sludge removal.

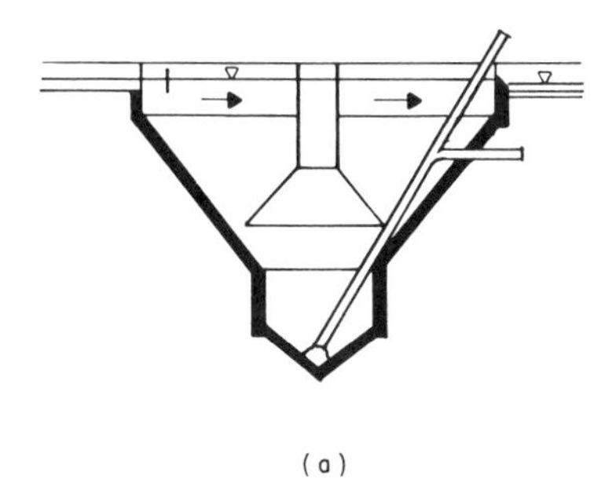

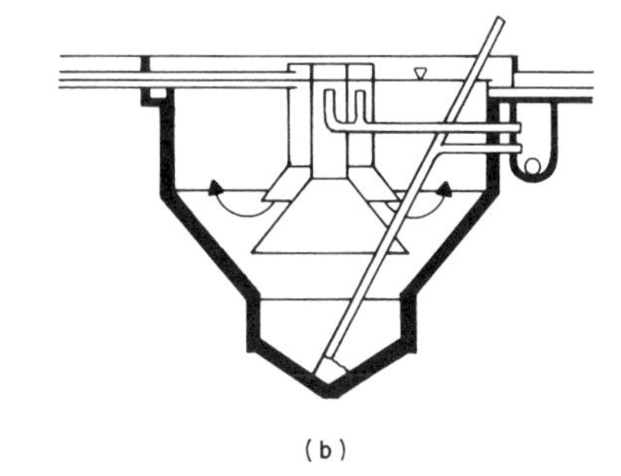

Figure 6.22 Examples of two-story clarifiers: (a) horizontal flow, (b) vertical and radial flow.

(35 000 ft^3). Between 1 000 and 2 000 m^3 volume, circular and rectangular tanks perform with about the same efficiency.

4. Tanks with movable sludge pipes. The rotary sludge-removing mechanism described under 3 can support sludge suction pipes or submersed pumps instead of scrapers. The orifice openings move along the bottom and sweep it clean if the sludge is light.
5. Two-storied tanks (Fig. 6.22). In these tanks, the lower compartment is for sludge storage and the upper compartment is for settling. Settled particles slide from the upper into the lower compartment. The two stories are so constructed that rising gas bubbles and sludge particles cannot enter the settling compartment. This arrangement represents the simplest and safest removal of decomposable solids since the sludge is not stored in the same volume as the cleaned supernatant and there is no need for cleaning mechanisms. The entire sedimentation chamber can be counted as an effective settling volume.

   If the sludge compartment is small, that is, if sludge storage is provided only for a few days of accumulated sludge, these type tanks do not differ substantially from conventional clarifiers. However, the sludge chamber can be designed to serve as a digestion chamber. In such cases, 2–3 months sludge storage volume should be considered. Settling tanks with the lower compartment designed as a digester are called Emscher* or Imhoff tanks (Fig. 8.8) if the wastewater flow avoids the digestion compartment. The name Travis tanks is used if part of the flow (e.g., $\frac{1}{5}$) is allowed to pass through the digestion volume. All of the flow can be directed through the digestion compartment when colloidal emulsions are to be degraded and broken up by anaerobic action.
6. Shallow settling basins. Such basins are flat and similar to sludge drying beds (Fig. 9.13). The basins perform alternatively as settling basins and after disconnecting the waste flow, as drying beds for sludge that has accumulated in the preceding period.

*Named after Emschergenossenschaft, a wastewater disposal agency located in Essen, Germany.

7. Settling ponds and lagoons are the simplest and cheapest unit operations. The accumulated sludge remains permanently in the pond or it must be periodically dredged. Their primary use is for controlling pollution by urban runoff and combined sewer overflows (Section 3.2).
8. Wastewater ponds and lagoons are usually designed for biological treatment. However, if the pond is not aerated and if the incoming wastewater contains settleable solids, the ponds also perform as settling basins. The efficiencies of ponds and lagoons are usually less and the area requirements greater than those for properly designed settling tanks and clarifiers.

In the design of all types of settling basins, consideration should be given to floating sludge or scum that accumulates on the surface. The scum can be odorous. The volume of scum can become particularly troublesome if shredded screenings are present in the influent to the basins. It can be reduced if most of the scum is removed in a preceding oil and grease separator or flotation unit. Even when these special pretreatment units are present, there will always be some scum present on the surface of settling tanks and skimming will be needed. The skimming device can easily be connected to the sludge-removing mechanisms as shown in Figures 6.18, 6.20, and 6.21.

The collected scum solids should be conveyed into a special container, dewatered, and pumped, for example, to a sludge digestion unit. The scum from the settling tanks is not the same as the floating sludge in the digestion units, however.

Technical performance of settling tanks is expressed using settleable suspended solids as a parameter. These are measured in Imhoff cones or graduated laboratory cylinders after 2 hr settling. A properly designed settling tank with, for example, 1.5 hr detention time, can remove between 85–98%—commonly 90–95%—of the settleable solids. The concentration of remaining effluent solids depends on several factors, the most important ones being the character of the waste, namely its freshness, and the concentration of influent suspended solids. Stale or anaerobic influent and/or low influent solids concentrations usually result in lower removal efficiency.

6.4 COAGULATION AND FLOCCULATION

The efficiency of all types of settling units can be increased and sedimentation accelerated if coagulation and flocculation of fine solids and colloids are used. This can be accomplished by adding coagulating chemicals. Examples of chemicals that can be added for this purpose are metal (iron and aluminum) salts and/or organic polymers. However, even when chemicals are not added, flocculation can occur if an industrial waste with coagulating properties is present in the influent to the treatment plant.

The formation of larger settleable flocs from fine particles and colloids is usually accomplished in two unit operations: (1) chemicals are first added to the wastewater in a rapid mixing unit (coagulation), and (2) the flocs are then formed and encouraged to grow in a flocculation basin or series of basins.

In the flocculation process, it is important to keep the flocs together and promote their growth. Therefore the velocity of wastewater in a flocculation basin should not exceed 0.3–0.4 m/s. If a series of basins are used, the turbulence in the last basin should be just sufficient to keep the flocs in suspension. Sand must be effectively removed in a grit chamber before flocculation as it would otherwise accumulate on the bottom of the flocculation unit.

Chemical flocculation of municipal wastewater is commonly employed for phosphorus removal or to enhance solids removal.

Chemical coagulation is specifically designed to remove colloids (usually with a negative surface charge) and very fine suspended particles from the wastewater. The chemicals that are added promote the formation of flocs and their growth by two major mechanisms, neutralization of the surface charge causing the particles to form larger flocs by electrochemical attraction and surface adsorption (perikinetic flocculation) and by precipitation of colloids (orthokinetic). They also can precipitate a number of chemical components in the wastewater such as phosphates and toxic metals.

Coagulant Chemicals. There are a number of available coagulant chemicals and their selection depends on the type of the colloids or compounds to be removed and also on the treatment economy. The most common chemicals used in treatment are

- Coagulant aids:
  - Ferrous sulfate
  - Ferric chloride
  - Aluminum hydroxide and salts
  - Organic polymers
- Phosphorus removal by chemical precipitation:
  - Ferric chloride
  - Aluminum sulfate
  - Lime
- Metal precipitation:
  - Caustic
  - Lime
  - Magnesium hydroxide
- Sludge conditioning and dewatering:
  - Lime
  - Ferric chloride
  - Organic polymers.

Wastewaters from iron pickling operations can provide the necessary iron, for example, for precipitation and removal of phosphorus.[22]

Ferric chloride, also called ferrichlor or chloride of iron, is marketed in three forms: as a solution (containing about 52% $FeCl_3$), in crystaline form (about 60% $FeCl_3$), and as an anhydrous powder (98% $FeCl_3$). All forms of ferric chloride are best fed into the wastewater stream as solutions in concentrations up to 45%.

Ferric chloride can also be formed by chlorination of ferrous chloride or on the site by passing a chlorine solution through a column of scrap iron. Roughly 2 kg of chlorine and 1 kg of iron will produce 3 kg of ferric chloride. Ferric chloride might be reduced to ferrous chloride if the wastewater is anaerobic and contains reducing substances. For example, about 1 kg of hydrogen sulfide can reduce 3.2 kg of pure ferric chloride. However, reduction of ferric compounds to ferrous is not easy. Nevertheless, additional chlorination is recommended to maintain the coagulating efficiency of this chemical when wastewater is anaerobic. The dosages are about 35 mg/liter of ferric chloride or 53 mg/liter of ferrous chloride and 8 mg/liter of chlorine when wastewater is relatively diluted. At higher concentrations of pollutants, the dosages are about 50% higher.

Ferrous sulfate ($FeSO_4 \cdot 7\ H_2O$) is cheaper since it is a byproduct. By chlorination, it can be converted to ferric sulfate as the ferric form is preferred in coagulation. This chemical is best fed in a dry form.

Aluminum compounds (mostly aluminum sulfate) can be purchased in a dry or liquid form. It is best fed as liquid alum.

Lime can be purchased as quick lime or hydrated lime. For most installations, hydrated lime in bags or bulk quantities is most attractive. It can be fed either in dry form or in a slurry, which requires certain precautions. The velocity in the pipe should be above 1.5 m/s and the pipes must be periodically cleaned.

Dosages of Chemicals. Chemical dosage is measured and controlled by a dry-feed (Fig. 6.23) or solution-feed apparatus, depending on the nature of the chemical. The most common are solution-feed systems in which, for example, 10% of anhydrous ferric chloride is diluted in water. The dosage should be flow proportioned in order to maintain optimum concentrations of the coagulating chemicals.

The optimum dosages of chemicals are best determined in a laboratory "jar test," in which samples of wastewater are placed in jars, dosed with different

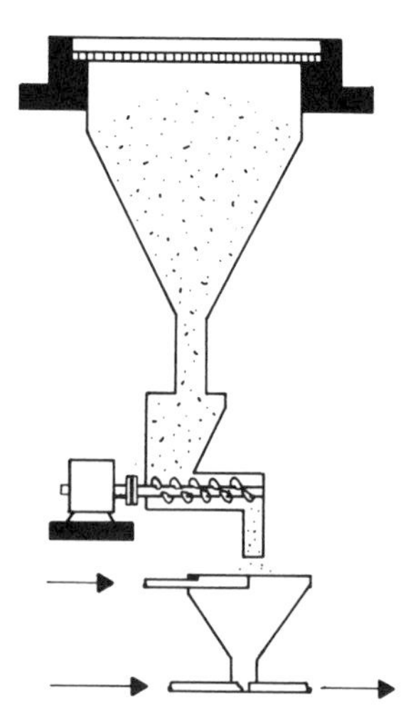

Figure 6.23 Dry chemical feed system.

concentrations of the chemicals, and rapidly and slowly mixed to promote and simulate mixing, coagulation, and flocculation processes. The samples are then analyzed for turbidity concentrations with the lowest value determining the optimum dosage. It should be noted that coagulation and flocculation processes are an "art" rather than a science and laboratory tests are indeed mandatory. The best dosages of chemical and/or coagulant aids for a particular wastewater can be found only by laboratory tests.

Another method of testing utilizes an instrument that measures the zeta-potential, which will approach zero at the optimum dosage.

As a result of chemical coagulation, the pH of wastewater often changes, therefore it should be measured throughout the course of the process and adjusted concurrently. For this purpose, electronic measurements by pH probes should be used. It should also be noted that sufficient alkalinity must be present to sustain the reactions. If the wastewater has very low or zero alkalinity, which might be the case with some industrial effluents, bicarbonate salts must be added with the coagulant.

The chemicals should be mixed with wastewater quickly and uniformly. Good mixing is obtained hydraulically by pumps or in turbulent mixing channels, or mechanically by rapid agitation of the wastewater–chemicals mixture by revolving paddles or by injected air. Aeration also helps to keep the wastewater fresh. Rapid mixing should be followed by slower agitation in a flocculation basin or a series of basins with a detention time of 10–20 min. This slow mixing period promotes the growth and buildup of the flocs. However, the flocs are relatively fragile and should not be destroyed in the process.

To improve flocculation and floc building, German practice recommends addition of brown coal slurry. In the United States, paper slurry used to be added to aid flocculation and also facilitate drying and incineration of the sludge; however, such practices are no longer used.

Sludge. After flocculation the wastewater is conveyed to a clarifier with 1–4 hr detention time. As stated previously, the sludge is of flocculant character and is best removed with funnel or circular upflow clarifiers.

As a result of the chemical addition and flocculation, sludge volumes are increased because (1) more suspended solids are removed, (2) colloidal and dissolved metals, phosphates, and other compounds will precipitate, (3) added chemicals are included in the flocs, and (4) the sludge itself is lighter and bulkier. In a typical application to a municipal wastewater, the sludge volume after chemical addition is about twice as much as that without addition. If lime is employed as a precipitant, alone or in combination with iron salts, the sludge volume is greater; however, lime helps to thicken the sludge. Brown coal or paper slurry addition increases both the sludge volume and its dry mass. An advantage of the use of polymers is that they do not significantly add to the sludge volume. In addition, they do not change the pH.

The increased sludge volumes do not pose a serious problem since chemical sludges readily pass through the digesters and improve mechanical dewatering properties of the digested sludge. However, the digester volume may be affected adversely by the addition of inorganic sludges inasmuch as the design of the digestion process is based on the solids retention time. Therefore, the digester volume should be increased by about 75% over that typical for plain sedimentation.

Efficiency and Design Considerations. As a treatment process, chemical precipitation accomplishes removal of solids and BOD about midway between plain sedimentation and complete biological treatment. With higher chemical dosages, efficiencies approaching those expected for biological treatment can be achieved under optimum conditions for wastewaters that have a lower soluble BOD content. The removal efficiency can be further enhanced if the effluent from the coagulation–flocculation process is treated by sand filtration and postchlorination.

Chemical treatment of municipal wastewaters is not common and it is not used in the United States. However, chemical precipitation is used for treatment of various industrial wastewaters in which toxic or otherwise objectionable compounds can endanger the operation of the biological treatment process and/or violate effluent standards. Residual dissolved metals present in the wastewater can precipitate at different optimum pH ranges and be subsequently removed by sedimentation. For the precipitation of metals the following optimum pH values are preferred:

Metal	Ferric Iron	$Chromium^{III}$	Copper	Zinc	Nickel
pH value	>4	8	7.5	8.8	9.5

When a wastewater contains a mixture of metals, a pH of 8–9 will precipitate all metals except nickel. When nickel is present, the pH has to be increased to 9.5–10.

Chemical treatment can improve the efficiency of overloaded primary and secondary clarifiers without expanding them or building new units, it enhances precipitation of metals and phosphates, it will prepare strong industrial wastes for biological treatment, and it is one of the components of tertiary (after biological) treatment unit operations if a highly purified effluent is required.

6.5 USE OF CHLORINE[23–26]

There is no other chemical that has more alleged uses in a typical wastewater treatment plant than chlorine. The uses of chlorine include

1. disinfection or destruction of disease-causing (pathogenic) microorganisms in the effluent and wastewater treatment bypasses and emergency discharges;
2. destruction or control of undesirable growths in sewage systems and treatment units such as filamentous organisms and infestation of trickling filters with the filter fly or fungus;
3. odor control;

4. improving efficiency of aerated skimming tanks;
5. improvement of clarification of suspended solids and colloids.

However, the use of chlorine has a drawback because in combination with organics present in the treated wastewater or in the water body receiving chlorinated effluents, the chlorine can form potentially carcinogenic organochlorine compounds. Also chloramines that are formed by chlorine replacing hydrogen in the ammonia ion are toxic to aquatic life. Therefore, regulations in a number of states of the United States require that after disinfection is completed, the effluent must be dechlorinated, resulting in a near zero residual chlorine measured at the outlet from the treatment plant. Dechlorination can be accomplished by reducing the chlorine with sulfur dioxide or similar sulfuric compounds in the +IV chemical stage. Activated carbon may also be used. Finger, Harrington, and Paxton[27] describe the instrument technology needed for reliable dechlorination. It appears that in the future, alternative means of disinfection will be sought and developed such as ozonization and ultraviolet rays, although the relative low cost of chlorine will probably make changes in present practice slow.[28,29]

If the chlorinated wastewater discharged into receiving waters is reused as potable water in the water–sewage–water cycle the chlorinated hydrocarbons (if present) should be removed in the water treatment process, typically by activated carbon adsorption.

Chlorine is toxic to organisms in relatively low concentrations. Therefore, caution must be exercised in storing and handling chlorine. Large plants usually use liquid chlorine gas that has been liquified under pressure and shipped and stored in steel cylinders. The storage of chlorine must always be in a separate and safe enclosed room with its own ventilation and accessibility from outside. For temporary or emergency chlorination, bleaching powder or calcium hypochlorite (containing about 70% of free available chlorine) is commonly employed. Chlorinated lime that contains from 25 to 37% available chlorine is used for temporary disinfection and odor control. A third, less frequently used source of chlorine is sodium hypochlorite (liquid bleach), which is primarily used for disinfection and algae control of swimming pools.

Mechanism of Chlorination. Chlorine is a strong oxidizing agent and acts as such in most of its chemical reactions. When added to water, chlorine assumes a number of active forms by hydrolysis, dissociation (ionization), and combination with nitrogen compounds that are present in the water. Important forms of active chlorine are hypochlorous acid (HOCl), hypochlorite ion (OCl^-), and chloramines ($NHCl_2$ and NH_2Cl). The content of HOCl and OCl^- measured after a certain reaction time (usually 15 min) constitutes free available chlorine, and that associated with chloramines constitute combined available chlorine. The chloramines are slower and weaker oxidants and disinfectants than the HOCl and OCl^- components. A combination of free and combined chlorine compounds then constitutes the total available chlorine.

Of the free available chlorine, hypochlorous acid is a more strongly disinfecting agent than hypochlorite ion. The concentration of HOCl relative to the total concentration of available chlorine is greatest at pH values below 7.5–8. Chlorination of sewage commonly will result in formation of chloramines. As a result of the destruction of ammonia and nitrogen compounds the break point chlorination required for drinking water chlorination is rarely achieved in wastewater treatment.

The dosage required for disinfection purposes depends on the amount of unoxidized organics and other reducing compounds (e.g., ferrous iron, nitrite nitrogen, or sulfides) present in the wastewater. These compounds are oxidized first. Oxidation of reduced substances produces so-called chlorine demand. After oxidation is completed, the remaining chlorine is available for disinfection.

The amount of chlorine that remains free is called the chlorine residual. The presence of residual chlorine for disinfection purposes should be ensured by laboratory analyses.

Dosages of Chlorine. The chlorine demand causes the chlorine dosage for disinfection purposes to vary with the strength of the wastewater. Typical chlorine dosages are given in Table 6.4.

Raw wastewater or primary effluents are chlorinated only when the biological treatment is put out of service because of damage caused by toxic spills. In such cases, it is beneficial to provide sufficient storage in the clarifiers and use chlorination only for a limited time period. During emergencies, the chlorination equipment can be installed as temporary.

If chlorine is used to control the formation of hydrogen sulfide in wastewater or in the sludge liquor from digestion tanks (to prevent odor or destruction of concrete), large amounts of chlorine may be needed. For this reason, ferrous chloride may be substituted with a subsequent binding of sulfur with iron.

Chlorine reduces the pH of the wastewater and makes it more corrosive and destructive to building materials. This adverse effect can be reduced if hypochloric acid is used instead of dissolved chlorine gas.

Presently, the need for chlorination of wastewater must be justified on the basis of the contact water use (swimming beaches) downstream. Several years ago, all municipal effluents in the United States had to be chlorinated.

TABLE 6.4 Recommended Dosages of Chlorine in Wastewater Treatment

Type of Effluent	Dosage (mg/liter)
Disinfection	
Raw municipal sewage	10–30
Primary effluent	5–20
Biologically treated effluent	2–20
Surface waters	0.1–1
Odor control	4
Destruction of fungus on trickling filters[a]	50
Control of filamentous growths in activated sludge	0.5–1.5% of returned mix. liquor volatile solids

[a] Only for a short time period.

REFERENCES

1. K. Imhoff, *Gesund. Ing.*, **60,** 599 (1937).
2. F. Zunker, *Gesund. Ing.*, **61,** 454 (1938).
3. W. W. Eckenfelder, Jr. and D. L. Ford, *Water Pollution Control—Experimental Procedures for Process Design.* Pemberton Press, Austin, TX and New York, 1970.
4. Metcalf & Eddy, Inc., *Wastewater Engineering: Treatment, Disposal, Reuse.* McGraw-Hill, 2nd ed., New York, 1979.
5. A. Hazen, *Transact. Am. Soc. Civil Eng.* **69,** 45 (1904).
6. T. R. Reynolds, *Unit Operations in Environmental Engineering.* Brook/Cole, Eng. Div., Monterey, CA, 1982.
7. K. H. Kalbskopf, *Komunalwirtschaft* **54** (9), 415 (1966).
8. L. Morales L. and D. Reinhart, *J. Water Pollut. Control Fed.* **56,** 337 (1984).
9. K. Imhoff and G. M. Fair, *The Arithmetic of Sewage Treatment Works.* Wiley, New York, 1929.
10. E. B. Fitch, *Sewage Ind. Wastes* **29,** 1123 (1957).
11. F. D. Prager, *Water Sewage Works* **97,** 144 (1950).
12. R. I. Dick, *J. Water Pollut. Control Fed.* **48,** 633 (1976).
13. V. D. Laquidara and T. M. Keinath, *J. Water Pollut. Control Fed.* **55,** 54 (1983).
14. R. Bettaque, *Vom Wasser* **27,** 60 (1960).
15. T. M. Keinath, *J. Water Pollut. Control Fed.* **57,** 770 (1985).
16. T. R. Camp, *Trans. Am. Soc. Civil Eng.* **111,** 895 (1946).
17. H. Abrose, E. R. Bauman and E. B. Fowler, *Sewage Ind. Wastes* **29,** 24 (1957).
18. F. Pöpel, *Gesund. Ing.* **70,** 241 (1949).
19. *Process Design Manual for Suspended Solids Removal.* U.S. Environ. Prot. Agency, Washington, DC, 1971.
20. W. J. Katz and A. Geinopolos, *Proceedings International Water Pollution Control Conference,* Pergamon Press, Oxford, 1962.
21. D. Groche, *Sttut. Ber. Siedlungswasserwirtschaft* **13** (1964).
22. *Process Design Manual for Phosphorus Removal,* U.S. Environ. Prot. Agency, Washington, DC, 1976.
23. *Chlorination of Wastewater,* Manual of Practice No. 4. Water Pollut. Control Fed., Washington, DC, 1976.
24. S. C. White, *Handbook of Chlorination,* 2nd ed. Van Nostrand-Reinhold, New York, 1986.
25. J. D. Johnson, *Disinfection-Water and Wastewater.* Ann Arbor Sci. Publ., Ann Arbor, MI, 1975.
26. *Municipal Wastewater Disinfection,* EPA Rep. No. 625/1-86/021. U.S. Environ. Prot. Agency, Cincinnati, OH., 1986.
27. R. E. Finger, D. Harrington, and L. A. Paxton, *J. Water Pollut. Control Fed.* **57,** 1068 (1985).
28. S. C. White, E. B. Jernigan, and A. D. Venosa, *J. Water Pollut. Control Fed.* **58,** 181 (1986).
29. G. Zukovs, J. Krollar, H. D. Monteith, K. W. Ho, and S. A. Ross, *J. Water Pollut. Control Fed.* **58,** 199 (1986).

CHAPTER 7

Biological Treatment Units

7.1 BASIC PRINCIPLES

Generally, pretreatment (screens, grit removal) and primary sedimentation, even with chemical addition, are not sufficient to provide an effluent that can be safely (nor legally in the United States) discharged into surface waters. Exceptions may include existing municipal sewage disposal systems in coastal areas where primary treated sewage is discharged into the sea by long outfalls (up to 8 km long in Passaic, NJ). Even these plants must demonstrate no adverse effects in order to avoid mandatory secondary treatment. In the vast majority of cases, wastewater treatment relies on unit operations employing microorganisms that decompose organics and purify the wastewater. The processes resemble natural purification of sewage in surface waters (selfpurification). These unit operations are called "biological treatment."

Anaerobic units in which anaerobic bacteria break down organic matter without oxygen represent a special case used primarily for pretreatment of high strength organic wastewaters (see McCarty[1]) or for stabilization of sludge. All other biological unit operations are aerobic. The microorganisms feed on the organics and nutrients in the wastewater and develop into large colonies in a form of slimy gelatinous layers on the surface of soil (land disposal unit operations) or filter media (trickling filters or rotating biological contact units) or as freely floating flocs in activated sludge units. The biological slimy layer or flocs absorb the fine or colloidal organic particles as well as dissolved organics that are then decomposed by the microorganism. The end products of decomposition are then transferred back into surrounding wastewater. The process may be described as follows:

$$\text{wastewater organics} + \text{microorganisms} + O_2 + \text{nutrients}$$
$$= CO_2 + H_2O + \text{microorganisms} + \text{residuals}$$

Obviously, dissolved oxygen must be present in the treated water. Oxygen is provided from the air–water interface as in lagoons, from air or pure oxygen injected in the units as in activated sludge units, or from passing wastewater droplets through the air as in biological contact units or land disposal systems.

At O°C and 1 bar atmospheric pressure, 1 m^3 of dry air weighs 1294 g; 1 m^3 of air also contains 209.4 liters of oxygen that weighs 300 g. Using this information and assuming the average wetness (vapor content) of the air, the weight of 1 m^3 of wet air is about 1 250 g and its oxygen content is approximately 280 g.

In biological treatment systems, oxygen is supplied in great excess over the actual demand. In an activated sludge system, only 5–15% of the oxygen available from the injected air is actually used, whereas the use of oxygen in a trickling filter with natural draft is 5% and with forced draft, it is even lower.

Selection of the System

The size of a biological treatment unit depends on the organic load. This is normally expressed in terms of mass of the 5-day biological oxygen demand (BOD_5) of the wastewater. This can be obtained from the influent flow (m^3/hr) times the influent concentration of the dissolved BOD_5 in mg/liter (= g/m^3). For example, if the flow is 1 000 m^3/hr and the dissolved BOD_5 concentration is 200 mg/liter, the plant load is $1\ 000 \times 200 = 200\ 000$ g/hr $\times$ 0.001 g/kg = 200 kg/hr.

When the information on the flow and BOD is not available (such as in preliminary pollution abatement studies), the treatment plant loading can be estimated from the population equivalent.[2] One population equivalent represents 40 g of BOD_5 per capita per day after primary sedimentation. The value of 40 g/cap-day presumes that all people living in the area are connected to sewers and the households are provided with flushing toilets. If flushing toilets are not common, as it is the case in many developing countries, the per capita load is reduced by 70% (to about 12 g/cap-day). Industrial contributions are expressed usually directly by a mass of BOD_5 or suspended solids per unit of a production output of the industry. In the Europe, industrial loadings in planning studies are converted to a number of population equivalents. The size of the various biological treatment units can then be approximately estimated from volumetric loadings given in Table 7.1.

From the information given in Table 7.1 one can see that activated sludge units require less volume than trickling filters. On the other hand trickling filters may require less land area than activated sludge, this being an advantage when land is expensive and/or sparse.

TABLE 7.1 Volumetric Loading of Biological Treatment Units

	Influent BOD_5 Load to 1 m^3 Volume of the Filter or Aerated Unit [g/(m^3-day)]	Population Number to 1 m^3 Volume of the Filter or Aerated Unit (Population/m^3)
Trickling filters		
Low rate (with nitrification)	200	5
Complete treatment (normal load)	400	10
High rate	800	20
Activated sludge		
With nitrification (low rate)	500	12
Complete treatment	1000	25

As specified previously, the term "nitrification" denotes the process of aerobic conversion of ammonia to nitrates, which is the second stage of the biological oxidation process (Fig. 5.3) performed by specific nitrifying bacteria. The term "complete treatment" implies the removal of most of the carbonaceous (first stage) BOD.

The loading values given herein can serve only for a preliminary sizing of the units. More accurate estimations, often based on laboratory and field treatability studies, may be required for a detailed design.

In special cases, it should be documented whether there are reasons to increase or reduce the volumetric loading values from those indicated here. The loading should be reduced when

1. a higher degree of treatment is required;
2. the influent contains higher concentrations of ammonia;
3. the wastewater is unusually diluted (and therefore the treatment time determined from the pollution load becomes shorter);
4. when significant wet weather contributions (stormwater and/or infiltration) are present;
5. pretreatment or primary treatment could fail or is not included;
6. the wastewater is unusually cold;
7. because of the type or operation of the treatment process, lower efficiency is anticipated;
8. the plant is relatively small with a small contributing area.

Correspondingly, higher volumetric loadings are possible when justified by treatment technology.

The performance of a treatment plant or unit is expressed as percentage removal computed from the difference between the influent and effluent concentrations divided by that in the influent. Removal efficiencies for three basic pollutants in various treatment units are given in Table 4.1. Of importance herein are the removal efficiencies for the BOD, which for most well-functioning biological treatment units can be at or above 90%. Such removal efficiencies can then be considered as complete treatment. One-third of the credit for the removal should be given to primary sedimentation in which most of settleable organics are removed and further disposed as sludge. It should be noted that these removal efficiencies are dependent on good design and competent operators. The best designed plant will show poor performance without trained operators.

System Configuration

Since a principal function of biological treatment is to convert nonsettleable (fine, colloidal, and dissolved) substances into settleable sludge, biological units must be followed by settling tanks that remove the sludge that builds up in the units.

The quantity of sludge that is removed by sedimentation for further treatment and disposal is called "sludge yield." This sludge requires treatment to reduce its volume and moisture, stabilization, and to minimize odor problems. A small portion of the produced sludge solids will escape over the weir of the settling tanks into the effluent. A massive loss of the solids is a result of clarification failure caused either by overloading the secondary clarifiers or by sludge bulking caused by overloading or underloading the biological treatment units. The solids that are discharged into the receiving water body with the effluent can settle onto the bottom and exert sediment oxygen demand.

Only a portion of the organics expressed as BOD reaching the biological treatment units is completely degraded to carbon dioxide, ammonia, nitrate, or other mineral residues. Much of the remainder is converted into sludge. The sludge yield depends upon the loading but for well-functioning treatment units it is typically

55% for activated sludge units
45% for high rate trickling filters
20% for low rate trickling filters.

The primary objective of biological treatment is to obtain an effluent that will not be damaging to receiving waters. In legal terms, this means that the effluent quality must meet effluent standards that are either based on the required performance or on the capacity of the receiving water body to assimilate the residual pollution discharge. Receiving waters suffering from excessive algal and macrophyte (aquatic weeds) growths may require the removal of nitrogen and/or phosphorus from the effluent since these two components are nutrients that may promote their growth. For example, the phosphorus concentration in effluents from municipal and industrial treatment plants located in the Great Lakes area of the United States and Canada must be at or below 1 mg/liter. It should be noted that most North American lakes have been found to be phosphorus limited, whereas estuaries and desert lakes and reservoirs may be nitrogen limited.

With regards to nitrogen, it appears in the effluent primarily in two forms: as ammonia if the treatment process has limited nitrification capability (usually when high rate treatment is used) or as nitrate. Small amounts of nitrite can be also present as will unoxidized proteinaceous organic nitrogen. In addition to being a nutrient for plants and algae, ammonia may also be toxic to fish if the pH of the receiving water body is on the alkaline side (greater than 7.5). In addition, oxidation of ammonia to nitrate in the receiving water body utilizes oxygen and has the same impact as carbonaceous BOD. This kind of oxygen demand is called nitrogenous biogical oxygen demand (NBOD).

Generally, control of algae growths in inland receiving water bodies is accomplished by reducing the phosphorus content of the effluent, nitrification is required for reducing toxicity of un-ionized ammonia (commonly when the pH of the receiving water body is above 7.5) and when the oxygen balance of the receiving stream could be disrupted by nitrification. However, nitrification may become a problem in shallow streams and in estuaries. Denitrification may be required if the nitrate content could adversely affect the downstream use of the receiving water body such as for drinking. Excessive algal growths of some estuaries can be controlled by denitrification of the effluents.

7.2 LAND DISPOSAL OF WASTEWATER[3–8]

To many people, land disposal of sewage is the most natural method of treatment and, consequently, the best method for protection of the quality of surface waters. Recently, land disposal methods received considerable attention in association with the "zero discharge" goal advocated in the United States in the Water Pollution Control Act of 1972. However, it must be recognized that by using land disposal, the wastewater discharge and its residual pollution effect are not completely eliminated but shifted to soil and groundwater resources.

Many soils have a very high capacity to receive and decompose pollutants contained in municipal and some industrial wastewaters; however, this capacity is reduced during winter months in colder climatic regions and is highly dependent on the soil characteristics. In arid regions of the United States and elsewhere, recycled treated effluents recharge groundwater and are also used for irrigation of crops and golf courses.[8] The nutrient content and the value of the water itself make the effluents particularly suitable for irrigation, whereas the use for groundwater recharge must follow strict hygienic guidelines.

Land disposal of wastewater in northern climates is more difficult and, in many cases, if the price of the land is high, or the effluent must be pumped for large distances, the economics may not favorable. In addition, the Western U.S. water laws consider wastewater as part of the water rights and cause a legal constraint that must be addressed.

In most cases, the use of wastewater for irrigation is an agricultural problem and systems must be designed using the guidelines of agronomy. Typically, the hydraulic loading and the nutrient loading are matched with the crop irrigation and nutrient requirements and hydraulic permeability of the soils. References 9 and 10 contain the guidelines for the design of the land disposal systems.

Agricultural lands can receive only a limited amount of wastewater, particularly in more humid climates when the annual rainfall is more than 600 mm (24 in.) and, as stated before, during winter months in colder climatic conditions.

Previously, land disposal of wastewater was thought feasible only for municipal effluents that are relatively free of toxic compounds. Recent advances of the state-of-the-art documented that disposal of both municipal and industrial effluents is feasible.

There are three common types of land disposal:

1. slow rate or crop irrigation (SR),
2. rapid infiltration (RI), and
3. overland flow (OF).

The key characteristics of these three types of land treatment are shown in Table 7.2 and in Figures 7.1–7.3.

TABLE 7.2 Typical Design Parameters for Land Treatment[a]

Feature	Processes		
	Slow Rate	Rapid Infiltration	Overland Flow
Application technique	Sprinkler or surface	Surface	Sprinkler or surface
Annual loading, m	0.5–6	6–125	3–20
BOD_5 loading, kg/(ha-day)	1–5	20–125	5.5–20
Field area, ha/(1 000 m^3-day)[b]	6–60	0.2–6	1.6–11
Population equiv./ha	25–125	500–3125	137–500
Typical weekly loading rate, cm	1.3–10	10–240	6–40
Minimum pretreatment	Primary[c]	Primary[c]	Grit removal[c]
Need for vegetation	Required	Optional	Required
Grade	<20%	Not critical	2–8%
Soil permeability	Moderate (loams)	Rapids (sands)	Slow (clays and silts)
Drainage	Surface	Tiles or wells	Surface
Depth to groundwater, m	0.6–1 (minimum)	1.5–3	Not critical
Climatic restrictions	Storage needed during freezing and high precipitation	None	Storage needed during freezing

[a] After U.S. EPA.
[b] To convert from ha/(1 000 m^3-day) to acre/mgd (million gallons per day) multiply by 9.34, to convert from kg/ha to lb/acre multiply by 0.9.
[c] With restricted public access, crops not for direct human consumption.

The direction of flow in the first two systems is primarily vertical, whereas in the overland flow system the flow is primarily horizontal.

The main objective of all three systems is the treatment of wastewaters, however, agricultural use is emphasized for the slow (SR) and overland flow (OF) systems. In these two systems, vegetation and the beneficial effect resulting from the uptake of nutrients (particularly of nitrogen) and filtration in OF systems are considered and included in the overall design. Vegetation is neither important nor needed in the rapid infiltration systems.

The limiting factors for the design of the three land treatment systems are given in Table 7.3.

Description of the Systems

Slow Rate (SR) Irrigation. Slow rate systems in more humid geographic zones are usually limited by the hydraulic capacity of the soils or by the nutrient (nitrogen) uptake of crops. In colder and humid regions (such as the northeastern portions of

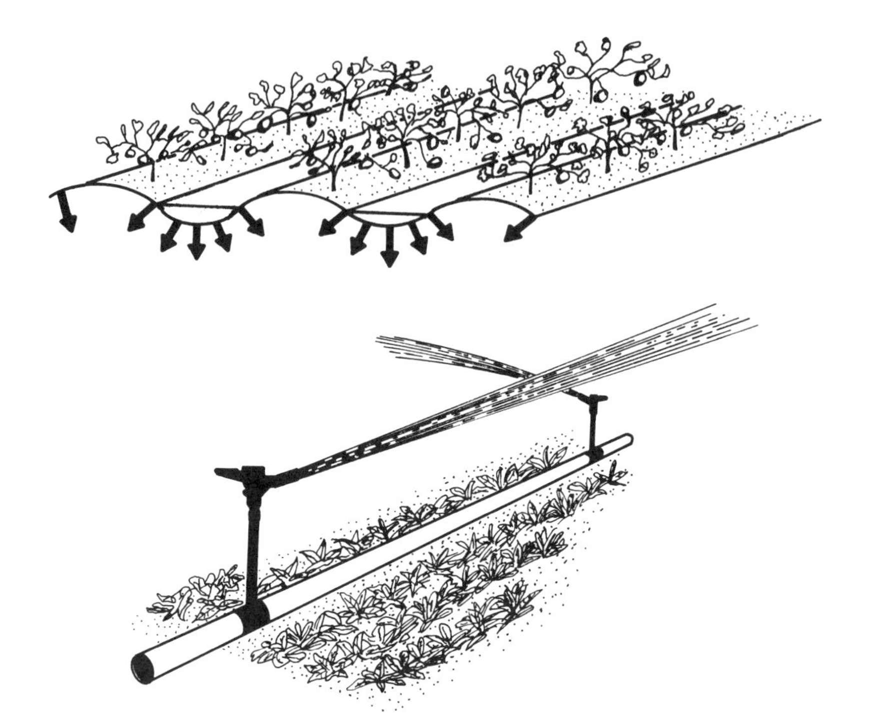

Figure 7.1 Slow rate land treatment.

the United States and most of Europe), storage may be needed as applications during high precipitation and during soil freeze periods are limited or not possible at all. The storage volume can be estimated as the largest number of consecutive days with the average temperature below freezing during a typical design year times the average wastewater flow. It is also important not to overload the fields with nitrogen as the excess nitrogen will oxidize and leach as nitrate into groundwater or return flow.

In more arid parts of the western United States and of the world, the water itself and not its nutrient content is the most valuable commodity. The application rates are then determined by the evapotranspiration rates and leaching requirements of the crops. The leaching requirement specifies the excess irrigation water over the evapotranspiration loss that must be applied to maintain the salinity of the soil

TABLE 7.3 Limiting Design Parameters

Slow rate (SR) processes	Hydraulic capacity of soils, salt buildup, nitrogen contamination of groundwater
Rapid infiltration (RI)	Hydraulic capacity
Overland flow (OF)	The longest flow length that provides the retention time for the removal of key pollutants specified by the discharge permit

water at acceptable levels. In such cases, the salinity of the wastewater and the salt tolerance of crops are the limiting factors as well as the drainage characteristics of the soils. Underdrains may be needed for systems that cannot accept the hydraulic loading without preventing soil waterclogging or salt buildup.

Rapid Infiltration (RI). The objective of rapid infiltration systems is wastewater treatment only and the removal of wastewater constituents is accomplished primarily by filtration, biochemical reactions of microorganisms residing in the topsoil, and adsorption. Rapid infiltration requires natural sandy soils. The term "Intermittent Sand Filtration" was previously used to describe similar systems developed in the early 1900s.[11] Winter operation is possible without interruption and storage may be unnecessary. A design example of a rapid infiltration system is presented in Section 12.2, Example 5. During operation of rapid infiltration systems, wastewater is applied in a 5- to 15-min time interval so that the water depth is about 5–15 cm. The water drains uniformly driving the air downward with it. The water usually disappears from the surface in approximately 1–2 hr followed by a drying period for the rest of the day. However, the timing of wet and dry periods is a function of the soil and wastewater characteristics, climatic conditions,

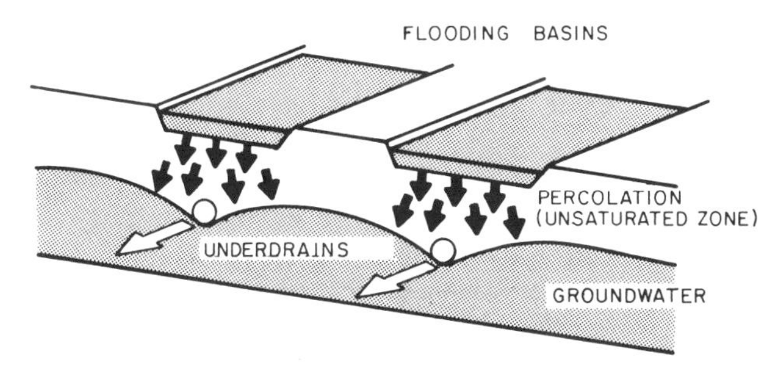

(a) RECOVERY OF RENOVATED WATER BY UNDERDRAINS

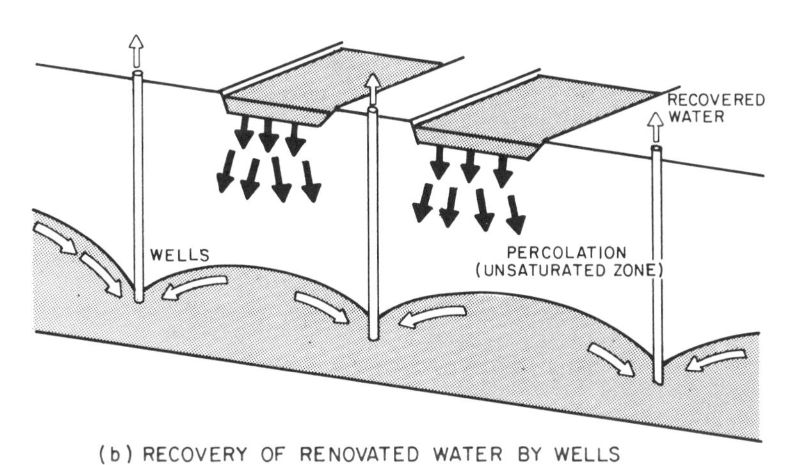

(b) RECOVERY OF RENOVATED WATER BY WELLS

Figure 7.2 Rapid infiltration.

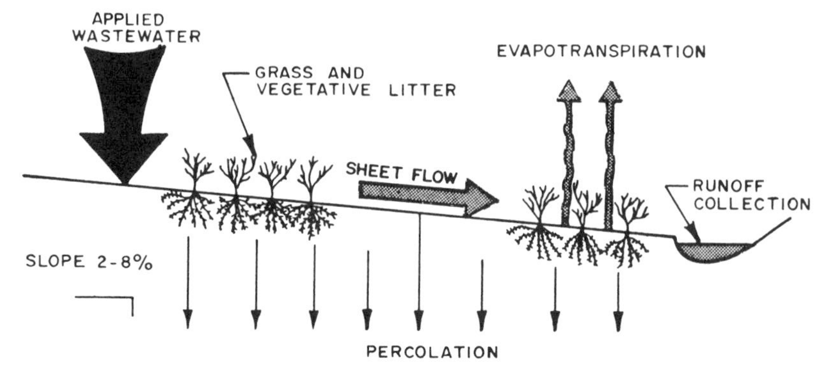

Figure 7.3 Overland flow treatment.

and effluent quality requirements. Problems with the RI systems usually stem from improper design and operation.[12]

Drainage pipes or drilled wells are used in RI system to collect the treated effluent. Vitrified clay or synthetic (PVC) perforated drain pipes with a minimum diameter of 15 cm are spaced 15 or more meters apart and are at depths of 2–5 m. In soils with high lateral permeability, spacing may approach 150 m. The pipes should be surrounded by gravel to keep sand from washing into the drains. Detailed information on drainage design can be found in references 13 and 14. For water harvesting and recovery, wells are more appropriate. Such systems have been installed in Phoenix, Arizona and Fresno, California. Systems on deep sand deposits used for recharging aquifers do not need underdrains.

Overland Flow (OF) Systems. In overland flow systems, wastewater is applied at the upper reaches of grass-covered slopes and allowed to flow over the vegetated surface to runoff collection ditches. The system is best suited for relatively impermeable soils or subsoils. The water is purified by the mechanical straining action of the grasses and biologically by microorganisms residing in a slimy layer on the top of the soil. The length of the overland flow and, hence, the residence time for the removal of the key pollution constituents (BOD_5, suspended solids, nutrients, or bacteria) are the primary design parameters for overland flow systems.

Current guidelines call for slope lengths in the range of 30–45 m for treatment of primary or raw municipal wastewater and a minimum of 45 m for treatment of oxidation pond effluent and for sprinkler distribution systems.[10] A model for the design of overland flow systems that may result in savings on land area was developed by Smith and Schroeder.[15] If properly designed, overland flow processes can be inexpensive solutions to municipal wastewater treatment problems.[16]

Treatment Performance

As shown in Table 7.4, the three land treatment systems introduced herein provide a high quality return flow that, in addition to irrigation, can be reused, with some limitations, for such beneficial purposes as recharge of groundwater or surface streams, replacement of potable water in park and golf course irrigation, and remedy of saltwater intrusion in coastal aquifers.

TABLE 7.4 Typical Effluent (Return Flow) Quality from Land Treatment Processes[a]

Constituent	Slow Rate	Rapid Infiltration	Overland Flow
BOD_5, mg/liter	2 (<5)[b]	5 (<10)	10 (<15)
Suspended solids, mg/liter	1 (<5)	2 (<5)	10 (<20)
Ammonia-N, mg/liter	0.5 (<2)	0.5 (<2)	4 (<8)
Total N, mg/liter	3 (<8)	10 (<20)	5 (<10)
Total P, mg/liter	0.1 (<0.3)	1 (<5)	4 (<6)
Total coliforms number/100 ml	0 (<10)	10 (<200)	200 (<2 000)

[a] After U.S. EPA.[9,10]
[b] The numbers in parentheses represent the upper range at the mid to low end of the loadings presented in Table 7.2.

There are a number of processes that are involved in the natural treatment. Solids are removed by straining in the soil and by vegetation, organics by bacterial decomposition by microorganisms residing in the soil and in the slimy active layer on the surface, and nutrients by plant uptake and also by adsorption on soil particles and organic matter (phosphorus, ammonia) and precipitation (some metals). Nitrification of the applied wastewater is essentially complete when appropriate hydraulic loading cycles are used. Very high nitrogen removals are achieved when the nitrogen application rate is matched with the plant uptake (SR and OF systems). Nitrogen removal in rapid infiltration systems is usually less than 50% unless specific operating conditions are established to maximize denitrification. These procedures include optimizing the application cycle, recycling portions of the renovated water that contain high nitrate concentrations, reducing the infiltration rate to enhance temporarily anaerobic conditions in the soil, and supplying an additional carbon source.

Pathogenic or coliform bacteria are effectively removed in the vertical flow systems (RI and SR) by filtration, adsorption, predation, and exposure to adverse conditions. Viruses are removed almost entirely by adsorption. The major mechanisms responsible for removal of microorganisms in OF systems include sedimentation, filtration through surface organic layer and vegetation, sorption onto topsoil particles, and predation. Generally, the removal efficiency for OF systems is less than that for SR and RI systems and disinfection may be required. It should be noted that the limiting factor in most instances is the cadmium content of the effluent, inasmuch as it has been documented to be taken up by fruit and edible portions of crops.

Application of Wastewater. The wastewater is applied either by surface flooding distribution systems (ridge and furrow or border applications) or by sprinklers. Flooding is usually intermittent and generally is cheaper than sprinkler systems.[3] As noted in Table 7.2, primary treatment is the minimal necessary requirement for most agricultural applications, golf courses, or in parks. Groundwater recharge requires secondary treatment and disinfection.

Sprinkler distribution systems may cause a number of problems. If the wastewater is brought to the fields by long pressure pipes, the wastewater becomes anaerobic and, as a result, odor is released when it is sprayed into the air. Also bacteria and parasites can be released into the air and onto the vegetation and, thus, impair its harvest and use. As a result, it may be necessary to provide secondary treatment and disinfection prior to application.

Stormwater runoff from surrounding areas should not enter fields irrigated or flooded with wastewater. Therefore, bypass conveyance drainage must be constructed to intercept surface runoff.

Use of Natural or Man-Made Wetlands

Wetlands, either natural or man-made, are a natural sink for nutrients, particularly during growing periods. Wastewater effluents have been applied to many wetlands and the interest in the use of wetlands for nutrient removal is increasing. In addition to nutrients, suspended solids and BOD can also be removed. Dubuc et al.[17] showed that in less than 1.5 km of flow, over 90% of the BOD, phosphorus, and nitrogen were removed by a peatland in a northern climate.

However, Nichols[18] cautioned against the use of wetlands for wastewater purification and stated that most of the nutrients retained during the growing season are released when the plants die. On the average, 1 ha of wetland can remove approximately 50% of the nitrogen and phosphorus from a typical municipal wastewater produced by about a 60 person population equivalent. A much larger wetland area is required for higher removal efficiencies. Application onto wetlands can be an effective method of wastewater posttreatment during warmer periods of the year if the wetland area is abundant and the population density is low.

In temperate zones in which seasonal dye-off of wetland plants and subsequent release of nutrients are not a constraint, man-made or natural wetlands can be designed to provide advanced treatment (effluent polishing) prior to discharge to lakes and other nutrient-sensitive water bodies in addition to further removal of BOD and suspended solids. The City of Orlando, Florida, discharges 1 m^3/s (20 mgd) of sewage effluent through 500 ha of former wetland that has been previously drained for agricultural purposes. The wetland has been replanted with desirable wetland vegetation.[19]

A state-of-the-art publication on the use of wetlands for treatment of municipal wastewater was prepared by Godfrey et al.[20] A design manual for wastewater disposal onto natural or man-made wetlands has been released by the U.S. EPA.[21] The report cautions against an indiscriminatory discharge of wastewater to wetlands.

7.3 ATTACHED GROWTH MEDIA BIOLOGICAL PROCESSES (FILTERS)[22–33]

Biological filters evolved from the intermittent sand filters (rapid infiltration) described in the preceding section. It was observed that the sand filter could biologically remove and degrade wastewater organics, however, aerobic conditions were required. The removal was not by filtration, but by biological contact. To reduce the surface area and increase the surface loadings, the active biological layer was made deeper, aerated more efficiently, and the filter was made from coarser grain materials (such as crushed rock, slag, anthracite coal, coke, and, as the latest development, synthetic materials).

Trickling Filters

The basic treatment operation of the trickling filters relies on the formation of a slime layer on the surface of the media. The thickness of the layer is about 2–3 mm, which is about the depth of penetration of the oxygen from the air. If the layer becomes thicker, part of it could become anaerobic and cause odor problems. The thickness of the slime layer is controlled by the hydraulic loading to the filter. By increasing the hydraulic loading the thickness of the layer can be kept under control. In order to balance the organic and hydraulic loading it is common practice to recirculate a portion of the effluent.

The typical detention time in a rock-filled trickling filter with a depth of 1.8–2 m is about 20–60 min. In a tall filter tower with plastic media, the detention time is about 20 min. The detention time in a treatment plant can be measured by adding a strong salt solution to the filter inlet. The average detention (residence) time occurs when one-half of the added mass of the salt is detected in the filter outlet.

Filter Types. Trickling filters are categorized into low-, normal- (intermediate), high-, and superhigh-rate filtration, which differ primarily by hydraulic loading.

Low-rate filters do not include recirculation. The classification of filters and their design parameters are given in Table 7.5.

Low-rate filters are commonly used for complete treatment, including nitrification. Because of the low loading, only a small amount of new biomass is formed. However, as the biomass in the form of a slimy layer on the media is increasing, the lower layer is being degraded and stabilized using oxygen from the air inside the filter. The biomass accumulated in the filter must be flushed from the filter, usually several times per year. The sludge is well stabilized, more granular than flocculant in character, and does not require digestion prior to final disposal. In

TABLE 7.5 Classification of Trickling Filters

Loading Parameter	Low-Rate Filter	Normal-Rate Filter	High-Rate Filter	Superhigh-Rate Filter
Hydraulic loading $m^3/(m^2\text{-day})^a$	1–4	4–10	10–30	30–50
		(includes recirculation)		
Organic loading, kg $BOD_5/(m^3\text{-day})^a$	0.08–0.32	0.24–0.48	0.48–1.0	0.8–1.6
		(organic load from recirculation is not included)		
Population equiv./m^3	2–8	4–12	12–25	20–40

[a] To convert from $m^3/(m^2\text{-day})$ to mgd/acre multiply by 1.07; to convert from $kg/(m^3\text{-day})$ to $lb/(1\ 000\ ft^3\text{-day})$ multiply by 62.

the United States and England, the sludge has been used as an additive to organic compost. The low-rate filters are very reliable and, in conjunction with primary and final settling tanks, the BOD removal is 75–85%.

Low-rate filters often become infested with a small mothlike fly, *Psychoda,* called the filter fly. Heavy infestation is associated with thick films and higher temperatures. The larvae of the fly burrows into the slime, covering the filter stone. After development into a fly, it may become extremely troublesome.

Low-rate trickling filters also harbor large numbers of aquatic earthworms and reddish sludge worms. During seasonal unloading periods, masses of worms can be disgorged, and they decay in secondary settling tanks or on drying beds, sometimes giving rise to objectionable odors.

Normal-rate filters are similar in design to low-rate units, however, they operate with recirculation. The hydraulic loading in this range and the hydraulic flushing rate result in a thinner biological layer being formed on the filter media. The excess of biological mass is carried by the flow to the bottom of the filter and in the effluent. The sludge has a more moist consistency and a larger mass and volume than that of low rate filters, and is not well stabilized. In many cases, the hydraulic loading is not sufficient to carry away the excess biomass and clogging of the filter can occur. This problem may be solved by increasing the size of the filter media to 75–100 mm.

High-rate filters are commonly used for pretreatment. The higher organic loading due to recirculation (which is commonly one-half to four times the influent flow) eliminates problems with clogging and filter flies. Because of the high load, more biomass is formed, however, the sludge is not stabilized and requires further digestion. The filter is relatively sludge free because of continuous flushing, however, flushing should not be considered a purely mechanical phenomenon, but rather a biological step in which the growth is balanced by faster removal of the biomass. As the biological mass on the higher rate filters is relatively young, little or no nitrification occurs and the effluent has less nitrate content.

Filter Media. The specific contact or effective area of the filter depends on the size of the filter media (Table 7.6). In the United States, the size of the filling media is typically between 4 and 8 cm and in England, between 2 and 5 cm. As the effective contact surface increases with the decrease in size, the smaller size media will increase efficiency but not without operating troubles. As shown in Table 7.6, crushed stone of a size from 4 to 8 cm will have an effective contact surface area of 45 m^2/m^3 of filter volume. Crushed stone with a size from 2.5 to 4 cm will have an effective area of 60 m^2/m^3. The advantage of increased contact surface by decreased grain size is offset to some extent by a reduction in the surface area of individual voids, which consequently causes greater resistance to air flow and hydraulic loading. Therefore, smaller sized filter material requires a shallower filter. As shown in Table 7.6 plastic media provide the highest specific contact surface. Such media are available from several manufacturers in two major types: bulk packed (made of polyethylene), consisting of small plastic shapes similar to short pieces of plastic tubing with internal fins, and modular (made of PVC), consisting of corrugated plastic sheets.

TABLE 7.6 Filter Media and Their Specific Contact Area

Medium	Nominal Size (cm)	Specific Surface Area (m^2/m^3)
Plastic (bulk packing)		115–200
Plastic (modular)		80–100
Redwood	120 by 87.5	50
Granite	2.5–7.5	60
Granite	10	40
Blast furnance slag	5–7.5	60

In addition to their higher specific surface area, the plastic media are 10–20 times lighter than crushed stone.

Filter Depth. The recommended filter depth for smaller media (3–4 cm in diameter) is about 1.8 m, whereas coarser filter media (5–8 cm in diameter) commonly have a filter height of 2–3 m in the United States. High rate trickling filters are shallower—typical depths range from 0.9 to 1.8 m. In addition to the size of the filter media, the depth of the filter also depends on the strength of the wastewater. Higher BOD concentrations in the influent require deeper filters, hence, the detention time and length of travel of wastewater in the filter are increased, resulting in a higher removal efficiency. Increased depth is possible as long as provisions are made for increased aeration. Very high (tower) filters up to 8 m high are feasible.[34]

The requirement of higher depth for increased removal efficiency and for treatment of high strength influents has lead to the development of superhigh-rate trickling filters with the synthetic media having large void space and high effective contact area. These filters have depths ranging from 3 to 12 m. The high depth filters are referred to as biooxidation towers.[35–37] Very high BOD loads, up to 5 kg/(m^3-day), are possible. Typical removal efficiencies vary with the BOD load: at 6 kg/(m^3-day) the removal efficiency is about 50%, at 2.5 kg/(m^3-day) it is about 65%, and at 1 kg/(m^3-day) about 80% removal efficiency can be expected.

The superhigh-rate tower filters are almost exclusively filled with light plastic media. Using conventional crushed stone would lead to structural problems.

Filter Arrangement. Although a single-stage filter may often be satisfactory and can achieve secondary treatment, an arrangement of two or more filters in series or in parallel is more reliable. The filters in series are used primarily for high strength organic wastes that would otherwise require very tall towers or for municipal wastewaters requiring nitrification. In the latter case, the first stage can be designed as a high rate filter followed by a low-rate filter. To reduce the sludge load on the second filter from the first one, an intermediate clarifier may be needed for treatment of high-strength wastewater.

As a result of a smaller depth of each individual filter, the two-stage process may not require forced air aeration as each filter will need proportionally less air.

However, if forced draft aeration is required in either case, one tower filter is cheaper than two shallower filters in series.

If the first stage is followed by a sedimentation unit, the second unit can be smaller and made of finer stone (5–8 cm) in order to create a larger surface area for the prefiltered and therefore weaker wastewater. It is also possible to alternate the first and second stage to provide time for the biomass in the highly loaded first filter to digest the accumulated organics, which helps to prevent clogging. This is called *alternate double filtration*. However, this type of filtration makes sense only if both units are of the same size, thus requiring a reduced load on the filters.

Filter Construction. The filter support must be constructed to allow free drainage of the water and no place for the sludge to accumulate. Furthermore, the air should be able to move freely upward from the bottom. This is accomplished with a double support in which the upper bottom provides support for the filter media and the bottom carries the filter effluent to clarification. Therefore, the bottom should have drainage channels. Between the upper bottom and the collected water is an air space and provisions should be made for flushing of the bottom channels. Typical designs of rock and plastic media filters are shown in Figures 7.4 and 7.5.

The bottom intake for the air must be so dimensioned that it will provide the required air supply to the filter. For example, for a normal-load filter that requires an air flow of 0.3 m/min = 5 mm/s and an air flow rate through the bottom openings of 1 m/s, the ratio between the air inlet openings and the total area of the filter is 0.005/1 = 0.5%. For a more conservative design, 1% is used.

Proper distribution of wastewater on the surface of the filter is essential. Sewage is sprayed on the surface from fixed or moving sprays (Fig. 7.6). Flows to stationary sprays or self-propelled rotary distributors are generally controlled by dosing tanks that also provide a hydraulic head of 1–2 m at the spray nozzle.

Rotary sprays that rotate slowly over the filter bed are more common than fixed sprays. They consist of two or four horizontal arms that rotate around a central hollow shaft. The wastewater flows through the shaft and is sprayed onto the bed

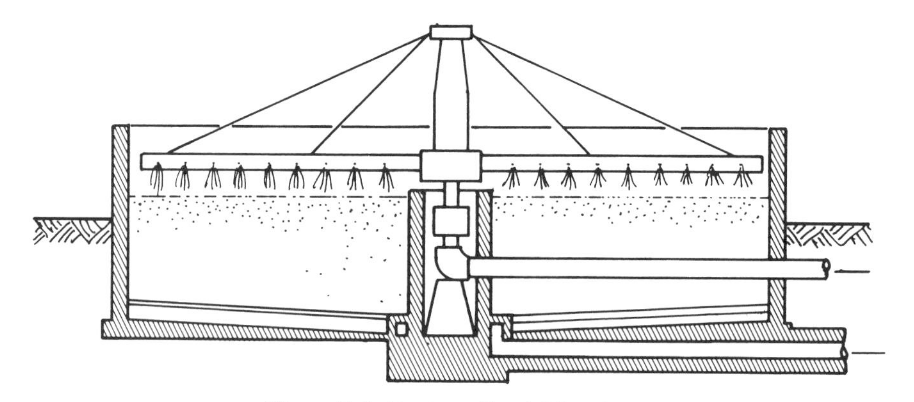

Figure 7.4 Rock media trickling filter.

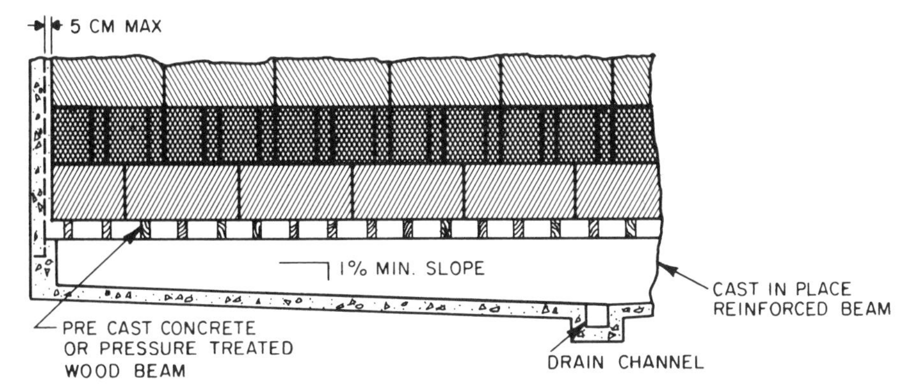

Figure 7.5 Plastic media filter.

through orifice nozzles along the side of the arms. Obviously, the shape of the filter must be circular. The required hydraulic head at the openings is less, about 0.5 m. The dosing of wastewater with rotary sprays is more uniform and since the orifice nozzles in the rotating arms are smaller and have a higher exit velocity they are also less prone to clogging. Furthermore, the dosing of wastewater above a particular section of the filter is intermittent and, therefore, allows more time for drying and less chance for pooling. The rotational speed is 0.3–5 rotations per minute.

Traveling distributors consisting of a perforated horizontal pipe supported from a truss spanning the entire width of the rectangular beds and running on rails on opposite walls have not been common in the United States but exist in Europe.

Whatever the method of distribution, oxygen is absorbed from the atmosphere by the sprays and the wastewater entering the filter is fresh and nearly saturated with air.

Ventilation and Air Supply. For very shallow filters less than 1 m in depth, air can be supplied from the surface. However, for all other filters it is necessary to create a draft through the filter, either upward or downward. The direction of the draft depends on the difference in densities between the air inside and outside of the filter. The temperature of the air inside the filter is about the same as that of wastewater. Therefore, if the outside temperature is less than that inside, the warmer and, hence, lighter air in the filter will rise. In the summer, when the outside temperature is warmer, downward drafts are common. However, the flow

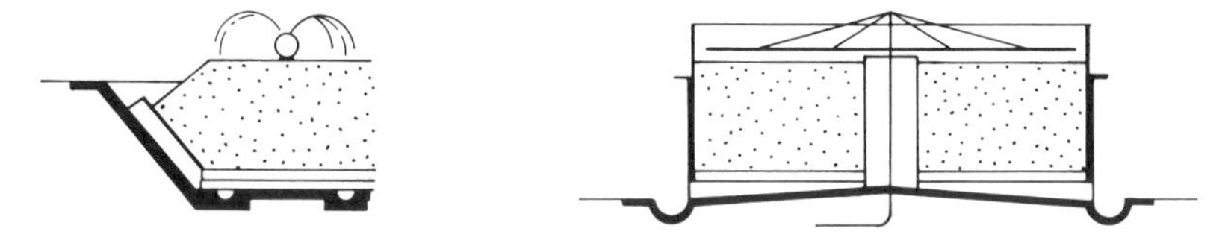

Figure 7.6 Fixed and moving spray distributors.

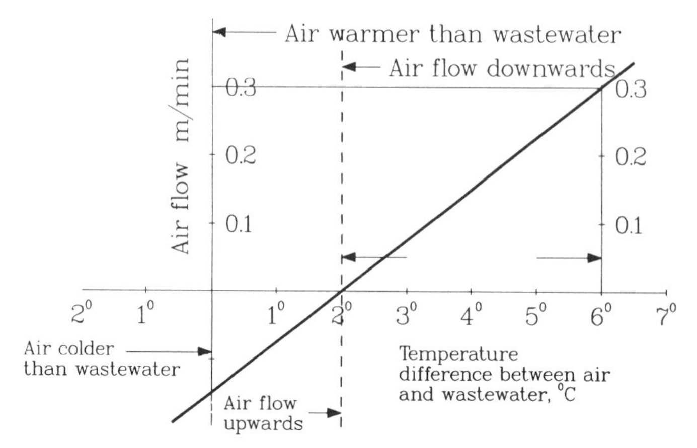

Figure 7.7 Airflow in trickling filters (after Halvorson) (to convert m/min to $ft^2/ft^2/min$ multiply by 3.3).

can be reversed upward during night hours. Thus, the direction of the draft during summer months can change twice a day. Halvorson and his co-workers[38] have estimated that if the air temperature difference between the inside and outside air is 6°C. the resulting draft is 0.3 $m^3/(m^2\text{-min})$ = 5 mm/s. The draft is reduced to almost zero if the temperature difference drops below 2°C. There is a straight line relationship in air flow related to temperature difference between and beyond these values (Fig. 7.7).

At an air flow of 0.3 m/min = 18 m/hr the oxygen input in the filter usually greatly exceeds the oxygen demand of the wastewater. This explains the fact that no mechanical draft is needed, even for superhigh-rate (tower) filters.

To guarantee adequate ventilation, the maximum water depth in the underdrains and wastewater collection channels should not be more than one-half the clearance between the upper and lower bottoms.

In some cases, mechanical draft filters may be needed. For forced ventilation, the underdrainage system must be sealed or bypassed and the beds completely enclosed. The air is usually blown on the top of the filter, creating a downward draft at a rate of about 0.3 $m^3/(m^2\text{-min})$ = 18 m/hr. Such filters are installed when low-temperature operation is required or to prevent fly nuisance. Covering the filters and using mechanical draft have no effect on the performance of the filter. It is therefore erroneous to call them "high-efficiency" filters."

Pretreatment. For pretreatment, primary sedimentation with approximately 1.5–2 hr detention should be used. Primary settling tanks with a larger area and submerged baffles are necessary when the wastewater contains mineral oils and grease that could severely damage the filter. Heavy organic loads such as those from industrial operations should be pretreated either by chemical precipitation, high-rate activated sludge treatment, or high-rate (roughing) trickling filtration. Pretreatment has an affect similar to dilution and the filter can therefore receive correspondingly higher hydraulic loading.

Secondary Clarification. As a rule, the majority of operations require secondary clarification following filtration. Clarification is mandatory for normal- and high-rate filters. The settling tanks should have a detention time of 1.5–2 hr. This detention time should be estimated from the total flow entering the sedimentation basin, including the return flow that is recycled back onto the filter. However, to reduce the size of the clarifier the flow can be recycled back onto the filter before the clarifier. The collected sludge is flocculant, light, and digestible. It is similar in character to activated sludge and can be handled in the same way, for example, by returning it to the primary tanks from which the excess settled mixed sludge is directed to digestion tanks. The volume of the tanks should be made large enough to accommodate the increased sludge input. Sludge volumes, water content, and organic constituents are greater in high-rate operations than in low-rate operations. The water content of sludge from low-rate operations is about 92% and for higher rate operation, it is about 96%, which represents a significant increase in volume.

Filter Design. In designing a trickling filter, both surface area and volume are important. The hydraulic surface load is a measure of the flushing power of the applied wastewater. The actual performance of the filter depends more on BOD loading. Typical hydraulic surface loadings and BOD volumetric loadings for different types of filters were given in Table 7.5. A load of 200 g of $BOD_5/(m^3\text{-day})$ is typical for a low-rate filter with nitrification, whereas 800 $g/(m^3\text{-day})$ is typical for a high-rate filter.

The organic loads can also be expressed in population equivalents (PE). Assuming that the population equivalent parameter of a municipal sewage after primary sedimentation is 40 g of $BOD_5/(\text{cap-day})$, the corresponding low-rate filter loading is $200/40 = 5\ PE/m^3$ and that for a high-rate filter would be $800/40 = 20\ PE/m^3$. Expressing the load in PE terms is advantageous when the PE loading for industrial contributions are known.

Plastic media filters provide full treatment at BOD_5 surface loadings up to 4 g/day per m^2 of the media surface. Therefore, their efficiency depends on the specific contact area that may be up to 200 m^2/m^3 (Table 7.6). Nitrification can be achieved at BOD_5 surface loadings up to 2 $g/(m^2\text{-day})$, expressed per surface area of the media. Plastic media filters are particularly well suited for arrangement in several steps.

In design, it must be determined whether the loading value should be at the lower or higher end of the ranges given in Table 7.5.

Recirculation. The hydraulic surface loading parameter is expressed in $m^3/(m^2\text{-day})$ [in the United States the equivalent is million gallons per acre per day = mgad = 0.935 $m^3/(m^2\text{-day})$]. It should include the recycled flow. This parameter is useful for normal- and high-rate filters where it expresses the flushing power of the filter operation, which, in turn, governs the thickness of the biological film on the contact media. According to Halvorson *et al.*,[38] the minimum hydraulic loading of high rate filters should be 0.8 $m^3/(m^2\text{-hr})$ = 0.8 m/hr. This is about 10 times more than that for low-rate filters.

The magnitude of the flushing power is greater than indicated by the average hydraulic loading because typical distribution systems result in higher instantaneous shock loads followed usually by a resting period during the passage of the sprayers. A typical rotating spraying mechanisms has several (2–6) arms rotating 50–400 times per hour. The flushing power then decreases with the number of arms and with the number of rotations per hour and it can be estimated according to the following formula:

$$S = \frac{q_f}{an}$$

where S = flushing power of the filter (mm/wetting)
q_f = surface overflow rate (mm/hr)
a = number of spray distributor arms
n = number of rotations per hour

Normal values of S should be between 2 to 6.

To obtain a desired hydraulic loading, it is often necessary to recycle a portion of the effluent. This increases the flow to the filter and reduces its strength. A recycle ratio of 3 means that the flow to the filter is composed of 1 part of sewage to 3 parts of treated recycle. For municipal sewage, the ideal BOD_5 concentration on the top of the filter should be about 100–150 mg/liter. That means that for typically weaker U.S. sewage, the recycle ratio should be about 0.5 to 1 whereas that for stronger, European sewage should be 2 to 3. The recycle can originate from the filter effluent before or after clarification. The former approach keeps the size of the clarifier smaller but it leads to the recycle of solids. The latter case requires a proportionally larger secondary clarifier to accommodate the increased flow.

Other advantages and benefits of recirculation include (1) reducing periods out of service to a minimum, (2) keeping self-propelled distributors turning by adjusting recirculation to influent flow, (3) limiting film thickness and fly breeding, (4) preventing odors by freshening the raw sewage, (5) seeding applied sewage with bacteria and other active microorganisms as well as nitrates, and (6) more effective loading and working of the deeper portions of the filter.

It may sometimes be desirable to mix the recycle ahead of primary treatment. This will freshen the incoming wastewater and initiate biological treatment in the settling tank. However, the size of the primary settling units must be made proportionally larger.

Figure 7.8 shows the typical arrangements of treatment units for filtration with a recycle.

Filter Maintenance and Operation. During operation, interruptions of flow to low-rate filters should be kept to a minimum. Frequent interruptions of 10–30 min of duration can adversely affect the bacterial decomposition process and also promote development of filter flies. The hydraulic loading of normal and high-rate filters should be essentially without interruptions.

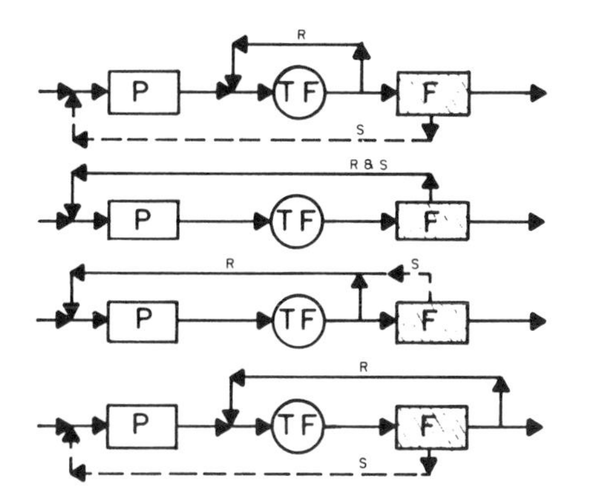

Figure 7.8 Typical recycle flow for trickling filters (R, recycle; S, sludge; P, primary clarifier; TF, trickling filter; F, final clarifier).

Clogging of filters should be prevented. It is indicated by an increase in organic settleable solids in the effluent and by ponding on the top of the filter. Ponding is also caused by hydraulic overloads or by algal or fungal growths on the filter surface or in the upper layers. Ponding can be controlled by chlorination (using concentrations of 20–50 mg of chlorine per liter), by increasing the recycle ratio to flush off the undesirable biomass, or by resting the filter for several days. Resting causes the microorganisms to die and become amenable to subsequent flushing. Another method available for filters with stone media is to dislodge the biomass in the upper 15 cm of the filter media and flush it away.

It should be noted that properly designed filters do not usually develop such problems if they have adequate flushing power, their filter media is not too fine, and they have adequate ventilation. If clogging does develop, the operating conditions should be rechecked and the hydraulic loading adjusted by recycle.

Odor problems are minimized by keeping the wastewater in the influent fresh by recycling and/or by aeration or prechlorination. Enclosing and ventilating the filters require that the exhaust air be deodorized.

Development of filter flies (*Psychoda*) can be controlled by periodically adding chlorine or insecticides to the influent. The application is particularly effective during the short breeding periods of the flies that at 20°C may last about 2 weeks. Again, high-rate filters have less problem with the flies than low-rate filters. The continuous flow of wastewater with relatively higher flushing power keeps the thickness of the biological layer in which the larvae live small, and also prevents the flies from escaping from the filter.

Trickling filters may become hydraulically overloaded during periods of rainfall. Up to 1.5 times the design flow can be treated on low-rate filters without significantly changing their efficiency. In high-rate filters, the recycle ratio should be reduced or the recycle stopped to accommodate the increased flows. Flows exceeding these loadings should be diverted to retention basins that may have to be installed to prevent overload. Also, the filters can be made larger. For example, in England, it is required that the filters be capable of accommodating and treating flows that are three times the dry weather flow, which makes the filters about 50–75% larger than those computed from typical population equivalent parameters or dry weather loadings.

Filters can perform satisfactorily even under winter operations. The temperature of the influent is warm and the heat loss from the filter is relatively small, maximally 4°C. High-rate filters perform better than low-rate filters during winter operating conditions since they accept more of the relatively warm influent flow. Heat losses can be minimized by elevating the top of the side walls above the filter surface to reduce the effect of cold winds and/or filters may be buried. The bottom openings for ventilation may be reduced during very low outside temperatures to the minimum size required for effective ventilation. Nevertheless, the removal efficiency of filters is less during cold temperatures. Pöpel has estimated that the performance of a filter treating 10°C cold influent is only about 62% of that at 20°C.

In extreme climates of the northern United States and Canada, and Scandinavia, it is advisable to cover the filters for winter operations.

Performance of Trickling Filters. Low- and normal-rate trickling filters with primary and secondary clarification can remove 65–95% of the incoming BOD load. A typical average removal efficiency for these trickling filters is 80%. The removal efficiency can be regulated by arrangement of the filters in series (several stages) or by recirculation. In general, the removal rate depends on the filter BOD loading. Figure 7.9 shows the BOD removal relationship established from

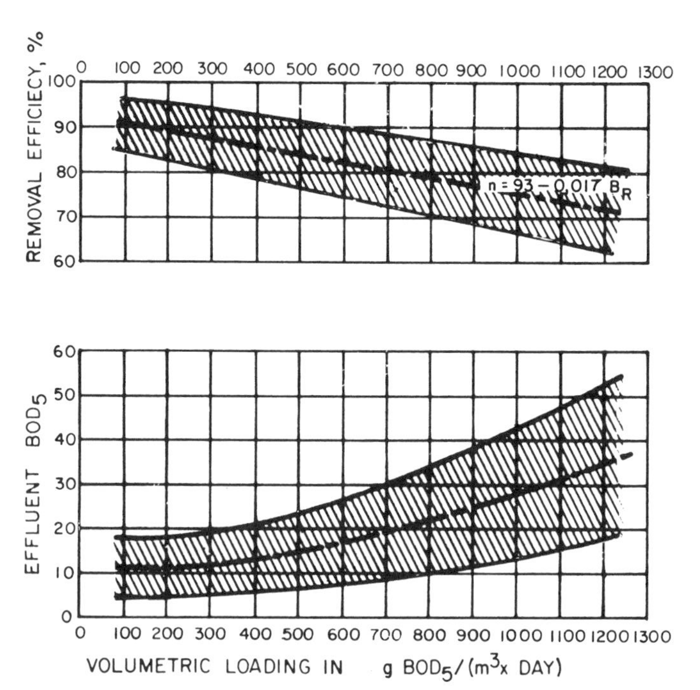

Figure 7.9 BOD removal in trickling filters [to convert from g/(m^3-day) to lb/ft^3/day multiply by 6.24×10^{-5}].

measurements of trickling filters in the Ruhr area of Germany. Imhoff and Fair[11] reported similar ranges for U.S. conditions. The best performance of filters in the Ruhr area was accomplished at loadings of 400 g/(m^3-day).

A very appropriate application of trickling filters is for pretreatment of wastewater (sometimes called roughing filters). Even when the load is highly variable, significant removals can be obtained.

The selection between low-rate and high-rate filters is somewhat subjective. The simpler low-rate filters are reliable and yield a more digested sludge. They also can oxidize ammonia to nitrate. The normal- or high-rate filters are smaller and, hence, cheaper. However, their removal efficiency is not as high and in order to achieve the removal efficiencies required in the United States, they require recirculation, resulting in a larger secondary clarifier. The sludge volume is larger and the sludge itself contains more water and is less stabilized. This results in an increased cost of digesters and sludge handling units and, as a result, the savings are not as high as one might expect.

As previously indicated, nitrification can be accomplished by low-rate filters or by two filters in series, the second one being a low-rate unit. Nitrification efficiency is related to the available contact surface area of the filter media,[39] however, often the design of trickling filters for nitrification requires parameters obtained from pilot treatability studies. An example of treatment plant nitrification performance is shown on Figure 7.10. Gulicks and Cleasby[40] have developed an empirical method of analysis for a trickling filter nitrification design that accounts for hydraulic loading, influent ammonia concentration, wastewater temperature, and specific contact area of the media. The method can also be used for design of filters with recycle.

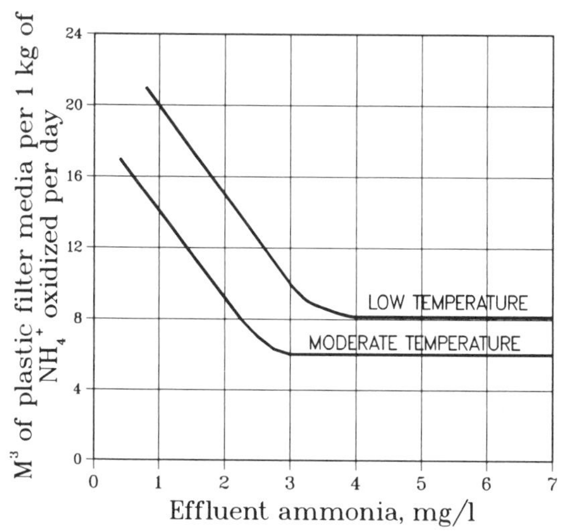

Figure 7.10 Nitrification in plastic media filters (to convert from m^3/kg to ft^3/lb multiply by 16). Adapted from Ref. 39.

The conclusions reached on the trickling filter performance by a U.S. Environmental Protection Agency survey[41] were as follows:

1. Trickling filters are capable of providing a high degree of treatment.
2. Of the plants with trickling filters 90% exhibited over 74% BOD removal.
3. Over 90% removal can be achieved by two-stage filtration.
4. The performance of the filters was not sensitive to load fluctuations within the ranges anticipated at each plant.
5. Recirculation had no significant effect on the filter performance.
6. Both rock and plastic media performed equally well.

Contact Aerated Beds

Contact beds, sometimes called Emsher beds, were developed in Germany shortly after World War I. They consist of submerged media in a tank through which wastewater moves in a horizontal direction. A number of media can be used such as crushed rock, coke, wood chips, aluminum plates, and synthetic materials similar to those previously introduced for trickling filters. Compressed air is blown through the contact media to maintain circulation and satisfy the oxygen requirements of the biological flocs. Such filters are not common in the United States.

Rotating Biological Contactors

These units have evolved from fixed contact units described in the preceding paragraph. In 1920s, Karl Imhoff and his co-workers developed and tested rotating drums filled with brushwood, but the results were not favorable. However, work at the University of Stuttgart by Hartmann and Pöpel resulted in elimination of most of the problems with mobile biological contactors and the first rotating biological contactors (RBC) were installed in Germany in the early 1960s.[42] Since 1972 they have been successfully marketed in the United States.

The units typically consist of a steel shaft (about 7–8 m long) with a series of mounted plastic discs placed in a tank. The bottom of the tank can be contoured (Fig. 7.11). The diameter of the disc is about 3–4 m and the discs vary in configuration, depending on the manufacturer. The media are available as standard (10 000 m^2/shaft) or as high density (15 000 m^2/shaft). The discs slowly rotate at a rate of 1–2 revolutions per minute and are about 40% submerged in the wastewater and 60% exposed to air. Although submerged in the water, the discs pick up a thin layer of wastewater from which the biomass that grows on the disc absorbs the organics and nutrients. During air exposure, the biomass can digest the absorbed pollutants and utilize oxygen. The bacteria form a slimy layer similar to that formed on trickling filters.

Excess biomass (sludge) is removed from the discs by the shearing forces created by the rotation of the disc in the wastewater. The rotation also creates mixing,

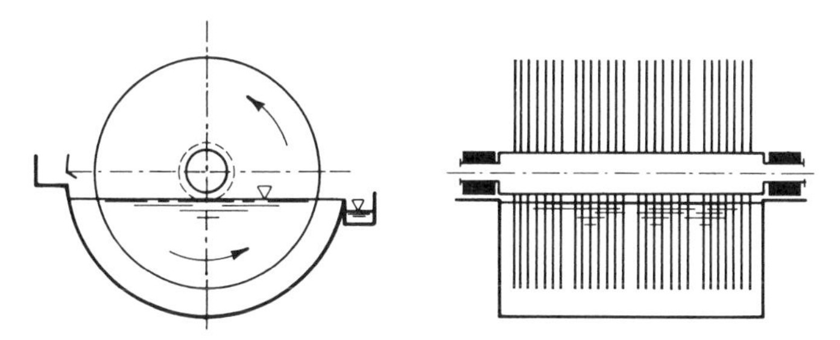

Figure 7.11 Rotating biological contractor.

which keeps the sludge solids in suspension so they can be carried to the final clarifier.

Usual installations employ four rotating shafts in series. The shafts are perpendicular to the direction of flow in the tank. Organic BOD is removed by the first two to three shafts whereas the last shaft often allows nitrifying organisms to develop.

RBC units are particularly suited for small plants. A design curve showing the effluent quality as a function of loading is shown in Figure 7.12.

RBC units are normally covered by fiberglass enclosures to prevent degradation of the plastic media by sun light and to eliminate the growth of algae. In cold regions, the enclosure eliminates freezing of the units.

For high-strength wastewaters, the loading to the first shaft must be reduced by recycle or supplemental aeration to prevent anaerobic conditions, however, recirculation of wastewater flow is normally not provided.

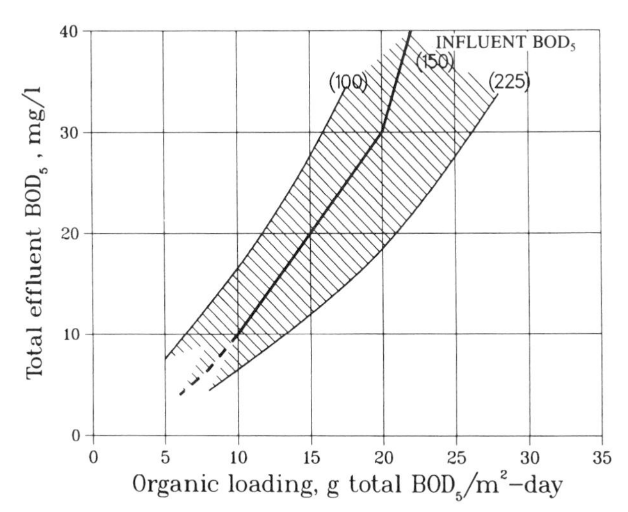

Figure 7.12 Effluent BOD_5 as a function of organic loading in RBC units (after U.S. EPA) (to convert from g/m^2 to lb/ft^2 divided by 4 885).

A check list for a trouble-free design of an RBC system includes the following criteria[43]:

1. The organic load to the first shaft should be below 30–40 g/(m^2-day) (6–8 lb/1 000 ft^2/day) of total BOD_5 and 12.5–20 g/(m^2-day) (2.5–4 lb/1 000 ft/day) of soluble BOD_5. Higher loading will create problems such as heavy and thick biofilm thickness and anaerobic conditions resulting in odor problems. The oxygen transfer efficiency is about 7.5 g/(m^2-day) (1.5 lb/1 000 ft^2/day).
2. High-density media should be used only in the later stages to avoid clogging with the biomass. In practice, this means that the high-density media should be used primarily for nitrification.
3. The bearing capacity of the shaft should be designed for a maximum film thickness of about 4–5 mm. The desired film thickness should be 1.2–1.5 mm.
4. Means for removing excess biofilm growth should be provided, including air or water stripping, chemical addition, or rotational speed control and reversal.
5. The shafts should have variable speed control, multiple treatment trains, removable baffles between the stages, potential for step feed, and recirculation of effluent for maximum operational flexibility.
6. Primary treatment should be provided prior to RBC treatment.
7. Supplemental aeration may be needed in the first stages of the RBC unit to handle occasional high loads or if high-strength wastewater is to be treated.
8. The average energy use for rotation is about 2 kW/shaft. With air-driven shafts the energy consumption is about 5 kW/shaft.
9. The RBC units should be covered or be located inside of a building.

The average BOD removal of the RBC units installed in the United States is about 82%, with a range of 75–90% and the average effluent BOD_5 is 18 mg/liter. The cost of operating an RBC unit is less than that for activated sludge but more than that for a trickling filter.

Nitrification can be achieved with RBC units when the bulk of the BOD is removed, that is, when the concentration of BOD_5 in the unit drops below 15–30 mg/liter. This is usually reached after three shafts.

7.4 SUSPENDED GROWTH BIOLOGICAL TREATMENT PROCESSES[44–48]

Activated Sludge Process

The activated sludge process for wastewater treatment originated when waste treatment engineers tried to purify sewage by simple aeration. It was observed that under a prolonged aeration, flocs made of living microorganisms have developed and settled in the tank after aeration and agitation stopped. When new fresh sewage was then introduced in the tank and the mixture aerated again the flocs were reactivated and purification proceeded at a much faster rate.

Essentially, activated sludge can be considered as an accelerated self-purification process in natural streams. The conditions and the processes are similar, only the concentration of the microorganisms and their density inside the flocs are much higher. The surface area of the biological flocs is about 2 000–10 000 m^2 per 1 m^3 of the aerated volume.[49] In order to maintain an aerobic condition necessary for the decomposition and removal of pollutants, artificial aeration is needed that can be provided by blowed compressed air, spray aeration, or injection of pure oxygen. By aeration the water in the tanks is also sufficiently mixed, which prevents the flocs from settling to the bottom and become septic.

Compared to trickling filter operations, the activated sludge process offers higher removal efficiencies, it is relatively free of odor, it has no fly nuisance, and works effectively under both winter and summer conditions. The highest efficiency is achieved at a temperature of 30°C. Disadvantages are that the process is mechanically more complex and more costly, and it produces more sludge that is less stabilized, hence, the size of the sludge digesters and sludge handling units is greater.

Pretreatment. Typically, the influent to an activated sludge process is pretreated by primary sedimentation as it is common for other biological treatment processes. The primary clarifier removes settleable suspended solids and the biological unit is then left to handle nonsettleable, colloidal, and dissolved pollutants. In some smaller plants, particularly when the influent suspended solids are mostly organic, the primary clarification can be reduced or even omitted, however, it should be kept in mind that inert suspended solids will pass the aeration units and will be removed by secondary clarifiers. The option of eliminating the primary sedimentation unit is not available to trickling filters and RBC units.

Types of Activated Sludge Systems

Categorization of activated sludge systems is primarily based (1) on the type of flow throughout the basin, (2) on the type of aeration or aeration devices, and (3) on the magnitude of the sludge load.

System Categorization Based on Type of Flow. There are basically two types of flow in an aeration basin:

1. *Plug Flow* basins are usually narrow elongated channels (length to width ratio of 5–50 to 1) through which particles from the influent pass as a group, or plug, at the same time. Hence, the detention time in the tank is about the same for all particles. Plug flow basins are constructed either as once through elongated channels (Fig. 7.13) or as circulation channels (Figs. 7.14 and 7.15). The "oxidation ditch" or "Carrousel tanks" are examples of circular plug flow reactors that differ by aeration systems.

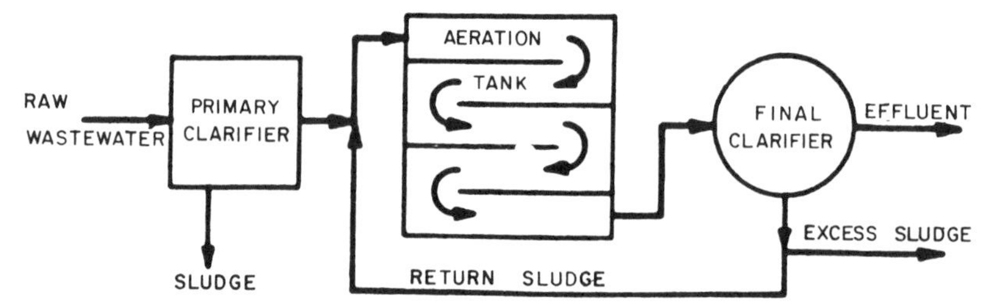

Figure 7.13 Conventional plug flow activated sludge plant.

Oxidation Ditches[49] employ a surface aerator (Kessener brush) located at one or more points above the surface of the channel, which also provides momentum for the horizontal movement of wastewater at a velocity of about 0.3 m/s. The oxidation ditch treatment is very popular for small treatment plants as it is very easy to operate with very little supervision and maintenance. The *Carrousel tanks* employ one or more vertical shaft surface aerators that also provide the propulsion for the oxidation ditch.

A modification of the oxydation ditch that uses part of the ditch for secondary clarification eliminates the need for a separate settling tank (Fig. 7.15). The settling occurs in an intrachannel clarifier or a ''boat'' settling unit located above the bottom of the ditch. As the flow passes below the clarifying channel, part of the mixed liquor enters through the clarifying channel bottom into a quiescent zone from which the clarified liquid is withdrawn into the effluent. In this way, the cost of the treatment can be reduced.[50]

Modifications of the purely plug flow reactor are tapered aeration and step aeration. In tapered aeration, the diffusers are spaced so that more air is supplied at the beginning of the tank where the oxygen demand is greatest and reduced along the length as the demand decreases. In the step aeration process (Fig. 7.16) the return sludge is mixed with a portion of the influent at the head of the tank but the influent is fed into the tank at different points along its length. The step aeration process was introduced by R. H. Gould[51,52] in the New York–Tallmans Island treatment plant. The step aeration process is a transition between the plug flow and completely mixed systems.

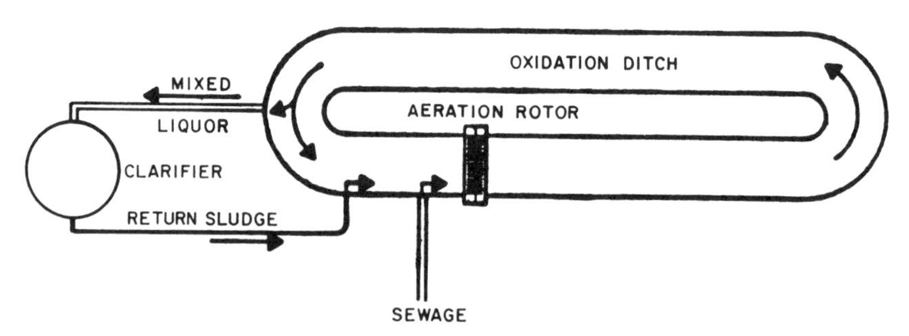

Figure 7.14 Circular activated sludge plant (oxidation ditch) with a Kessener brush for aeration and separate secondary clarification.

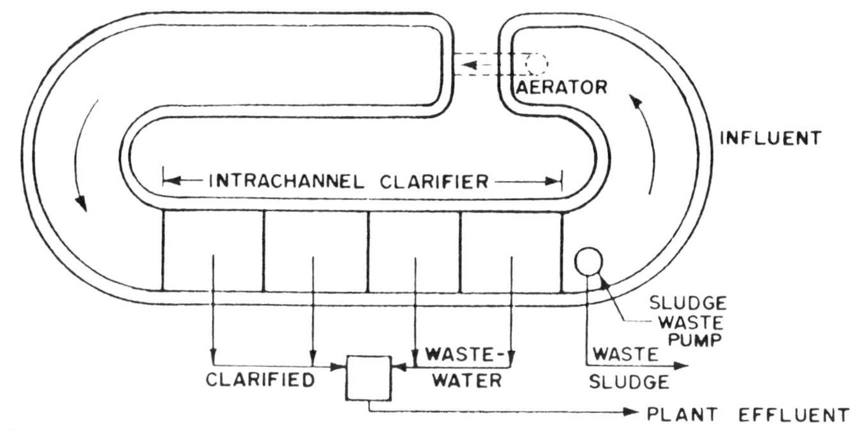

Figure 7.15 Circular activated sludge plant (carrousel) with intrachanel clarifier (after Christoper, *Civ. Eng.*, May 1983).

With these two modifications of plug flow reactors, overloading and insufficient oxygen supply at the beginning of the reactor will be avoided.

2. *Completely Mixed Basins* are usually square or rectangular basins with a low length-to-width ratio. Mixing by the aerators results in a uniform distribution of solids throughout the basin. As a consequence, the solids reside in the basin with a different residence time and some influent solids will reside in the tank for a very short time. Figure 7.17 shows examples of completely mixed basins. It should be noted that several completely mixed tanks in a series will resemble plug flow systems.

Categorization Based on Type of Aeration. There are numerous types of aeration systems; however, their main purpose is as follows: (1) to provide enough oxygen for the removal of the biodegradable organics and keep the biota in the tank aerobic, and (2) to provide mixing that will keep the solids in suspension.

Aeration systems can be divided basically into the following[53–56]: (1) Diffused aeration, (2) mechanical aeration, and (3) pure oxygen.

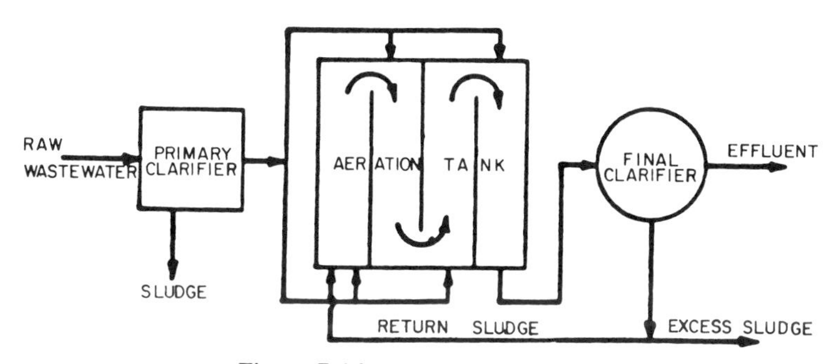

Figure 7.16 Step aeration plant.

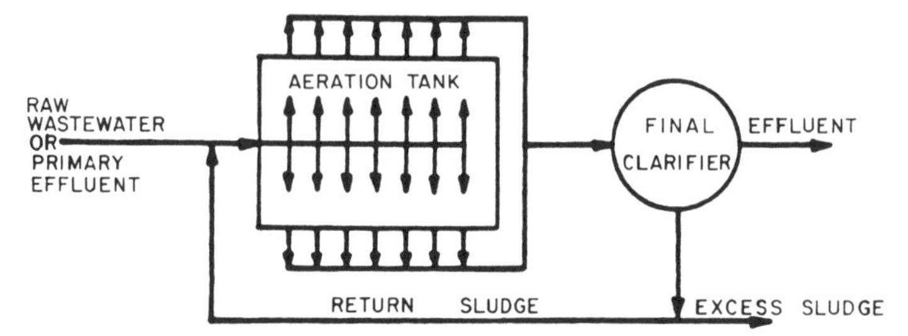

Figure 7.17 Complete mix plant.

In Diffused Aeration Systems, compressed air is blown into the tank through diffuser plates, domes, or a pipe located at or near the bottom of the tank (Fig. 7.18). The tanks are elongated with a relatively high length-to-width ratio, resulting in near plug flow conditions. The depth of the tanks is usually 3–5 m and ridges and furrows provide for aeration symmetrically from the floor, whereas spiral flow tanks have the aeration plates located near the wall. The air is introduced either as fine bubbles from porous plates or pipes, or medium or coarse bubbles from perforated pipes. In the INKA process shown in Figure 7.19, low-pressure air is introduced through the stainless steel or plastic pipes with a perforation on the bottom of the pipe that is located at a depth of about 0.8 m below the surface.

Mechanical Aerators work on the principle of breaking the surface of the wastewater in the tank and exposing it to the air. The Simplex system shown in Figure 7.20 uses a rapidly rotating impeller that exposes the water to the air located in a single square or rectangular tank broken into square compartments. The wastewater, mixed with the activated sludge flocs, flows from the bottom upward through a centrally mounted draft tube to the surface. After aeration, the liquid with the sludge returns back to the bottom. The contents of the tank are turned over about once in 20 min. Surface spray aeration works on the same principle but without the draft tube.

The Kessener brush systems achieve circulation and aeration by an elongated

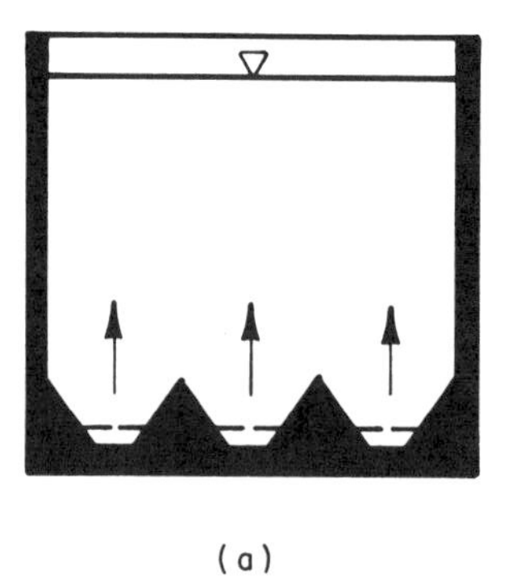

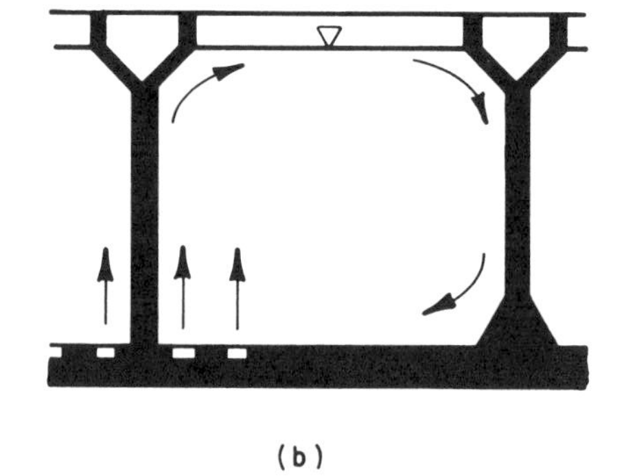

Figure 7.18 Diffused air activated sludge system: (a) ridge and furrow, (b) spiral flow.

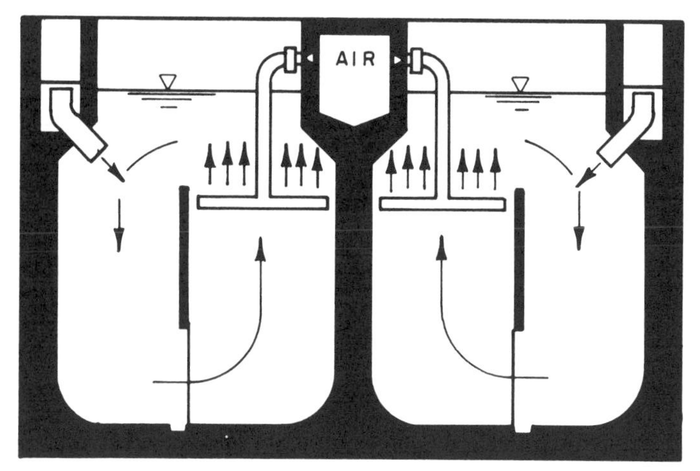

Figure 7.19 INKA aeration tank.

brush mounted at the wastewater surface along the side of the tank (in contrast to oxidation ditches in which the brush is mounted perpendicular to the direction of flow). The brush consists of stainless steel combs that rotate 40–120 times per minute and the wastewater lifted by the brush is sprayed over the surface of the tank.

Pure Oxygen Systems were first developed by M. Pirnie in 1946. D. A. Okun[57] reported on laboratory pilot studies using pure oxygen for treatment. However, the popularity of these systems increased only when oxygen became cheaper and more readily available. In one system manufactured in the United States, the oxygen is made at the treatment plant by separating nitrogen from the air.

With pure oxygen systems, the saturation value of oxygen in water is much higher (about five times) than that for air, and since the oxygenation capacity of

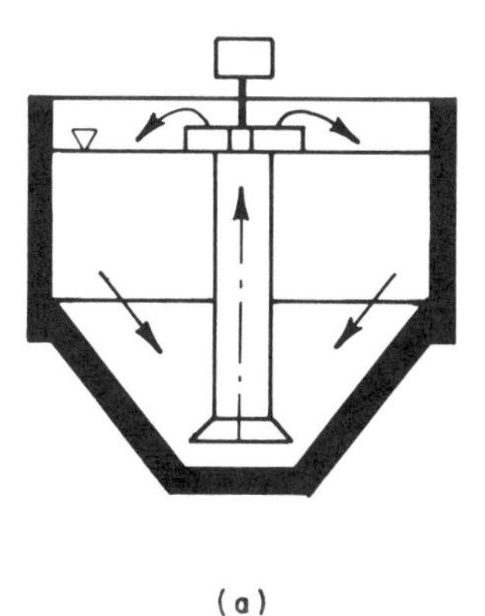

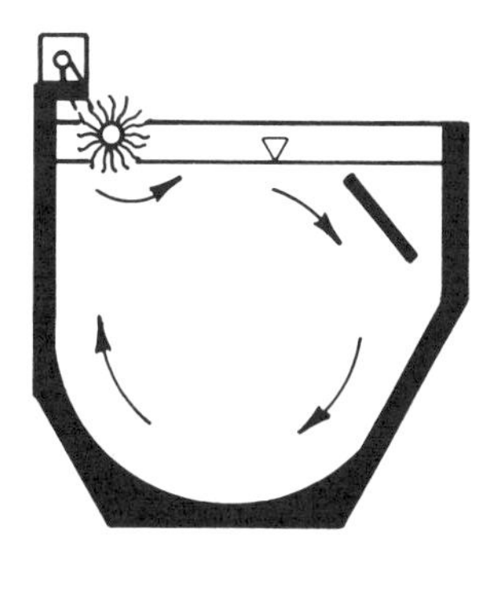

Figure 7.20 (a) Simplex activated sludge unit; (b) Kessener brush activated sludge system.

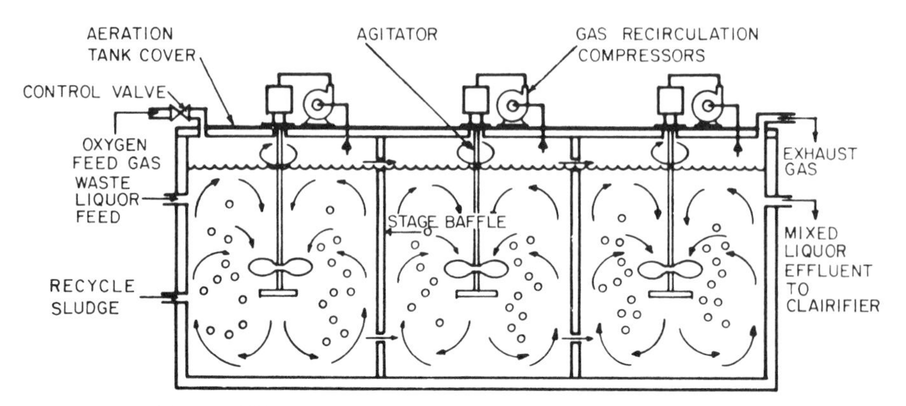

Figure 7.21 Pure oxygen ("UNOX") activated sludge unit.

aeration systems is proportional to the difference between the oxygen saturation value and the DO concentration in the tank, more oxygen can be supplied. This enables the maintenance of higher concentrations of activated sludge in the tanks and improves the treatment efficiency and/or reduces the size of the aeration tank. The economy of the process is improved if the treatment is performed in several enclosed tanks in series, where the off gas is collected and recycled inasmuch as only a small portion of the oxygen is used in each step (Fig. 7.21). Because the aeration units are completely enclosed no odors are released to the environment.

Categorization Based on Sludge Loading. The systems can be divided into conventional, high-rate, and low-rate activated sludge treatment.

Conventional Activated Sludge Systems usually consist of one or several elongated (plug flow) or square (completely mixed) reactors that are basically designed for the removal of carbonaceous BOD (Fig. 7.22). The overflow from the primary clarifier is mixed with the return sludge from the underflow of the secondary clarifier and the combined flow then enters the aeration basin. The detention time for a typical municipal wastewater is 4–5 hr. A portion of the returned sludge that equals the daily sludge production in the system (sludge yield) is usually diverted to the primary clarifier and to the sludge treatment jointly with the primary sludge.

BOD removal efficiencies should average 85–95%, resulting in total BOD_5 concentrations in the municipal sewage effluents of less than 30 mg/liter.

High-Rate Activated Sludge Systems have a shorter detention time, usually 2 hr, and commonly do not employ primary sedimentation. Obviously, the treatment efficiency is less than that for conventional treatment, usually about 75%, which means that such systems alone would not provide the treatment levels required in the United States. High-rate systems can be effectively used in combination with other systems such as trickling filters.

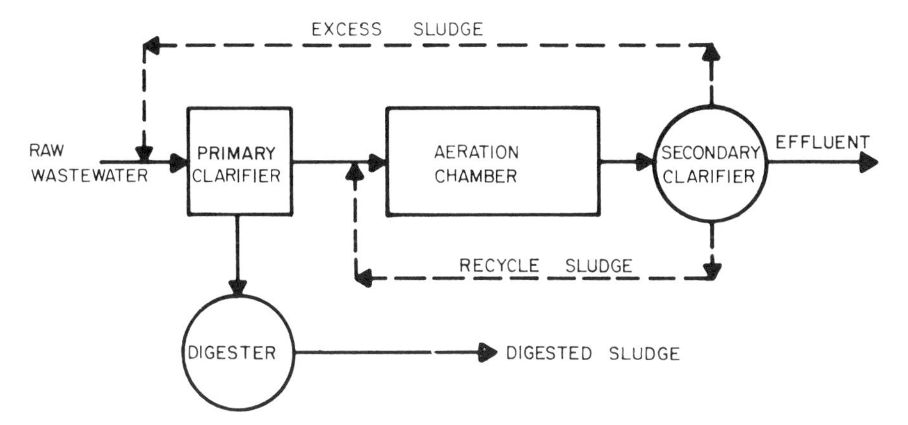

Figure 7.22 Flow schematic for simple stage activated sludge system.

There are several types of high-rate systems:

1. Contact stabilization units are particularly suitable for wastewaters that have an appreciable amount of BOD in suspended and colloidal solids. The highly adsorptive properties of the activated sludge are used to adsorb as much of the solids as possible in a short contact period and digest them in a peripheral side basin located on the sludge return (Fig. 7.23). The aeration tank has a relatively short detention time of 1–2 hr. The stabilization tank for the bacteria has a longer detention time (about 12 hr), however, the flow rate is smaller since only return flow with high solids concentrations are stabilized.[58]

   Contact stabilization units also provide safety against the effect of toxic spills that could adversely affect the biota. As the toxic spill can be contained in the aeration basin only, the damage to the biota is minimal inasmuch as the sludge stored in the stabilization basin will provide fresh and active biomass immediately after the toxics are removed from the aeration basin.
2. Units with a reduced concentration of activated sludge in the aeration basin, resulting in a high load on the sludge. A short period of aeration is combined with a relatively low concentration of the sludge in the aeration tanks (about 650 mg/liter). This process, first introduced in the New York–Jamaica treatment plant, is called *modified aeration*. The aeration requirement is reduced to about 50% of that needed in a conventional activated sludge process. The sludge is highly unstabilized, but it has relatively good settleability characteristics and lower moisture content.
3. Units with a high concentration of activated sludge and increased aeration. Such units were installed in Germany (Wuppertal). The volume of the aeration units is about one-third of that for a conventional process. The removal efficiency and sludge characteristics are similar to the preceding system.[59]

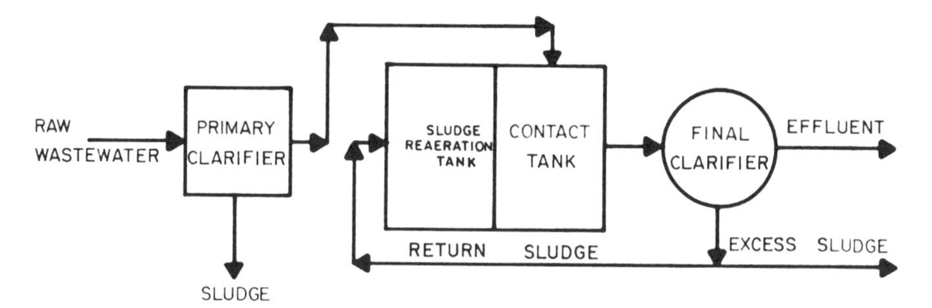

Figure 7.23 Contact stabilization activated sludge system.

In Extended Aeration Systems, the loading is so low that the microorganisms are starved and undergo partial autooxidation, which results in a more stabilized sludge and, often, nitrification. The detention time in the aeration units is greater than 24 hr. The process has been applied mainly to the treatment of domestic sewage from small communities, housing developments, and recreational areas. The oxidation ditches previously mentioned are the best known examples of extended aeration.

The basic design characteristics of activated sludge systems based on the sludge load will be given in subsequent sections.

Activated Sludge Process Characteristics

Activated Sludge Flocs. The biological decomposition process in activated sludge units proceeds in two steps. In the first step, part of the organic matter is oxidized and new cellular mass is developed to gain energy. Dissolved as well as colloidal organic matter is utilized. In the second step the bacteria cluster together by extracellular polymers, which occurs after the rate of bacterial growth declines. The polymers form bridges and connections between the individual microorganisms.[60,61] The second phase occupies the major portion of the aeration time.

In addition to the bacteria that form the flocs, higher organisms such as protozoa predate on the free flowing microorganisms. The flocs also consist of a slimy matter in which the bacteria and protozoa reside. The microorganisms of the activated sludge flocs grow spontaneously provided sufficient amounts of nutrients are present. Typical ratios of carbon, nitrogen, and phosphorus needed for optimum growth are 100:10:2. This can also be expressed as 3–8 kg of nitrogen and 1–1.5 kg of phosphorus needed for each 100 kg of BOD removed. If nitrogen and phosphorus are not present in sufficient amounts in the wastewater itself—a common occurrence in many industrial wastes—they must be added.

There are two levels of structure in the activated sludge flocs.[61] The microstructure is imparted by processes of aggregation and bioflocculation, essentially by one microorganism sticking to another. The macrostructure of the flocs is formed by filamentous organisms that form a network or backbone within the floc.

Since the detention time in the aerated tanks is relatively short (several hours) it does not provide sufficient time for the flocs to fully develop. Thus, a significant portion of the flocs is returned from the secondary settling tanks back into the aeration units as a sludge return. The remainder of the flocs, which equals the biomass produced in the system during a certain period of time, is wasted. This makes the residence time of the microorganisms in the activated sludge process much longer than the detention time of wastewater in the units. The amount of sludge formed and wasted daily from the secondary clarifiers is called *sludge yield.*

The average solids residence time (SRT), reflecting approximately the time the microorganisms spend in the system (also called the sludge age), is an important parameter in the operation and design of the activated sludge process. The SRT parameter is estimated as

$$\mathrm{SRT} = \frac{\text{quantity of sludge in the system (kg)}}{\text{sludge yield (kg/day)}}$$

It is more appropriate to estimate the SRT (sludge age) using the volatile rather than the total solids contents of the sludge. The longer the SRT (sludge age), the more stabilized sludge is obtained. However, an optimum SRT is about 2–4 days. For nitrification, the SRT should be longer.

The settleability and volumetric measure of the sludge after settling are expressed by the Zone Settling Test and as the Sludge Volume Index (SVI).

Sludge Volume Index (SVI) is the volume 1 g of sludge occupies after 30 min of settling expressed in ml. A good settling sludge would have an SVI between 50 and 100 ml/g. Bulking sludge, with poor settling characteristics, could have an SVI up to 400 ml/g. Daigger and Roper[62] have shown that the SVI, a parameter that can be easily and quickly measured, can be correlated to the settling velocity of the solids in secondary clarification.

Poor settling or bulking of the sludge is caused either by overloading or underloading the units. This results in a failure of flocs. The types of floc failures and their symptoms and remedies have been described by Jenkins, Richard, and Daigger.[61] The following types of floc failures resulting in poor settling and sludge bulking have been listed by the authors:

Dispersed growth is caused by a microstructure failure in which microorganisms do not stick to each other. Dispersed growth produces nonsettling sludge and turbid effluent.

Slime or jelly formation (viscous bulking) is caused by microstructure failure in which too much extracellular material is produced. The sludge has low compactibility, which may result in the thickening failure of the secondary clarifiers.

Pinpoint flocs formation is the result of a macrostructure failure caused by a shortage of filamentous microorganisms. The large aggregates of the sludge settle well (low SVI), however, the effluent is turbid because of the presence of small aggregates.

Sludge bulking is a macrostructure failure in which too much filamentous organisms growth is present. The filaments then interfere with compaction and settling of the sludge, which frequently results in the thickening failure of the clarifier. If thickening failure does not occur the effluent is relatively clear, however, the SVI is very high.

In addition to sludge flocs failure, foam formation in the aeration units and secondary settlers can severely impede the activated sludge process. Foam formation is caused either by nonbiodegradable surfactants and/or by the presence of a microorganism *Nocardia* sp. The bacterial (*Nocardia*) foam is viscous, stable, and often chocolate colored and contains large quantities of activated sludge flocs. Foaming may occur when the activated sludge is nutrient limited.

Sphaerotilus and other filamentous bacteria can develop into unsightly stringy growths that attach themselves to walls, gates, and baffles of channels and tanks.

Sludge Production. The excess sludge (sludge yield) has to be diverted to subsequent stabilization and disposal. In small plants, it is advantageous to divert it into primary settling tanks and provide joint sludge handling units for both primary and biological sludges. The size of the sludge handling units must then be increased to accommodate the increased sludge volumes.

Conventional activated sludge units produce about 0.5–0.8 kg of sludge per kg of BOD_5 removed. By increasing the sludge age, the flocs have more time to digest the absorbed organics and, thus, less sludge is produced. This process is called extended aeration and the sludge yield can be reduced to about 0.3 kg/kg of BOD_5 removed. On the other hand, high-rate plants that have a short sludge age (less than 2 days) can have sludge yields as high as 1 kg of sludge/kg of BOD_5 removed.

Loading to Activated Sludge Units. The decision whether to employ a low-rate, conventional, or high-rate system depends on the economics, desired effluent quality, nitrification requirement, and other factors. Normally, the size of the tank is expressed as detention time during an average 24 hr dry weather flow. Conventional systems have a detention time of 8–12 hr, high rate as low as 4–5 hr, and extended aeration system may have a detention time up to 24 hr. As it is with trickling filters, it is also possible to express the loadings per aeration volume in g of $BOD_5/(m^3\text{-day})$. The following loadings are typical for common completely mixed systems (lower loadings should be applied for conventional, plug flow units):

For removal of carbonaceous organics only	800 g $BOD_5/(m^3\text{-day})$
Oxidation of nitrogen compounds (nitrification)	<500 g $BOD_5/(m^3\text{-day})$

These parameters can also be converted to their population equivalents (assuming 40 g of BOD_5 of settled sewage per capita per day) as

Full treatment of carbonaceous organics	$800/40 = 20\ PE/m^3$
Nitrification	$500/40 = 12\ PE/m^3$

Besides the detention time and volumetric loading, the load can be expressed as sludge load. This is the most appropriate parameter that expresses the amount of food delivered to a unit mass of microorganisms per time, hence, it is called *the food to microorganism ratio* (F/M). The units are kg of BOD_5 delivered to the aeration basin in 1 day per total mass of active biological solids present in the aeration basin [usually expressed as Mixed Liquor Volatile Suspended Solids (MLVSS) or Mixed Liquor Suspended Solids (MLSS), which is the weight of the dried residue of the sludge in the tank before ignition (MLSS) or loss by ignition (MLVSS) divided by the volume of the sample]. Mathematically:

$$\frac{F}{M} = \frac{\text{kg of BOD}_5/\text{day in the influent}}{\text{kg of MLVSS or MLSS in the aeration volume}}$$

Conventional activated sludge treatment units operate at an F/M (MLSS) ratio of 0.2–0.4. Lower values are used when nitrification is desired, whereas higher values are used for high load systems. Very high F/M ratios (over 1.0) may result in filamentous growths and sludge bulking, whereas very low F/M ratios (below 0.1) may result in dispersed growths.

Figure 7.24 shows the percentage BOD removal versus loading of F/M (MLSS) ratio, sludge yield, and sludge age. It includes the effect of secondary clarification. At higher loadings the performance deteriorates and the resistance to shock loadings becomes worse. Also at higher loadings the removal efficiency is more or less a function of the influent load, whereas at loadings below $F/M < 0.3$, the effluent

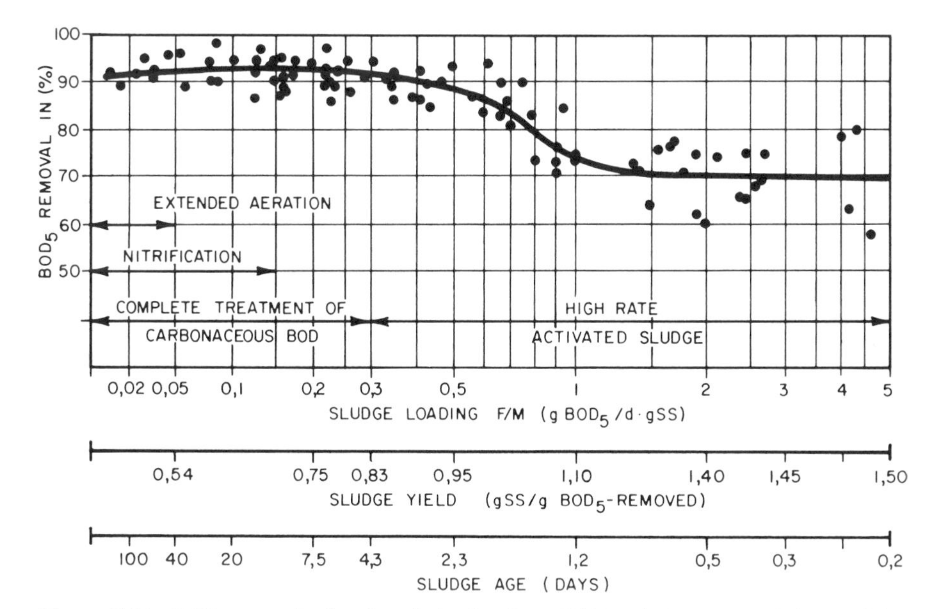

Figure 7.24 BOD removal related to sludge loading (F/M), sludge age, and sludge yield parameters obtained from the operational data of German treatment plants.

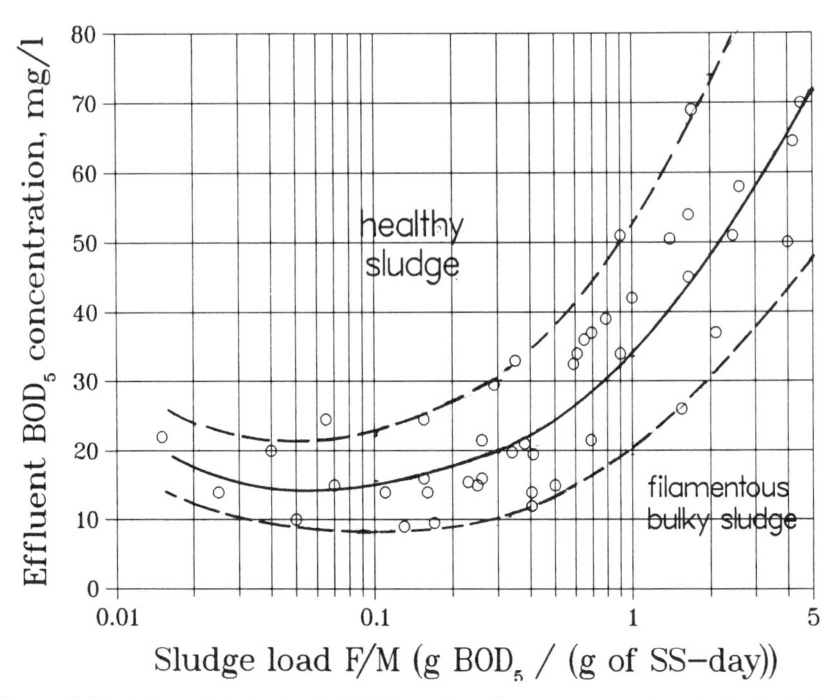

Figure 7.25 Effect of sludge load (F/M) on the effluent concentrations in the final clarifier.

BOD is almost independent of the load. The sludge growth values (yield) are shown in Figure 7.24, which presume pretreatment with a primary clarifier and a typical municipal wastewater. Figure 7.25 shows the relationship of the effluent BOD concentration (both dissolved and suspended) to the F/M ratio.

Air Supply to Aeration Systems. The air supply to the aeration units yields dissolved oxygen concentrations between 1 and 3 mg/liter at all compartments of the aerated units. Higher concentrations are neither needed nor economical. Most of the oxygen is utilized near the influent portion of the units, the rate of DO use being smallest near the effluent. However, this depends on the type of aeration unit. In completely mixed tanks (tanks in which the level of mixing and type of flow provide nearly uniform distribution of the wastewater and air), the aeration requirement is more or less evenly distributed. In plug flow reactors, that is, reactors that are elongated, the air supply rate is higher near the influent (tapered aeration) or the wastewater inflow is distributed along the tank (step aeration). Also the oxygen supply can be automated with the help of oxygen probes that measure the DO concentrations throughout the tank.

Oxygen is supplied in the tank either by blowing compressed air or pure oxygen from the bottom or the side of the tank (diffuse aeration—Fig. 7.18) or by mechanical aerators (Fig. 7.20).

TABLE 7.7 Oxygenation Capacity Requirement for Various Activated Sludge Processes

Type of the Activated Sludge Process	Oxygenation Capacity Requirement per BOD Loading (g of O_2/g of BOD_5)
Extended aeration	2.5–3
Single stage with nitrification	1.8–2.5
Conventional (removal of carbonaceous BOD)	1.2–2.0
Partial treatment (high rate)	≤ 1.0

The air supply requirement can be computed from the oxygen demand of the incoming wastewater. The higher the sludge age and the lower the sludge loading (F/M ratio), the higher the oxygen requirement because more of the incoming BOD is removed in the tank and the sludge is more stabilized. Furthermore, the aeration system should be able to handle sludge and higher loads. For these reasons, the aeration equipment is dimensioned for the maximum BOD load, whereas the size of the aeration unit itself is based on the average load. German practice recommends that the aeration capacity is dimensioned for 1.5 times the average BOD load. The oxygenation capacity (OC) requirement for different activated sludge processes is given in Table 7.7.

The values of the oxygenation capacity requirement in Table 7.7 account for the daily variation in the BOD load.

The oxygenation capacity (OC) or oxygen transfer values of aerators are expressed in terms of kg of O_2/kW-hr ($= 0.61 \times$ lb/hp-hr). The higher the aerator OC capacity, the more efficiently the aerator performs. The OC value of aerators is determined by pilot experiments performed under standard conditions at 10° or 20°C temperature using tap water containing no oxygen. Zero oxygen concentrations can be obtained by saturating the water with nitrogen gas or by adding oxygen-consuming chemicals such as sulfites with a cobalt catalyst. Dividing the oxygen required by the aerator by the oxygen transfer efficiency adjusted to the situation in the aeration tank, one can obtain an estimate of the size of the required aerator. This will be shown in the following simple example:

Example: For partial treatement of a wastewater from 100 000 people (population equivalent) the BOD load is approximately 100 000 $\times$ 40 g of BOD_5/(cap-day) = 4 000 kg/day, assuming primary treatment of the wastewater. The oxygenation capacity requirement for this type of treatment is from Table 7.7 about 1 g of O_2/g of BOD_5. Hence, the oxygenation requirement is 4 000 kg of BOD_5/day $\times$ 1 kg of O_2/kg of BOD = 4 000 kg O_2/day $\times$ (1/24) = 167 kg O_2/hr. If the oxygenation capacity of the selected aerator is 1.5 kg O_2/kW-hr (a typical value) the power requirement for aeration is 167 (kg O_2/hr)/1.5 (kg O_2/kW-hr) = 110 kW. Two aerators with 55 kW power capacity each can be selected.

TABLE 7.8 Air Supply Parameters for Diffuse Aeration Systems

		Air Requirement (m^3)	
System Type	Oxygen Utilization (%)	Per kg of BOD_5 Removed	Per Population Equivalent (PE) per Day
Fine bubble	11	48.7	1.7
Middle size (1.5–3 mm diameter)	6.5	82.4	3.0
Coarse bubbles	5.5	97	3.5

For diffused aeration systems, the amount of air required can be calculated by assuming the percentage oxygen removed in the tank. This is a function of the aeration system (bubble size), as shown in Table 7.8. For a basic reference, see Eckenfelder and McCabe.[63]

The following example shows the estimation of the air requirement for fine bubble diffused aeration:

Example: As shown in Table 7.8 only about 11% of the available oxygen is used in a fine bubble aeration system. Therefore, assuming that 1 m^3 of air contains about 280 g of oxygen, and that only about 11% of that is used then the air requirement for the removal of 1 kg of BOD_5 is $1.5 \times 1\,000$ g of $BOD_5/(280 \times 0.11) = 48.7$ m^3 of air. The multiplier of 1.5 converts the 5-day BOD to the ultimate carbonaceous BOD demand. Assuming that the efficiency of the BOD removal in the tank is 90% and that the PE equivalent for sewage receiving primary treatment is 40 g/(cap-day), the air flow requirement for 1 PE is

$$1.5 \times 0.9 \times 40 \text{ (g/PE-day)}/(280 \times 0.11) = 1.75 \text{ m}^3/\text{(PE-day)}.$$

The air requirement values given in Table 7.8 are based on an aeration depth of 3 m. Smaller depths will result in lower aeration efficiencies whereas higher depths will improve the efficiency. However, the energy requirement for oxygenation remains about the same.

The efficiency of middle and coarse size bubble aeration is not as good as that for fine bubble diffusers. It can be improved if the bubble is broken in the tank by an agitator that can be attached directly above the orifices from which the air is released. Pasveer and Sweeris[64] have shown that the diffused aeration systems perform best at an optimum horizontal velocity of the wastewater above the aerator of 0.5 m/s.

To maintain adequate mixing, the maximum width of aeration tanks should be approximately twice the depth. This width can be doubled by placing the diffuser line along the center line of the tank.

For the design of activated sludge aeration systems, the following parameters are useful:

Fine Bubble Diffused Aeration	
Depth-to-width ratio of the tank is 1 : 1 for spiral flow (max 1 : 2)	
Depth of aeration entry	3–6 m
Air intake	1–3 $m^3/(m^3$ of tank volume-hr)
(Air volume related to 0°C and 1 atm pressure)	
Oxygen intake	30–200 g $O_2/(m^3$-hr)
Air flow	5–15 m^3 of air/(m^2-hr)
	20–60 m^3 of air/(m^2 of plate area-hr)
Diffusers distance	30–50 cm
Energy use	6 W-hr/(m^3 of air × m of aeration depth)
Power level	> 10 W/m^3 (> 50 hp/mg)

Tanks with larger depths exhibit more energy saving. On the other hand, detergents present in the wastewater have a detrimental effect on aeration efficiency.

Surface Aerators	
Depth-to-width ratio	1 : 3–1 : 8
Depth	2.5–4 m
Power requirement	15–30 W/m^3 of tank volume (75–150 hp/mg)
Aerator diameter	0.5–2.5 m
Impeller velocity, v_u	3–5 m/s (perimeter)
Splash distance	$0.3v_u^2$

The energy consumption and oxygenation capacity increase with the second power of the diameter of the impeller and the third power of the impeller velocity. These can be regulated by changing the rotational velocity (range 50–100% of OC).

In tap or rainwater, higher oxygenation capacity can be expected, whereas the presence of detergents and other similar compounds reduces OC. Table 7.7 provided the typical ranges of the oxygenation capacity. A typical oxygenation rate per unit of energy consumption is 1.5 g O_2/kW-hr.

At the bottom of the tank, the velocity of the liquid must be greater than 0.2 m/s to prevent formation of sludge deposits. This is accomplished by the mixing energy input of the aerators, which depends on the size of the tank. In general, the power requirement for keeping the solids in suspension is[65]

Aeration Tank Volume	Power Requirement
500 m^3 (0.13 mg)	20 W/m^3 (100 hp/mg)
1,000 m^3 (0.26 mg)	15 W/m^3 (75 hp/mg)
2,000 m^3 (0.53 mg)	10 W/m^3 (50 hp/mg)

Good mixing can be achieved if the direction of rotation of the impeller is opposite to the direction of water near the aerator. Erosion caused by aerator action could damage the bottom of the reactor.

Surface aerators are noisy and noise control measures should be implemented if the distance from residential areas is not sufficient.

Diffuser plates and diffuser pipes submerged in wastewater should always provide a free passage of the air. Rust and clogging can occur if the process is interrupted by breaks or stopping the air supply. Therefore, the air should be filtered and corosion-resistant materials should be used. Replaceable diffuser pipes provide an advantage of an uninterrupted process when cleaning or replacement is needed.

Sludge Separation

Secondary Settling. Secondary clarification is required in order to maintain the needed, relatively stable concentrations of the biomass (flocs) in the aeration tanks. This is an integral part of the process in which the sludge is separated from the liquid and thickened, and, subsequently, a part of the sludge is returned back into the reactor and the remainder is directed to sludge handling and disposal. The clarifier must separate most of the solids since larger concentrations of solids in the clarifier effluent could cause a violation of the effluent standard. Also, the solids are highly organic and biodegradable and exert biochemical oxygen demand. A massive loss of solids from the clarifier would indicate clarifier failure.

There are certain rules and rule-of-thumb design parameters that should be followed in the design of secondary clarification. The sludge should not reside in the settling tank for too long (maximum 6 hr) since anaerobic conditions almost immediately develop, resulting in nitrogen gas production by denitrification. This can upset the clarification process. The best types of settling tanks are the funnel or other circular upflow clarifiers (Figs. 6.10, 6.12, and 6.13), however, horizontal flow clarifiers of the type shown in Figure 6.18 have also been used. The detention time of the liquid in the clarifier should be around 1.5–2 hr, however, tanks with longer detention times of 3–6 hr have been used in Germany. The weir overflow velocity should be small, below 15 $m^3/(m\text{-}hr)$. This is because the flocs are relatively light. The effluent concentration of solids in the overflow should be below 30 mg/liter. The suspended solids concentration in the clarifier influent is usually around 3–5 g/liter and that in the clarifier underflow is around 8–15 g/liter. The depth of the clarifiers is usually 2–3.5 m.

Practitioners would normally use a surface overflow rate parameter, in addition to the detention time and weir overflow rate, for sizing the secondary settling tank, similar to the primary settling tank design. Such rule-of-thumb overflow rates have been suggested in ranges between 1.0 and 2.5 $m^3/(m^2\text{-}hr)$ (600–1500 gpd/ft^2), depending on the mixed liquor suspended solids concentrations. Overflow rates of 0.5–1.5 $m^3/(m^2\text{-}hr)$ have been recommended in Germany, whereas higher rates are typical for U.S. treatment plants. More concentrated solids in the aeration tank effluent require a larger secondary clarifier.[11]

However, recent advances in the development of the so-called "solids flux theory'' showed that the surface overflow rate that indicates the particle settling velocity of individual flocs is not as relevant in the clarifier design as was thought.[66,67]

The clarifier, as was previously elucidated, has two functions: solids separation (clarification) and solids thickening. Laquidara and Keinath[68] have shown that only the thickening criterion must be considered in secondary clarifier design, since the clarification failure criterion given by the settling velocity of the individual flocs is automatically satisfied. Thickening failure occurs when the surface of the thick solids blanket reaches the overflow of the clarifier as opposed to clarification failure, whereby an excessive amount of solids (dilute blanket) reaches the overflow. The solids flux theory appears to provide the basis for secondary clarifier design.

The solids flux is defined as the mass of solids passing a unit area of a horizontal plane of a clarifier in a unit time. In a batch test in a laboratory or pilot cylinder-settler, the downward movement of the sludge solids is caused only by the settlement velocity. The solids from activated sludge reactors undergo flocculent settling and interact with each other. At higher concentrations (above 500 mg/liter), the flocs settle more or less in a unified mass with a distinct solids–liquid interface, that is, a zone. Thus, the *zone settling velocity* (the velocity of the settling of the solid–liquid interface) is then considered as the primary settling characteristic of the sludge (analogous to the settling velocity of individual particles in granular settling). A good quality sludge should have an initial zone settling velocity (settling velocity corresponding to the solids concentration in the aeration tank) of about 2.4 m/hr. As stated in the previous section, the Zone Settling Velocity is correlated to the Sludge Volume Index (SVI).[62]

At lower solids concentrations, zone settling is either uniform or may even increase because of flocculation. However, as the solids concentration increases, the particle interaction slows down the settling velocity and settling becomes hindered. The floc particles coalesce and the mass settle as a blanket (diluted blanket), forming a distinct solid–liquid interface. The solids that reach the bottom collapse and a thickened layer (thick blanket) is formed, the surface of which rises until the thick blanket reaches the solid–liquid interface. At this moment settling practically stops, however, the solid–liquid interface can still decrease as a result of compression of the solids in the thick blanket because of gravity thickening. This phase of zone settling is called compression settling.

In a continuous unit, such as a secondary clarifier, two factors contribute to the downward movement of the sludge solids: (1) settlement or the action of gravity (zone settling), and (2) the removal of thickened sludge in the underflow. This combined settling process is shown in Figure 7.26. Mathematically, the total downward flux of solids, G_T, is given by

$$G_T = G_u + G_g$$

where

$$G_u = CV, \qquad G_g = CU, \qquad \text{and } V = Q_u/A$$

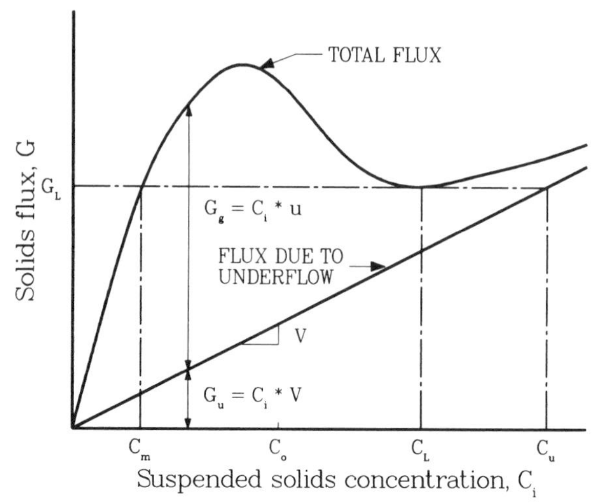

Figure 7.26 Solids flux concept for design of secondary clarifiers.

in which G_u = solids flux caused by underflow [$g/(m^2\text{-hr})$],
G_g = solids flux caused by gravity [$g/(m^2\text{-hr})$],
U = zone settling velocity in m/hr at the solids concentration in the clarifier of C (mg/liter),
V = underflow velocity (m/hr),
Q_u = flow withdrawn from the clarifier in the underflow (m^3/hr)
A = cross-sectional area of the clarifier (m^2).

The course of the total solids flux curve in Fig. 7.26 shows that there is a certain minimum in the total solids flux rate denoted as G_L that represents the largest area of the clarifier needed to transmit a given load of solids to the clarifier. As long as the concentration of the solids entering the clarifier is below the limiting concentration, C_L, and the applied solids flux is below the limiting solids flux, G_L, thickening, and, consequently, clarification failure will not occur.

For further references and design examples of secondary clarification design, see references 69–71.

A more simplified concept used in Germany employs a volumetric sludge loading parameter defined as the hydraulic surface loading of the sludge volume in the influent to the clarifier. In order to maintain the effluent concentration of the activated sludge solids below 30 mg/liter the sludge volumetric loading should be below 0.3 $m^3/(m^2\text{-hr})$. It is computed as the product of hydraulic loading times the volume of the sludge in the influent [Sludge Volume Index (ml/g)/10^6 times the concentration of sludge in mg/liter]. If the SVI is, for example, 125 ml/g and the influent dry solids concentration of sludge is 2 000 mg/liter, then the sludge volume in the influent is $125 \times 2\,000/10^6 = 0.25\ m^3/m^3$.

For an SVI of 100 ml/g = 0.1 m^3/kg, the sludge volumetric loading of 0.3

m^3/m^2 corresponds to a surface loading of suspended solids of 0.3/0.1 = 3 kg of dry suspended solids/(m^2-hr). Consequently, for an SVI of 200 ml/g, the limiting surface loading would be 1.5 kg/(m^2-hr).

Sludge Return. As previously stated, the sludge withdrawn from the clarifier underflow is either returned to the aeration tank or diverted to sludge handling and disposal. The amount of sludge diverted should equal the sludge yield. The relative amount of the returned sludge is identified as a return flow ratio, which is a percentage of the returned sludge volume or the percentage ratio of the volume of returned sludge to the volume of the influent wastewater. It can reach values as high as 50–100%. In older treatment plants, the return flow ratio was much less, about 10–30%.

The return flow ratio depends on the concentration of the solids in the aeration tank and in the clarifier underflow. This relationship is shown in Figure 7.27. As previously specified, a good estimate of the underflow solids concentration of a sludge consisting of healthy flocs is 1–1.5% (10–15 g/liter).

The purpose of the sludge return is primarily to maintain high sludge levels in the aeration units. Thus, the return flow should be related to the required MLSS

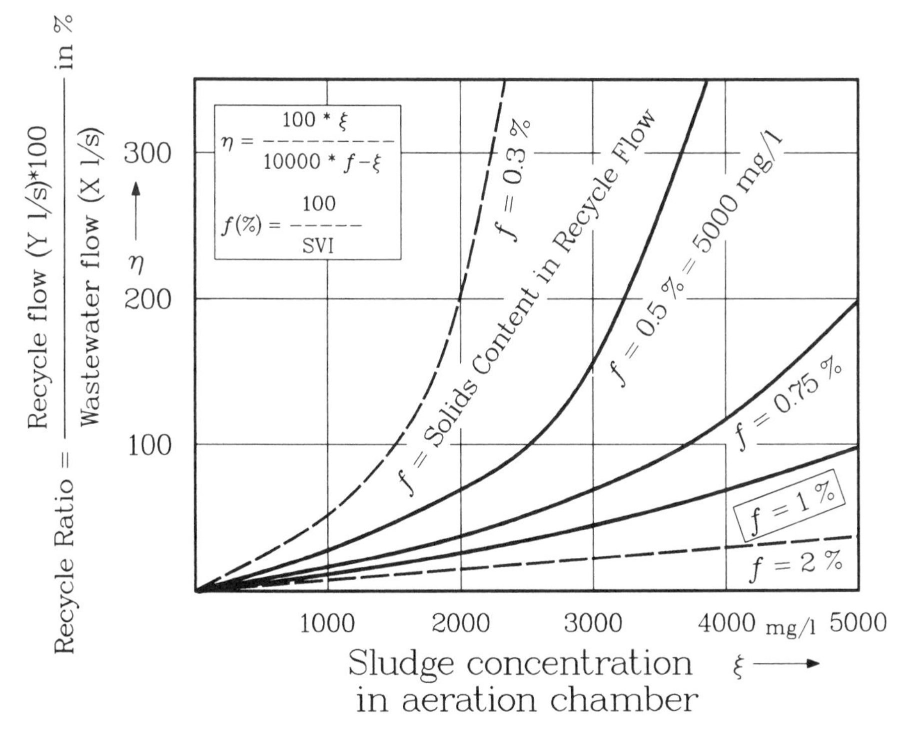

Figure 7.27 Recycle flow ratio as a function of sludge concentration in the aeration unit and in the recycle flow (SVI = sludge volume index, ml/g).

concentrations in the basin. For example, to maintain a mixed-liquor concentration in the tank of 2 500 mg/liter (0.25%) when the SVI is 125 ml/g ($f = 100/125 = 0.8$), the return flow ratio is (using the relationship from Fig. 7.27)

$$100 \times 2500/(10\,000 \times 0.8 - 2\,500) = 45\%.$$

The return flow has an effect similar to that in high-rate trickling filters. In both cases, it is possible to increase the efficiency of undersized overloaded plants to the level of properly sized and loaded plants by regulating the return flow. However, the sludge must be healthy.

Waste Sludge. The waste sludge, equaling the sludge yield, represents the daily increase of the sludge mass that must be removed and diverted to sludge handling. The amount of sludge wasted is about 5 liters/(cap-day) if the sludge is wasted from the mixed-liquor line or 2 liters/(cap-day) if the sludge is wasted from the return sludge line (see Table 9.1). Assuming that the typical U.S. per capita wastewater volume is about 400 liters/day then the waste sludge flow represents 0.5 or 1% of the total wastewater flow, depending on the location of the waste withdrawal. The most common practice in the United States is to thicken the activated sludge in the secondary clarifier and to waste from the return sludge line (Fig. 7.22). The waste sludge is then discharged to the primary tank or to a thickening tank for further handling. As previously specified, the mass of the wasted sludge is about 0.5–1 kg/kg of BOD_5 removed, depending on the sludge age.

If the wasted biological sludge is digested with the primary sludge the size of the digester units must be increased and there will be more gas produced in the units. Generally, the size of the digesters must be at least doubled because the mixed digested sludge has a higher water content than digested primary sludge alone.

If the excess sludge is directed to a separate thickener, the load on the thickener should be about 50 kg/(m^2-day) (10 lb/ft^2/day). This results in detention times of the sludge in the thickener of 12–24 hr. The separation of the excess water from the sludge is facilitated by slow mixing with vertically mounted rotating bars that tend to dislodge gases from the sludge. By thickening, the solids content in the sludge can be increased from 1–1.5% to about 4% (= moisture content 96%). Thickening prepares the sludge for mechanical dewatering by vacuum filtration or other treatment. Commonly centrifugation is an alternative to gravity thickening.

Raw thickened sludge can be dewatered by filter belt presses and vacuum filters. The sludge cake can then be heat dried and sold as a soil conditioner. In smaller plants it is more economical to mix the biological sludge with the primary sludge, raw or digested, and mechanically dewater the mixture. For information on dewatering of raw sludges the reader is referred to a paper by Trubnick et al.[72] However, it should be noted that dewatering of sludges without digestion is not a preferred alternative.

Operation and Maintenance of Activated Sludge Plants

An activated sludge plant is not free of problems. The most important rule is to operate the plant within the ranges specified for each type of operation. Both overloading and underloading can upset the process and result in poor performance. Typical operating and design parameters are given in Table 7.9.

The activated sludge process can become overloaded if the influent wastewater is abnormally strong or the time of aeration is too short. Deteriorating sludge is characterized by a stale odor, poor separation from the mixed liquor, a tendency to rise in the final settler, by foam formation, and by the presence of filamentous organisms. Sludge bulking is the swelling of the sludge volume caused by filamentous microorganisms. It is characterized by a high SVI (up to 400 ml/g as compared to 50–100 ml/g for a healthy flocculent sludge). Essential corrective measures include[61,73,74] (1) reducing the amount of sludge returned and wasting more sludge in order to build up a fresh supply of sludge, (2) increasing the air supply, (3) bypassing part of the settled sewage to reduce the load, (4) chlorination of the returned sludge, (5) adding hydrogen peroxide in lieu of chlorine, (6) adding nutrients if nitrogen and phosphorus are insufficient, (7) changing the mode of operation to step aeration, (8) adding coagulating chemicals such as ferric chloride

TABLE 7.9 Typical Operating and Dimensioning Parameters for Activated Sludge Systems

Parameter	Type of Process: Conventional Completely Mixed	Type of Process: With Nitrification	Type of Process: Extended Aeration
Average effluent quality, mg BOD_5/liter	20	15	15
Volumetric loading, g BOD_5/(m^3-day)	1 000[a]	500	250
Mixed-liquor suspended solids, mg MLSS/liter	3 300	3 300	5 000
Sludge loading (F/M), g BOD_5/(g MLSS-day)	0.3	0.15	0.05
Sludge yield, g of solids/g BOD_5 removed	0.8	0.7	0.5
Aeration time, hr	3.6	7.2	14.4
Sludge age, days	4	10	25
Oxygen utilization, g O_2/g BOD_5	2.0	2.5	2.5
Clarifier loading, g solids/(m^2-hr)	2 500	2 000	2 500

[a] Complete mix systems, lower volumetric loadings (about 25%) should be used for conventional, plug flow systems.

to the final sedimentation tank, and (9) temporarily ceasing the discharge of the digester supernatant to the aeration tank.

Clarifier can also malfunction if the DO drops to zero and denitrification sets off. The fine bubbles of N_2 can give a rise to a sludge blanket, resulting in thickening failure of the clarifier and overflow of solids.

Jenkins, Richard, and Daigger[61] recommend estimating the chlorine dose for control of bulking based on the total suspended solids inventory, including the solids in the secondary clarifier. For the reported cases, the chlorine dosages ranged from 3 to 6 kg of Cl_2 per 1 000 kg of SS. Use of hydrogen peroxide requires somewhat higher dosages, however, hydrogen peroxide also adds oxygen to the supernatant clarifier, which is important if anoxic conditions exist in the clarifier. Specchia and Gianetto[75] suggest addition of powered activated carbon to increase treatment efficiency if poorly biodegradable organics are present in the wastewater. The powdered activated carbon is gradually added to the aeration basin to a concentration of about 1 000 mg/liter and kept at this level by replacing the amount removed each day with the wasted sludge.

If the secondary clarifier becomes temporarily overloaded, Keinath[76] suggests increasing the recycle rate or changing to step feed of the influent (change the influent feed location from the first to last compartment). Tube settlers that can substantially increase the efficiency of primary clarifiers are ineffective for secondary clarifiers.

Underloaded treatment plants (a common occurrence when designers are too generous in estimating future loads) can be easily remedied by disconnecting excess aeration capacity or by designing the plant capacity for present conditions with a possibility of easy future expansion.

The start up period of an activated sludge plant can begin with simple aeration of sewage in the aeration units. The sludge buildup will last about 7 days. In the initial stage, foam will accumulate on the surface of the wastewater in the tanks. When enough sludge is present the foam will disappear as long as enough sludge is recycled. Persistent foaming problems caused by detergents in the influent can be controlled by antifoaming chemicals or by spraying the surface of the aeration tank.

The Two-Stage Activated Sludge Process

This process is shown in Figure 7.28. It is rarely used in municipal treatment plants, however, it is advantageous for high-strength industrial wastes (e.g., dairy or meat packing or slaughtering operations). The first step can be designed as a high load aeration unit whereas the second step is for achieving the effluent limits and for stabilization of the sludge. Each stage has its own clarifier. Excess sludge from the first clarifier is discharged for stabilization to the second unit. A trickling filter can be used as an alternative for the second stage.

A more common application of the two-stage process is for plants requiring nitrification. The first stage has biomass removing primarily carbonaceous BOD

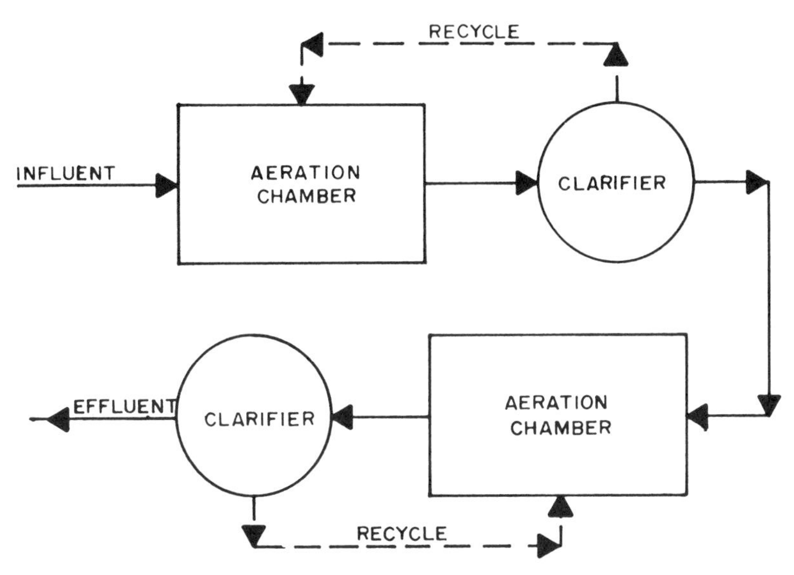

Figure 7.28 Two-stage activated sludge process.

whereas the second stage contains nitrifying organisms that oxidize ammonia to nitrate. Each stage can be optimized independently of each other.

Sequencing Batch Reactors (SBR)

Batch treatment has been used in several ways throughout the history of wastewater treatment. For example, the early oxidation ditches in some small communities were operated in batches. However, the sequencing batch treatment systems are relatively new. Depending on the wastewater flow the SBR systems may consist of one to four batch reactors in parallel operation.[77,78]

The operating sequence for the system is shown in Figure 7.29.

In the first period of the operation cycle, the tank containing sludge from the previous cycles is filled with raw wastewater. After completion of the fill period the wastewater is diverted to the next batch reactor. If only one tank is available wastewater must not be diverted, therefore, storage may be needed. However, one-tank operations are used for establishments in which wastewater is produced only for a portion of a day.

After the fill period, the wastewater and sludge are aerated and allowed to react. It should be noted that the reaction has already commenced during the fill period.

During the settling period, the air input is discontinued and the solids are allowed to settle. This period should not be too long in order to prevent sludge rising. During the idle period, the excess sludge is withdrawn from the tank.

A full operating system consists of the headworks, several reactors, an aeration device, a mechanism to withdraw the wastewater and sludge, and a control system.

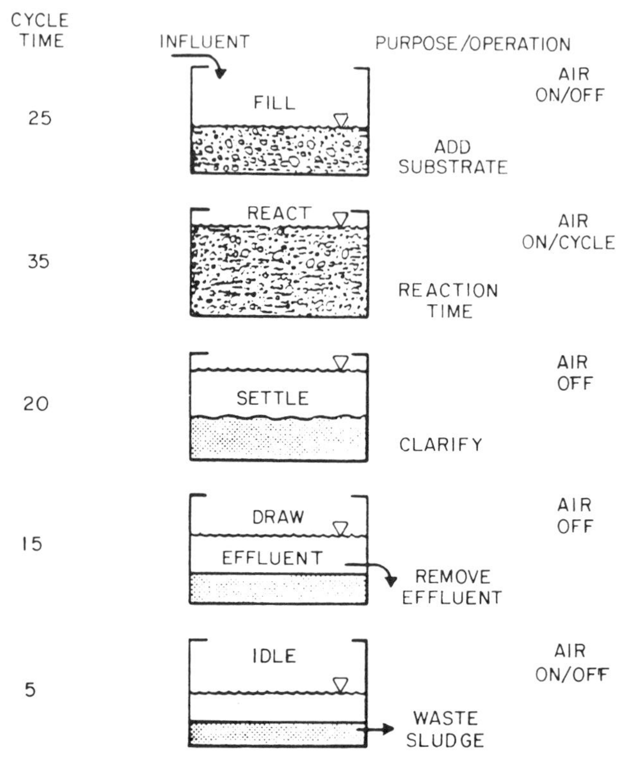

Figure 7.29 Typical SBR operation for a complete cycle in one tank (after U.S. EPA).

The SBR systems can be designed to achieve nitrification/denitrification by repeating the cycle without aeration and using primary sludge as a carbon source as shown by Abufayed and Schroeder.[79]

A demonstration project for this technology was conducted in Culver, Indiana.[80] The system had two reactors and the operation cycle in each reactor lasted 6 hr. The capital and operation costs are lower than those for a conventional activated sludge plant, but comparable to the cost of an oxidation ditch.

7.5 LAGOONS AND PONDS[81–85]

Lagoons and ponds are units that are easily constructed, economical, and easy to operate. Their major drawbacks are the large area requirement and lower treatment efficiency, particularly during low winter temperatures. For this reason, they are not attractive for colder climates. The units can be categorized into the following groups:

1. Aerobic lagoons or ponds,

2. Facultative lagoons or ponds, and
3. Aerated lagoons.

Aerobic Lagoons or Ponds are designed to have a positive dissolved oxygen concentration throughout the lagoon. Without mechanical aeration, this is accomplished by algae that develop in the basin. The lagoons are relatively shallow (0.2–0.5 m) to allow for light penetration and optimum conditions for algae development. The treatment efficiency depends on the temperature and climatic conditions. For the climatic conditions of the southern United States, the detention time should be about 3–5 days, resulting in an area requirement of 3–5 m^2/cap or BOD loading of 100–300 kg of BOD_5/(ha-day). For colder conditions typical of central Europe (Germany) or the northern United States and southern Canada, the detention time must be much longer (about 20 days) and, subsequently, the area requirement is ≥ 10 m^2 per capita or population equivalent. Mixing the contents for a few hours each day is essential to maintain aerobic conditions throughout the basin.

Facultative Lagoons or Ponds are deeper than aerobic ponds, hence, they are divided into an aerobic layer at the top and an anaerobic layer on the bottom. The aerobic layer is created by algae that produce oxygen by photosynthesis. The bottom deposits undergo anaerobic decomposition, producing methane and other gases. As a result, an anaerobic layer at the bottom is created. The depth of the pond is usually 1–2.5 m. By proper design and applying acceptable loads (see Table 7.10) the oxygen demand should not exceed the supply. Therefore, acceptable treatment can be achieved without creating odor problems.

The overflow from the aerobic and facultative ponds is either directly discharged

TABLE 7.10 Design and Operation Parameters for Lagoons[a]

Parameter	Aerobic	Facultative	Aerated
Depth, m	0.2–0.5	1–2.5	2–5
Detention time, days			
Warmer climate[b]	2–6	7–50	2–10
Colder climate[c]	10–40	50–180	7–20
BOD_5 loading			
In kg/(ha-day)			
Warmer climate	90–120	30–80	—
Colder climate	60–90	10–30	—
In kg/(m^3-day)			0.01–0.3
Percentage BOD removal	80–95	70–95	80–95
Algae concentration, mg/liter	100	10–50	Nil

[a] Adapted from references 82 and 83.
[b] Average winter air temperature above 10°C.
[c] Average winter air temperature at or below 10°C.

or flows through a sedimentation pond. The effluent can contain significant quantities of algae.

The sludge accumulated on the bottom of these ponds is well stabilized. Periodically, the sludge deposits should be dredged and disposed. It is convenient to operate two or several ponds in parallel configuration. In this case, one pond at a time can be drained and the accumulated sludge dried in place and subsequently removed with a front end loader.

Low-rate aerobic ponds can sustain fish populations and even be used for fish production, a practice that would be very attractive in developing countries. However, fish ponds have been in existence in central Europe for centuries. Such ponds again have a larger area requirement and their operation does not allow for continued treatment since in the Fall, the pond water volume is substantially reduced, fish are harvested, and the pond is refilled by high water runoff in the Spring. About 500 kg of fish, mostly carp, can be produced from 1 ha of the pond. As treatment is not an objective in the design of fish ponds, it is recommended that the wastewater receive treatment before it is discharged into the pond.

Ponds Employing Water Plants. These effective treatment units have emerged in the 1980s.[21,86–88] From a number of aquatic plants, the water hyacinth has been shown to be most efficient. Water hyacinths are tropical plants and hence do not survive in colder climates. However, an experimental system of water hyacinth ponds located in a greenhouse with an area of 2 ha was built in Austin, Texas. Three parallel ponds, each 275 m long, have proved to be effective in reducing pollutants all year round to well below the limits required by the state discharge permit.

In addition to significant and high BOD and suspended solids removals, water hyacinth ponds are also very effective for nitrogen control as documented by Hauser.[87] The ponds are commonly plug flow, elongated, and relatively shallow (0.5 m deep) basins with a dense cover of plants at the surface. A review on water hyacinths for wastewater control along with the design parameters for such systems was presented by Reddy and Sutton.[88] Most of the present uses are for nutrient removal, primarily nitrogen control, and for effluent polishing.

For mosquito control the ponds can be stocked with mosquitofish (*Gambusia affinis*). A design manual for ponds employing water plants has been published by the U.S. EPA.[21]

In Aerated Lagoons, oxygen is supplied by mechanical surface or diffuse aeration similar to the activated sludge units previously introduced. Essentially, the only difference between the aerated lagoon and activated sludge systems is that in the lagoon systems, sludge is not returned. As a result, the microorganism concentration in the lagoon is smaller. Typical depths of lagoons range from 1.8 to 3.6 m (6–12 ft). The aeration level in the basin is determined from either the requirement of (1) uniform oxygen mixing and maintaining all solids in suspension (aerobic basin), or (2) uniform oxygen distribution but insufficient mixing to keep the solids in suspension. In the latter, a portion of the lagoon near the sludge deposits can become anaerobic (aerobic–anaerobic lagoons). Uniform oxygen distribution can be accomplished at aeration power levels of 3–4 W/m^3 (0.015–0.02 hp/1 000 gal). Such aeration levels will not maintain all solids in suspension, however.

Unlike activated sludge units that are not very sensitive to slow temperature variations, the lagoons efficiency is highly dependent on the temperature of the wastewater in the lagoon, which, in addition to the heat inputs from the wastewater, is a response to solar and atmospheric radiation, wind, and aeration spray area. Novotny and Krenkel[89] developed a model for the computation of temperature and evaporation of aerated lagoons based on the heat balance concept.

Aerated lagoons require secondary treatment, sedimentation ponds (usually about 1–2 days detention, see Chapter 8 for design specifications), or clarifiers. The design of secondary clarifiers for lagoons should presume flocculent discrete or zone settling, depending on the concentration of suspended solids in the basin.

7.6 ANAEROBIC TREATMENT[1,90–93]

Anaerobic fermentation is a viable method of treating high-strength organic wastewater such as that from meat packing and slaughtering, food processing, alcohol distillery, and beer fermentation or leachate from landfills. In addition, on-site household wastewater disposal systems, septic tanks, and soil adsorption systems are essentially anaerobic. These processes are similar to the anaerobic digestion that has been used for stabilization of sludges for more than 80 years.

Degradation of organic matter in anaerobic processes is accomplished in two steps. In the first step, complex organic compounds are converted into simple soluble organic acids, aldehydes, ketones, and alcohols. In the second step, these mid-products of decomposition are converted to the final products, carbon dioxide, methane, and water. Small amounts of hydrogen sulfide (H_2S) are also produced, which causes problems in utilizing the gas. The two reactions can be schematically represented as follows:

$$\text{organics} + \text{microorganisms} + \text{nutrients} \xrightarrow{K_1} \text{organic acids}$$

$$\text{organic acids} \xrightarrow{K_2} CO_2 + CH_4 + H_2O.$$

The Anaerobic Contact Process. This unit's design is quite similar to the activated sludge process. It consists of a completely mixed reactor with internal and external solids separation. Regular solids wasting is required. Steffen and Bedker[95] obtained better than 90% BOD removal with 12–13 hr retention at BOD loading rates of 2.8 $kg/(m^3\text{-day})$ in a meat packing wastewater treatment plant. Mixing is critical for a successful operation.

The anaerobic filter employs a packing onto which the anaerobic organisms adhere and treatment occurs as the wastewater passes upward through the filter. Most of the present applications use plastic media. The retention time of wastewater ranges from 5 to 10 days. Eckenfelder[96] reported better than 85% removals for rice processing and wheat starch wastes at volumetric filter loadings of less than 2 kg of $BOD_5/(m^3\text{-day})$. Another application of the fixed media reactor for low-strength wastewater was demonstrated in Knoxville, Tennessee.[97] Effluent limits of 30 mg/liter of BOD_5 and suspended solids were met at loading rates of 0.25 kg of $BOD_5/(m^3\text{-day})$ and hydraulic retention times of 9–10 hr. The anaerobic trickling filter has also been successfully used in treating a high-strength ($BOD_5 > 30{,}000$ mg/liter) leachate from a landfil.[98] In this facility, better than 90% BOD removals are achieved with average detention time of 7.4 days and BOD loadings of 7 kg/(m^3-day). In addition to the high organic removals, more than 80% of the copper, lead, and zinc are also removed from the effluent and retained in the formed biomass.

The Upflow Anaerobic Sludge Blanket System (UASB). This system utilizes an upward hydraulic flow similar to an Imhoff tank (Fig. 9.8). The UASB reactors have been developed in Holland and are available in the United States under a licence. Raw wastewater enters the unit in the lower, digester compartments. Effective sludge and wastewater contact is achieved by the mixing resulting from gas formation in the sludge bed through which the wastewater is passing. The solids are separated from the effluent in the upper, clarifier compartment. Reported design loadings for food processing wastes such as sugar beet refineries, breweries, potato processing, and dairies ranged from 12 to 25 kg of COD per m^3 of the digest compartment per day. Lettinga[94] conducted tests in a UASB reactor with raw domestic sewage and obtained 65–85% COD reduction at temperatures in the range of 8–20°C and detention times of 12 hr.

The Anaerobic Expanded Fluidized Bed Reactor utilizes a solid media that is dispersed and retained in suspension by recirculation of the fluid around a tall column. The incoming water is introduced at the bottom of the reactor and moves upward. The microorganisms grow on the surface of the solid media and, as a result, the specific density of the solid enveloped with microorganisms will change. When the growth becomes excessive, the particles are carried to the top of the reactor where they are collected and separated from the effluent. The biological growth is then washed off the surface of the media and clean media are returned into the reactor. A few commercial systems of this type of treatment have been recently applied to some industrial wastewater streams.

REFERENCES

1. P. L. McCarty, *Anaerobic Waste Treatment Fundamentals, Public Works,* **95,** No. 9, 107–112, No. 10, 123–127, No. 11, 91–95, No. 12, 95–99, (1964).
2. K. Imhoff, *Sewage Works J.* **17,** 409 (1945).
3. R. C. Loehr et al., *Land Application of Wastes,* Vols. 1 and 2. Van Nostrand Reinhold, New York, 1979.

4. S. C. Reed and R. W. Crites, *Handbook of Land Treatment Systems for Industrial and Municipal Wastes.* Noyes Data Corp., Park Ridge, NJ, 1984.
5. M. R. Overcash and D. Pal, *Design of Land Systems for Industrial Wastes-Theory and Practice.* Ann Arbor Sci. Publ., Ann Arbor, MI, 1979.
6. K. W. Brown, J. B. Evans, Jr., and B. D. Frentrup (eds.) *Hazardous Waste Land Treatment.* Butterworth Publ., Ann Arbor, MI, 1983.
7. W. E. Sopper and S. N. Kerr, *Utilization of Municipal Sewage Effluent and Sludge on Forest and Disturbed Land.* Pennsylvania State Univ. Press, University Park, 1979.
8. G. S. Pettygrove and T. Asano (eds.), *Irrigation with Reclaimed Municipal Wastewater: A Guidance Manual.* Lewis Publ., Chelsea, MI, 1985.
9. *Application of Sludges and Wastewaters on Agricultural Lands: A Planning and Evaluation Guide,* EPA WH-546, Pub. 255. U.S. Environ. Prot. Agency, Cincinnati, OH, 1978.
10. *Process Design Manual for Land Treatment of Municipal Wastewater,* EPA 625/1-81-013. U.S. Environ. Prot. Agency, Cincinnati, OH, 1981. Supplement on ''Rapid Infiltration and Overland Flow,'' 1984.
11. K. Imhoff and G. M. Fair, *Sewage Treatment,* 2nd ed. Wiley, New York, 1956.
12. S. C. Reed, R. W. Crites, and A. T. Walace, *J. Water Pollut. Control Fed.* **57,** 854 (1985).
13. *Drainage Manual,* U.S. Bureau of Reclamation, Dept. of Interior, Washington, DC, 1978.
14. Am. Soc. Agron., *Drainage for Agriculture,* Manual No. 17. Madison, WI, 1974.
15. R. C. Smith and E. D. Schroeder, *J. Water Pollut. Control Fed.* **57,** 785 (1985).
16. R. C. Smith and E. D. Schroeder, *J. Water Pollut. Control Fed.* **55,** 255 (1983).
17. Y. Dubuc, P. Jaanneteau, R. Labonte, C. Roy, and F. Briere, *Water Resour. Bull.* **22,** 297 (1986).
18. D. S. Nichols, *J. Water Pollut. Control Fed.* **55,** 495 (1983).
19. J. Jackson, P. K. Feeney, R. A. Morrel, and J. D. Click, *Wetlands Creation and Reclamation through Land Application of Wastewater.* Paper presented at 59th Conference of Water Pollution Control Federation, Sec 45, Los Angeles, CA, October 1986.
20. P. J. Godfrey, E. R. Kaynor, S. Pelczarski, and J. Benforado, *Ecological Considerations in Wetlands Treatment of Municipal Wastewaters.* Van Nostrand-Reinhold, New York, 1985.
21. Design Manual-Constructed Wetlands and Aquatic Plant Systems for Municipal Wastewater Treatment, EPA Rep. No 625/1-88/022, U.S. Environ. Prot. Agency, Washington, DC, 1988.
22. *Process Design Manual-Wastewater Treatment Facilities for Sewered Small Communities,* EPA Rep. No 625/1-77-009. U.S. Environ. Prot. Agency, Cincinnati, OH, 1977.
23. *Upgrading Trickling Filters,* EPA Rep. No. 430/9-78-004. U.S. Environ. Prot. Agency, Cincinnati, OH, 1978.
24. K. Imhoff, *Sewage Works J.* **9,** 91 (1937).
25. H. D. Halvorson, *Water Works Sewerage,* **83,** 307 (1936).
26. C. J. Velz, *Sewage Works J.* **20,** 607 (1948).
27. K. Imhoff, *Sewage Works J.* **10,** 350 (1938).

28. W. W. Eckenfelder, Jr., *Trans. Am. Soc. Civ. Eng.* **128,** 371–398 (1963).
29. *Handbook of Trickling Filter Design.* Public Works Journal Corp., Ridgewood, NJ, 1970.
30. G. Rincke, *GWF, das Gas-und Wasserfach* 108, 667 (1967).
31. *Grundsätze für die Bemessung von einstufigen Tropfkörpern (Fundamentals of Design of Single Stage Trickling Filters)* ATV Manual A 135, St. Augustine, FRG, 1983.
32. J. Oleszkiewicz, *J. Water Pollut. Control Fed.* **52,** 2906 (1980).
33. Y. C. Wu and E. D. Smith (eds.), *Fixed-Film Biological Processes for Wastewater Treatment.* Noyes Data Corp., Park Ridge, NJ, 1983.
34. K. Imhoff, *Gesund. Ing.* **74,** 41 (1953).
35. E. Sarner, *Plastic Packed Trickling Filters.* Ann Arbor Sci. Publ., Ann Arbor, MI, 1986.
36. P. N. Chipperfield, *J. Water Pollut. Control Fed.* **39,** 1860 (1967).
37. E. H. Bryan in *Fixed Film Biological Processes for Wastewater Treatment* (Y. C. Wu and E. D. Smith, eds.). Noyes Data Corp., Park Ridge, NJ, 1983.
38. H. O. Halvorson and G. M. Savage, *Sewage Works J.* **9,** 888 (1936).
39. *Process Design Manual for Nitrogen Control,* EPA Technology Transfer 4-64. U.S. Environ. Prot. Agency, Washington, DC, 1975.
40. H. A. Gulicks and J. L. Cleasby, *J. Water Pollut. Control Fed.* **58,** 60 (1986).
41. *Upgrading Trickling Filters,* EPA 430/9-78-004. U.S. Environ. Prot. Agency, Washington, DC, 1978.
42. K. R. Imhoff, *History and Development of the Rotating Biological Contactor Process,* Int. Symp., Felbach 1983. Eur. Common Market Secretariat, St. Augustin, FRG, 1983.
43. *Design Information on Rotating Biological Contactors,* EPA Rep. No. 600/52-84-106. U.S. Environ. Prot. Agency, Cincinnati, OH, 1984.
44. P. L. Bush and W. Stumm, *Environ. Sci. Technol.* **3,** 49 (1968).
45. D. L. Ford and W. W. Eckenfelder, Jr., *J. Water Pollut Control Fed.* **39,** 14 (1967).
46. J. J. Ganczarczyk, *Activated Sludge Process, Theory and Practice.* Dekker, New York, 1983.
47. R. E. McKinney and M. P. Hanwood, *Sewage, Ind. Wastes* **24,** 117 and 280 (1952).
48. Committee on Water Pollution Management, *J. Environ. Eng. Div., Am. Soc. Civ. Eng.* **105,** 283–296 (1979).
49. M. G. Mandt and B. A. Bell, *Oxidation Ditches in Wastewater Treatment,* Ann Arbor Sci. Publ., Ann Arbor, MI, 1982.
50. S. Christopher and R. Titus, *Civil Eng. (NY)* **53**(5), 39 (1983).
51. R. H. Gould, *Civ. Eng. (NY)* **19**(1), 30 (1949).
52. R. H. Gould, *Sewage, Ind. Wastes* **22,** 997 (1950).
53. A. L. Downing, *J. Inst. Publ. Health Eng.* **59**(4), (1960).
54. *Aeration in Waste Treatment,* Manual of Practice No. 5, Water Pollution Control Federation, Washington, DC, 1971.
55. *Oxygen Activated Sludge Wastewater Treatment Systems,* EPA Technol. Transfer Semin. Publ., U.S. Environ. Prot. Agency, Washington, DC, 1973.
56. A. A. Kalinske, *J. Water Pollut. Control Fed.* **48,** 2472 (1976).
57. D. A. Okun, *Civ. Eng. (NY)* **18** (5), 32 (1948).

58. T. R. Haseltine, *J. Water Pollut. Control Fed.* **33,** 946 (1961).
59. F. Pöpel, *GWF, das Gas-und Wasserfach* **96,** 533 (1955).
60. M. W. Tenney and W. Stumm, *J. Water Pollut. Control Fed.* **37,** 1370 (1965).
61. D. Jenkins, M. G. Richard, and G. T. Daigger, *Manual on the Causes and Control of Activated Sludge Bulking and Foaming.* Water Res. Comm., South Africa, and U.S. EPA, Ridgeline Press, Lafayette, CA, 1986.
62. G. T. Daigger and R. E. Roper, *J. Water Pollut Control Fed.* **57,** 859 (1985).
63. W. W. Eckenfelder, Jr. and J. McCabe, *Diffused Air Oxygen Transfer Efficiencies,* Adv. Biol. Waste Treat., Pergamon Press, New York, 1963.
64. A. Pasveer and S. Sweeris, *J. Water Pollut. Control Fed.* **37,** 1267 (1965).
65. E. Knop and K. H. Kalbskopf, in *Advances in Water Pollution Control,* Proc. 3rd Int. Conf., Pap. II-11. IAWPR 1969, (1970).
66. R. I. Dick, *J. Water Pollut. Control Fed.* **48,** 633 (1976).
67. R. I. Dick and M. T. Suidan, in *Mathematical Modeling for Water Pollution Control* (T. M. Keinath and M. P. Wanielista, eds.). Ann Arbor Sci. Publ., Ann Arbor, MI, 1975.
68. V. D. Laquidara and T. M. Keinath, *J. Water Pollut. Control Fed.* **55,** 54 (1983).
69. M. D. Mynhier and C. P. L. Grady, *J. Environ. Eng. Div., Am. Soc. Civ. Eng.* **101,** 829 (1975).
70. L. D. Benefield and C. W. Randall, *Biological Process Design for Wastewater Control.* Prentice-Hall, Englewood Cliffs, NJ, 1980.
71. Metcalf & Eddy, Inc., *Wastewater Engineering: Treatment, Disposal and Reuse.* McGraw-Hill, New York, 1979.
72. E. H. Trubnick and P. K. Mueller, *Sewage, Ind. Wastes* **30,** 1364 (1958).
73. H. Wagner, *Water Sci. Technol.* **16,** 1 (1984).
74. Y. C. Wu H. N. Hsieh, D. F. Carey and K. C. Ou, *J. Environ. Eng. Div., Am. Soc. Civ. Eng.* **110,** 472 (1984).
75. V. Specchia and A. Gianetto, *Water Res.* **18,** 133, (1984).
76. T. M. Keinath, *J. Water Pollut. Control Fed.* **57,** 770 (1985).
77. R. L. Irvine and J. Bender, *Technology Assessment of Sequencing Batch Reactors,* EPA Rep. No. 600/52-85/007. U.S. Environ. Prot. Agency, Cincinnati, OH, 1985.
78. R. L. Irvine and A. W. Bush, *J. Water Pollut. Control Fed.* **51,** 235 (1979).
79. A. A. Abufayed and E. D. Schroeder, *J. Water Pollut. Control Fed.* **58,** 387 (1986).
80. R. L. Irvine and L. H. Ketchum, *Full Scale Study of Sequencing Batch Reactors,* EPA Rep. No. 600/52-83-020. U.S. Environ. Prot. Agency, Cincinnati, OH, 1983.
81. W. W. Eckenfelder, Jr., *Water Quality Engineering for Practicing Engineers.* Barnes & Noble, New York, 1970.
82. E. F. Gloyna, *Basis for Waste Stabilization Pond Design,* Adv. Water Qual. Improv., Vol. 1. Univ. of Texas Press, Austin, 1966.
83. *Design Manual—Municipal Wastewater Stabilization Ponds,* EPA Rep. No. 625/1-83-015. U.S. Environ. Prot. Agency Cincinnati, OH, 1983.
84. E. J. Middlebrooks, N. B. Jones, J. H. Reynolds, M. F. Torpy, and R. P. Bishop, *Lagoon Information Source Book.* Ann Arbor Sci. Publ., Ann Arbor, MI, 1978.
85. E. J. Middlebrooks, C. H. Middlebrooks, J. H. Reynolds, G. Z. Watters, S. C. Reed,

and D. B. George, *Wastewater Stabilization Lagoon Design, Performance and Upgrading.* Macmillan, New York, 1982.

86. R. Stowell, R. Ludwig, J. Colt, and G. Tchobanoglous, *J. Environ. Eng. Div., Am. Soc. Civ. Eng.* **107,** 919 (1981).
87. J. R. Hauser, *J. Water Pollut Control Fed.* **56,** 219 (1984).
88. K. R. Reddy and D. L. Sutton, *J. Environ. Qual.* **13** (1), 1–9 (1984).
89. V. Novotny and P. A. Krenkel, *Evaporation and Heat Balance in Aerated Basins. Water 1973.* Am. Inst. Chem. Eng., New York, 1973.
90. R. E. McKinney, *TAPPI Environ. Conf. Proc.*, p. 163 (1984).
91. *Technology Assessment of Anaerobic Systems for Municipal Wastewater Treatment,* EPA Rep. No. 600/2-82-004. U.S. Environ. Prot. Agency, Cincinnati, OH, 1982.
92. M. S. Switzenbaum, *Environ. Technol. Lett.* **5,** 189 (1984).
93. A. Y. Obayashi and J. M. Gorgan, *Management of Industrial Polutants by Anaerobic Treatment.* Lewis Publ., Chelsea, MI, 1985.
94. G. Lettinga, R. Roersma, and P. Grin, *Biotechnol. Bioeng.* **25,** 1701 (1983).
95. A. J. Steffen and M. Bedker, *Proc. Ind. Waste Conf. Purdue Univ.* **16,** 423 (1961).
96. W. W. Eckenfelder, Jr., *Principles of Water Quality Management.* CBI Publ. Co., Boston MA, 1980.
97. J. C. van den Hennel, R. J. Zoetemeyer, and B. J. Boelhouwer, *Biotechnol. Bioengin.* **23,** 2001 1981.
98. P. E. Schafer, G. C. Woelfel, and J. L. Carter, *Proc. Ind. Waste Conf.*, Purdue Univ. **41,** 383–389 (1987).

CHAPTER 8

Advanced (Tertiary) Wastewater Treatment

In industrialized countries, the term primary treatment implies separation of the solids from raw wastewater and secondary treatment implies biological removal of colloidal and dissolved organics by activated sludge or trickling filters. Primary and secondary treatment normally removes about 90% of the BOD and suspended solids, which is usually sufficient. In the United States, all municipal wastewater discharges must receive at least secondary treatment as mandated by law. The exceptions apply to existing plants located in coastal areas that have very long ocean outfalls. In these cases, primary treatment may suffice, however, it must be demonstrated that the discharge is not harming the marine environment and the plants must obtain a discharge variance from the Environmental Protection Agency.

The term tertiary or advanced wastewater treatment (AWT) is used to describe treatment processes that go beyond secondary treatment.[1–5] Situations requiring tertiary treatment may arise when (1) the effluent discharge is located on a smaller stream with a low waste assimilative capacity, (2) the stream is already overloaded by existing wastewater discharges, (3) the effluent is directed to a lake or a stratified reservoir, (4) the stream or lake requires additional protection (such as streams designated as sources of drinking water), and (5) the effluent or a substantial portion of it is to be reused, for example, in recharging an aquifer, or, as an extreme, as a source of potable water. In such cases, the removal of suspended solids, biodegradable organics (BOD), or nutrients (nitrogen and phosphorus) from the effluent in excess of secondary treatment limits may be required.

The most common advanced treatment processes are[1–5]

Advanced Treatment of	Unit Process
Suspended solids and biodegradable organics	Polishing pond, microsieves, sand filter
Colloidal materials or color	Coagulation, flocculation, adsorption
Phosphorus	Simultaneous precipitation, postprecipitation, algal or water plants pond
Nitrogen	Ammonia stripping, nitrification, denitrification, ion exchange, water hyacinth pond, breakpoint chlorination, wetlands
Heavy metals	Ion exchange, precipitation–flocculation, adsorption
Organic chemicals	Adsorption

Tertiary treatment is neither a substitution nor a remedy for malfunctioning primary and secondary treatment units. For the processes to be economical, the wastewater must receive full primary and secondary treatment.

Expected effluent quality from different advanced unit treatment processes are given in Table 8.1. The basic secondary treatment effluent values in Table 8.1 are based on the following design parameters:

F/M ratio for activated sludge	0.3 kg BOD_5/(kg MLSS-day)
Trickling filter loading	400 g/BOD_5/(m^3-day)
Waste stabilization pond	10 m^2/PE
Aerated lagoon	20–30 g BOD_5/(m^3-day)

TABLE 8.1 Typical Effluent Characteristics of Advanced Treatment Units[a]

Process	BOD_5 (mg/liter)	COD (mg/liter)	P_{tot} (mg/liter)	N_{tot} (mg/liter)	N-NH_4^+ (mg/liter)	SS (mg/liter)
Basic biological treatment	20	90	10	30	25	20
1. Biological treatment with nitrification	15	80	10	25	10	20
2. Two-step biological treatment with nitrification	12	75	10	25	5	15
3. Activated sludge with biological denitrification at 200% recycle	15	80	9	10	12	20
4. Biological treatment followed by a polishing pond	12	75	8	30	25	12
5. Biological treatment with nitrification followed by a polishing pond	10	70	8	25	10	10
6. Biological treatment with nitrification followed by microsieves	10	70	10	30	10	10
7. Biological treatment followed by rapid sand filtration	10	68	10	30	25	10
8. Biological treatment with nitrification and rapid sand filtration	7	60	10	25	10	7
9. Activated sludge with simultaneous precipitation	15	75	1	30	25	25
10. Activated sludge with nitrification and simultaneous precipitation	12	65	1	28	12	20
11. Activated sludge with simultaneous precipitation and filtration	5	45	0.2	28	25	5
12. Activated sludge with nitrification simultaneous precipitation and filtration	5	40	0.2	25	10	5
13. Activated sludge followed by aqueous plants lagoon (water hyacinth)[b]	<10	65	<8	<5	<4	<10

[a]Adapted from Hegemann et al.[6]
[b]After U.S. EPA.

For basic biological treatment with nitrification the following design parameters were employed:

F/M for activated sludge	0.15 kg BOD_5/(kg MLSS-day)
Trickling filter	200 g BOD_5/(m^3-day)

Polishing (Maturation) Ponds[8,9] (Fig. 8.1) have the purpose of retaining residual solids from the secondary effluent and allowing the wastewater to stabilize. The optimum detention time is about 1–2 days and the depth is between 1 and 2 m. With these design parameters, algae growth will be limited. Sludge accumulation rates are very low and cleaning may be required only once in several years. It is advisable to divide the required volume into two or more basins in series and to aerate the overflow from each basin on weirs or cascades. A 2-day detention time maturation pond will also remove about 50% of the residual BOD_5 from the secondary clarifier and about 30% of the residual COD.

Microscreening[10] has been used as a tertiary treatment step since the early 1950s. A microscreener (also called a microstrainer or microsieve) consists of a motor-driven rotating drum with a microscreen medium, mounted horizontally in a rectangular channel (Fig. 8.2). Feed water enters the drum through the open end and passes radially through the screen with a concomitant deposition of solids on the inner surface of the screen. The size of openings in the screen ranges from 25 to 35 μm. At the top of the drum, pressure jets of effluent water are directed onto the screen to remove the mat of the deposited solids. The backwash flow, which amounts to about 2–5% of the total waste flow, is directed to the inflow of secondary clarifiers. The sizing of the microsieves is based on hydraulic loading,

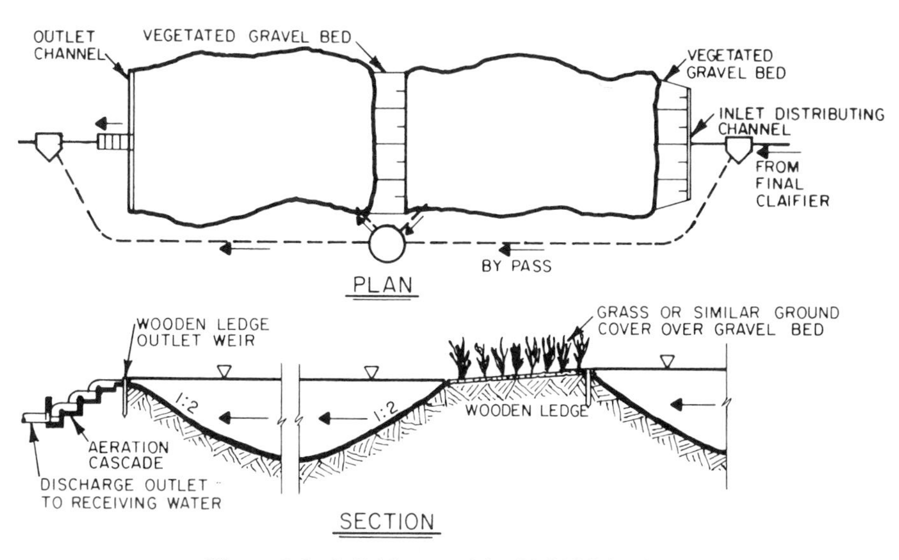

Figure 8.1 Polishing pond for 20,000 inhabitants.

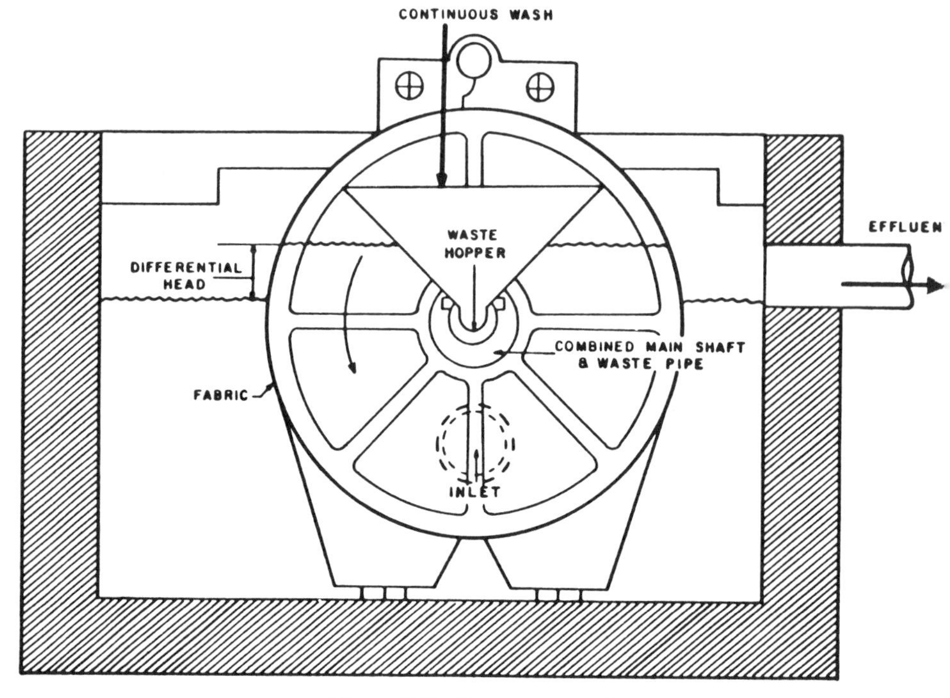

Figure 8.2 Microscreen unit.

typically 15 $m^3/(m^2\text{-hr})$ (6 gpm/ft^2), or solids loading, typically 0.4 $kg/(m^2\text{-hr})$ (0.08 $lb/ft^2/hr$) of submerged screen area. The removal of suspended solids is about 50–80% and the BOD removals were reported in ranges from 30 to 70%. The BOD removal efficiencies depend on the BOD content of the filterable solids and, hence, on the sludge loading of the preceding biological units. The hydraulic head loss through the sieve is about 7.5–15 cm.

Filters.[10,11] Conventional filters are made of silica sand with 1–3 mm grain size and with a filter depth of 0.5–2 m. Water passes through the filter either in an upflow or downflow direction, however, filter applications for tertiary treatment of wastewater effluents are "rapid" (outside the United States called "American") downflow gravity filters made of sand. The hydraulic loading is typically 5–10 $m^3/(m^2\text{-hr})$ (2–4 gpm/ft^2), however, filters were successfully operated at hydraulic load rates up to 25 m/hr (10 gpm/ft^2).

The maximum headloss through the filter is about 1–2 m. When the headloss caused by clogging of the filter by trapped solids and microorganisms exceeds this limiting value, the filter must be backwashed. During the backwashing process, the filter effluent water moves in an upward direction and dislodges the filter layers and the trapped solids that are then washed toward the inflow to the secondary clarifier. However, because of very large backwash flow rates occurring during a relatively short time period, an equalization basin may be needed to prevent upsets in the clarifier function. The backwash flow amounts to 3–6% of the total flow with a backwash flow rate as high as 100 $m^3/(m^2\text{-hr})$. To facilitate the liberation of trapped solids and microorganisms from the filter during the backwash cycle, air flow at rates about 100 $m^3/(m^2\text{-hr})$ is sometimes used. Air is also used for scouring of the solids in the initial stages of the backwash cycle.

During the backwash cycle, the filter bed expands about 20%. After the backwash is completed, the filter media, which was fluidized during the backwash, will resettle with the heaviest media fractions on the bottom and the lightest on the top as in a conventional sand filter. This is not effective because the "work" is done in the top few millimeter of the filter. Multimedia filters, made of larger but lighter anthracite coal granules (specific gravity about 1.5), and smaller but heavier garnet (specific gravity 4) and silica (specific gravity 2.65) sand particles, are more effective and allow longer filter runs.

The treatment efficiency of filters is better than that of microsieves. Reported removal efficiencies range from 70 to 90% for suspended solids and 50 to 70% for BOD.

Phosphorus Removal.[12] Phosphorus in wastewaters may exist in three forms: (1) organic phosphorus that is associated with organic molecules, (2) orthophosphate (PO_4^{2-} form) that exists as an anion, and (3) polyphosphates, mostly from detergents. Only orthophosphate can be chemically precipitated, however, most of the organic phosphorus and polyphosphates are converted to the orthophosphate form during biological treatment.

Phosphorus removal by biological treatment alone may reach about 30%. Most of the phosphorus removed in the biological units is incorporated into the waste sludge. If the waste sludge and supernatant from the sludge digestion units are disposed of on land and not returned to the system, more phosphorus can be removed in the biological units. However, most of the phosphorus removal methods use some kind of precipitation of the phosphates from the wastewater. The precipitating chemicals include iron and aluminum salts (ferric chloride and aluminum sulfate) and lime, although lime is the least preferred. The chemicals can be added (1) to the influent of the primary settling tank, (2) to the influent of the activated sludge units (simultaneous precipitation), (3) to the influent of the secondary clarifier, or (4) to the influent of separate tertiary precipitation and sedimentation or filtration units (postprecipitation).

In a Simultaneous Precipitation system, iron or aluminum phosphate flocs become a part of the return sludge and are wasted with it. This saves about 50% of the amount of chemical dosage. The recommended dosages in Switzerland were about 10 mg Fe/liter in the form of ferric chloride and in Germany, 1 g of Fe per 1 g of PO_4^{2-} (or 3 g of Fe per 1 g of P). These dosages provide about 85% phosphorus removal. Recommended dosages in the United States range from 1.7 to 3 g of metal to 1 g of P in the wastewater to achieve 1 mg/liter of P in the effluent or about 90% removal of P. pH correction is not generally needed. The precipitated phosphate is not released into the supernatant in the sludge digester units and the sludge volume is not increased, however, its dry content will rise by about 15% and the settling characteristics of the sludge are generally improved. The greatest advantage of coprecipitation is minimum capital cost and savings on chemicals. Waste pickle liquor ($FeSO_4/H_2SO_4$ mixture) from steel mills is a cheap source of iron that has been successfully used for simultaneous precipitation of phosphates from wastewater.[12]

In the Tertiary Phosphorus Removal (postprecipitation) chemicals are added to the effluent from secondary clarification, however, this requires addition of treatment units (mixers, flocculation basins, and tertiary clarifier or filter). Alum is the most preferred chemical since iron leaves a residual concentration in the effluent causing a slight color problem and lime is not as effective.

Biological Phosphorus removal is a relatively new process. Several full-scale operations have been reported to operate satisfactorily. Activated sludge that has been deprived of oxygen for a period of time has an increased uptake of phosphorus upon reintroduction into the aeration basin (luxury uptake). Other operational parameters that can be optimized to achieve a higher phosphorus uptake are F/M ratio, hydraulic retention time, and the solids retention time—sludge age. At higher F/M loads and lower solids and hydraulic retention times, the phosphorus removals are higher.[13,14]

Algal or water hyacinth ponds designed for effluent polishing remove phosphorus and nitrogen by uptake by the aquatic plants. These ponds are relatively shallow with a detention time of about 3–5 days with a design area of up to 2 000 population per 1 ha of the pond. Such ponds are best suited for warmer climates. Their removal efficiency for phosphorus is low but it can be improved by harvesting the produced biomass.

Nitrogen Removal.[3,15] Nitrogen in raw municipal wastewater is principally in the form of organic nitrogen, both soluble and particulate, ammonia, and some nitrate. Primary sedimentation removes a portion of the particulate organic matter, resulting in about 20% reduction of the nitrogen entering the plant. Conventional biological treatment (without nitrification) will remove more particulate and soluble organic nitrogen and decompose a portion of it to ammonium and other inorganic forms. Thus, the total nitrogen removal in a conventional primary–secondary facility is about 25–35%.

The amount of nitrogen removed in the biological units is proportional to the mass of sludge produced and wasted from the system. The overall percentage of the removed nitrogen depends on the sludge stabilization and disposal methods. Sludge hauled away from the plant contains about 5–10% of nitrogen. In anaerobic digesters, a substantial portion of the sludge nitrogen is released into the supernatant and, subsequently, the nitrogen will reappear in the effluent. Therefore, digested sludge has a lower nitrogen content than sludge that is mechanically dewatered and heat treated. Consequently, plants with anaerobic digesters have a lower overall nitrogen removal efficiency.

Further biological removal of nitrogen is possible only in low-rate biological treatment units when the residual organic and ammonium nitrogen is converted to nitrate by nitrification. It should be noted, however, that nitrification alone does not remove nitrogen from the effluent. It converts it to a nitrate form that does not deprive receiving waters of oxygen, as does ammonia, which has an oxygen demand of about 4.3 g O_2/g of NH_4^+. Both nitrate and ammonia forms can be damaging to receiving waters.

The major processes for nitrogen removal are biological nitrification–denitrification, breakpoint chlorination (superchlorination), selective ion exchange for ammonium ion removal, and air stripping for ammonium removal (ammonium stripping). Ion exchangers are relatively expensive in comparison to biological techniques.

Ammonia Stripping involves raising the pH of water by lime to very high values (to pH > 11) when most of the ammonium ion is converted to ammonia gas. Subsequently, the gas is stripped by using spray towers that are very similar to mechanical draft cooling towers. The effluent is then neutralized. Ammonia stripping is not reliable in cold weather. For this reasons, the North Lake Tahoe plant (California) successfully uses an enclosed air stripper in conjunction with clinoptilolite (ion-exchange resin selective for NH_4^+) for NH_4^+ removal. A consideration in chosing air stripping is that if the ammonia is not converted to nitrate, denitrification may not be necessary.

In Biological Nitrogen Removal Systems, nitrification by a low-rate or two-step aerated biological treatment is followed by denitrification in an anoxic unit. Two-step biological treatment–nitrification systems have a higher efficiency of nitrification. The key design factor for single-stage nitrification in activated sludge units is the sludge age, which should be long enough to allow the nitrifying microorganisms to develop and remain in the system. The sludge age should be at least 10 days. For trickling filters, the loading should not exceed 200 g/(m^3-day).

In the anoxic unit, anaerobic and facultative microorganisms use the nitrate as a source of oxygen, converting it to nitrogen gas that can escape from wastewater. Basically, it is the same biochemical reaction of decomposition of organic matter as that occurring in aerated units. Therefore, if the denitrification process follows biological treatment that includes nitrification, a carbon source supplement such as methanol must be added because most of the organics from the influent have been already removed in the preceding steps.[16,17] Both fixed growth (trickling filters) and suspended growth (fluidized bed or mixed activated sludge) reactors can be used for tertiary denitrification. Methanol (wood alcohol) is the cheapest and most common source of carbon. With methanol-suspended growth systems, the removal rate of nitrate measured by Sutton et al.[18] is about 0.1 kg NO_3^- removed/(kg MLVSS-day) at 10°C temperature in activated sludge reactors. At 20°C the removal rate is about 0.2 kg NO_3^-/(kg MLVSS-day).

In low-rate biological treatment systems with recycle, nitrification and denitrification can be accomplished without adding supplemental chemicals. One system consists of two biological units. In the first unit (anoxic), the nitrate recycled from the second unit is denitrified by microorganisms using the organic carbon from the influent. No nitrification of the influent nitrogen takes place in the first unit. In the second low-rate aerated unit, organics are further decomposed and ammonia is converted to nitrate. With a high recycle, the nitrate is then returned to the first

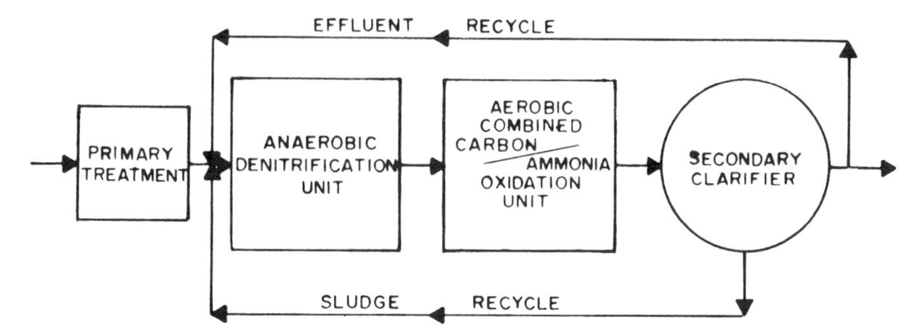

Figure 8.3 Biological (aerobic–anaerobic) nitrification–denitrification process.

anoxic unit for denitrification (Fig. 8.3). The disadvantage of this system over the post- (tertiary) denitrification using methanol is a relatively low rate of denitrification, which is about one-half of that typical for postdenitrification. This results in a large volume requirement for the first denitrifying reactor. Also the nitrogen removal is lower than in methanol-based systems. In some treatment systems, different compartments can be designated as aerobic and anoxic units. A comparison of one- and two-step nitrogen removal systems has been published by Kayser et al.[19]

A modification of the activated sludge process designed for both nitrogen and phosphorus removal was developed in South Africa.[20,21]

Aquatic Plant lagoons have been successfully used for effluent polishing and nutrient removal in warmer climatic regions. Design procedures were presented in a manual by the U.S. EPA[7]. Typical hydraulic loading for tertiary treatment in a nonaerated water hyacint pond is 600 m^3/ha-day and that in aerated ponds is up to 1000 m^3/ha-day. Other plants that have been tested include duckweed and pennyworth[7].

In the Breakpoint Chlorination process, oxidation of ammonia by chlorine leads to end products composed predominantly of nitrogen gas. A laboratory scale study using Blue Plains (Washington, D.C. area) secondary effluent indicated an optimum chlorine dosage of 8 g Cl_2/g NH_4^+ within an optimum pH range of 6–7. More than 90% of ammonia nitrogen removal can be achieved.[22]

Soil Bacteria can both nitrify and denitrify the effluents.[23] The process is relatively similar to that occurring in the treatment plant. Nitrification occurs when the soil is aerated whereas denitrification requires flooded conditions. In addition, ammonia is effectively adsorbed by clay and soil organic particles. Soil adsorption systems (such as those used for disposal of household sewage from septic tanks, land disposal, and groundwater recharge systems) usually very effectively nitrify and remove the ammonium. However, nitrate removal is usually low unless alternate flooding and drying periods in soil systems are implemented. Conventional soil adsorption system designs do not allow for such practices. Higher nitrate levels of groundwater are commonly a problem if soil adsorption systems are used for full or tertiary treatment of wastewater in highly permeable soils (sand or sandy loam soils).

Wetlands, natural or man-made, have also been used for effluent polishing. The problem in colder climates is the release of nutrients during the dormant season.

REFERENCES

1. *Tertiary Treatment,* Manual of Practice No. 3. Inst. Water Pollut. Control, London, 1974.
2. R. L. Culp, G. M. Wesner, and B. L. Culp, *Handbook of Advanced Wastewater Treatment.* Van Nostrand Reinhold, New York, 1978.
3. D. J. De Renzo (ed.), *Nitrogen Control and Phosphorus Removal in Sewage Treatment.* Noyes Data Corp., Park Ridge, NJ, 1978.
4. M. P. Wanielista and W. W. Eckenfelder, Jr., *Advances in Water and Wastewater Treatment: Biological Nutrient Removal.* Ann Arbor Sci. Publ., Ann Arbor, MI, 1978.
5. *Manual for Water Renovation and Reclamation.* Natl. Inst. Water Res., Pretoria, South Africa, 1978.
6. W. Hegemann et al., *Korresp. Abwasser* **31,** 311 (1984).
7. Design Manual. *Constructed Wetlands and Aquatic Plant Systems for Municipal Wastewater Treatment*, EPA/625/1-88/022, U.S. Environ. Prot. Agency, Washington, DC, 1988.
8. K. R. Imhoff, *Water Sci. Technol.* **14,** 189 (1982).
9. K. R. Imhoff, *Water Sci. Technol.* **16,** 285 (1984).
10. *Process Design Manual for Suspended Solids Removal,* EPA 525/1-74-003a. U.S. Environ. Prot. Agency, Cincinnati, OH, 1974.
11. P. C. C. Isaac and R. L. Hibberd, *Water Res.* **6,** 464 (1972).
12. *Process Design Manual for Phosphorus Removal,* EPA 625/1-76-001a. U.S. Environ. Prot. Agency, Cincinnati, OH, 1976.
13. J. L. Barnard, *J. Water Pollut Control Fed.* **47,** 143 (1975).
14. H. A. Nichols, D. W. Osborn, and A. R. Pitman, *Water Sci. Technol.* **17** (11–12), 73–87 (1985).
15. *Process Design Manual for Nitrogen Control,* EPA 625/1-75-007. U.S. Environ. Prot. Agency, Cincinnati, OH, 1975.
16. M. C. Mulbarger, *J. Water Pollut Control Fed.* **43,** 2059 (1971).
17. J. L. Barnard, *J. Water Pollut. Control Fed.* **45,** 705 (1973).
18. P. M. Sutton, K. L. Murphy, and B. J. Jank., *Prog. Water Technol.* **10** (1/2), 241–248 (1978).
19. R. Kayser and G. Ermel, *Gewässerschutz-Wasser-Abwasser* (*GWA*) **59,** 275. Gesellschaft f. Forderung des Siedlungswasser-wirtschaft, Aachen, FRG, 1983.
20. J. L. Barnard, in *Advances in Water and Wastewater Treatment-Biological Nutrient Removal* (M. P. Wanielista and W. W. Eckenfelder, Jr., eds.). Ann Arbor Sci., Ann Arbor, MI, 1978.
21. J. L. Barnard, *Water Wastes Eng.* **11** (7), 33 (1974).
22. T. A. Presley, D. F. Bishop, and S. G. Roan, *Environ. Sci. Technol.* **6,** 622 (1972).
23. V. Novotny and G. Chesters, *Handbook of Nonpoint Pollution: Sources and Management.* Van Nostrand-Reinhold, New York, 1981.

CHAPTER 9

Sludge Handling

After approximately 10 hr treatment in a typical plant, the wastewater has lost most of its suspended and decomposable colloidal and dissolved pollutants and has become purified and relatively harmless and ready for disposal to receiving water bodies. The pollutants that have been removed from the wastewater have for the most part been converted to sludge. Its volume amounts to about 1% of the total wastewater flow received by the treatment plant and may contain pathogenic microorganisms and toxic materials. In a short time it can turn into an odiferous decomposing and objectionable material. Its economical and hygienic handling is the topic of this section.

Figure 9.1 provides an overview of possible alternative processes in sludge handling and disposal.[1] The goal is to dispose of the final sludge residuals safely onto land or into storage lagoons, or into the sea.

The task of planners and designers is to select the most economical and feasible combination of the processes. In the majority of cases, raw sludge is thickened, aerobically or anaerobically stabilized in a digester, then dried and directly disposed onto land, into a landfill or incinerated. Other disposal schemes employ mechanical dewatering of the thickened stabilized sludge followed by composting or incineration. In Milwaukee, Wisconsin, heat-dried sludge is bagged and sold successfully as a low-nutrient content organic fertilizer.

Figure 9.2 shows the volume and mass of sludge in different stages of the sludge-handling process.[2]

9.1 SLUDGE CHARACTERISTICS

By observing the appearance and color, and sensing the odor of the sludge, it is possible to make a judgment on its state.

Raw sludge produced by plain sedimentation is gray and has an objectionable odor. Fecal matter, papers, and vegetable residues are evident. It is very difficult to dewater and even then generally gives off foul odors. The sludge liquor is turbid and smelly. Anaerobically digested sludge is black and has a tarry odor. It can be deposited onto the sand beds in layers 20 cm thick and becomes forkable in 1 or 2 weeks. If well-digested sludge is poured out of a white enamel container, it leaves white lines on the surface of the container that have been traced by rivulets

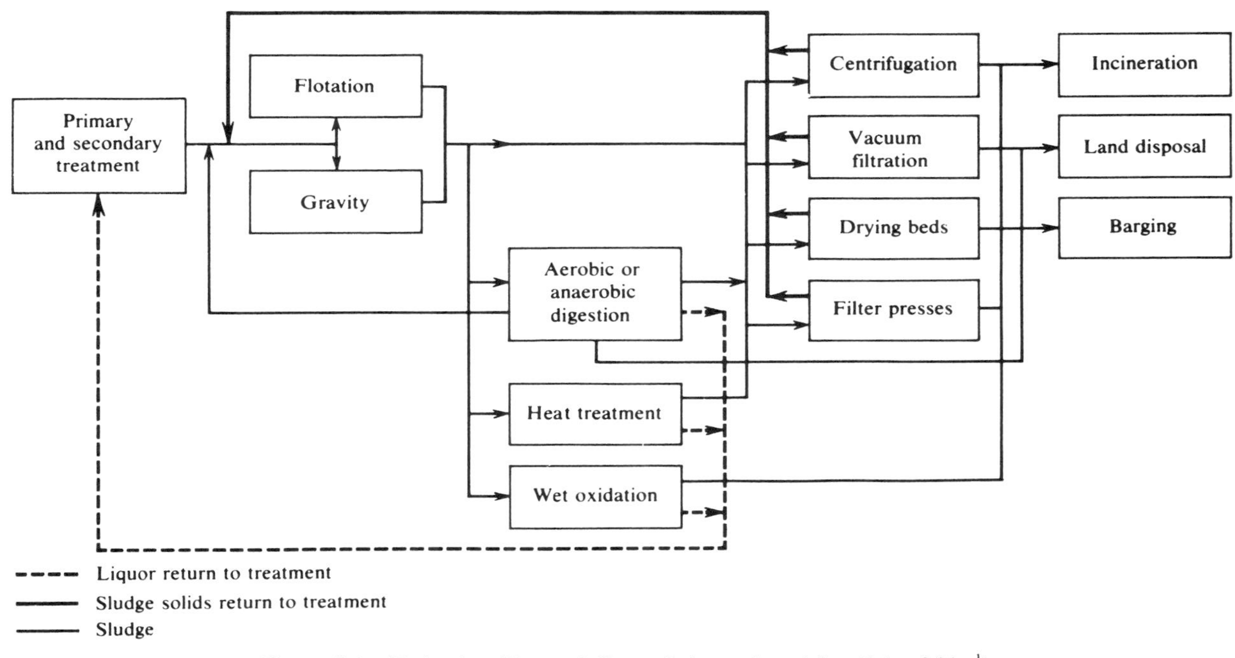

Figure 9.1 Sludge handling and disposal alternatives (after Eckenfelder[1]).

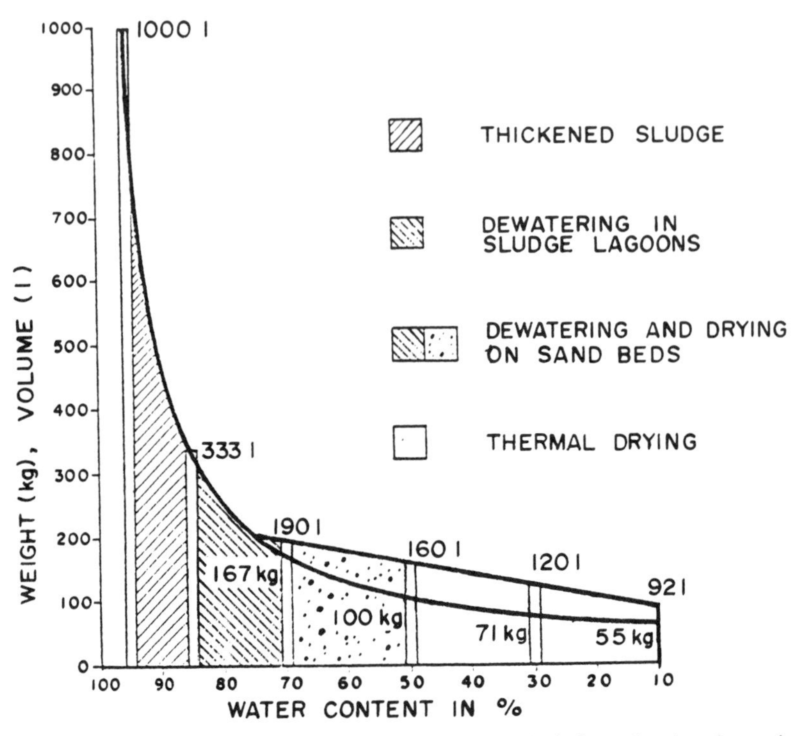

Figure 9.2 Reduction of volume and weight of digested sludge related to its moisture content.

of water separating from the granulated sludge particles. This quick analysis of the sludge can be used by the operators to check on the digestion stage of the sludge.

Aerobically digested sludge is brown and has an earthy smell. During summer weather, it can be dried on sandbeds in about 2 weeks.

Fresh secondary sludges from both activated sludge and trickling filter processes are brown and flocculent. They do not dry readily even in thin layers.

The *water content* of sludge is a very important characteristic that is determined in the laboratory by drying the sludge sample in a drying oven at 103°C. The weight loss upon drying represents the moisture content, where

$$\text{percentage moisture} = 100 \times \text{loss in weight}/\text{initial weight}$$

For example, when the weight loss difference before and after evaporation is 90%, the water content is 90% and the dry weight of the sample is 10% (= 100 g/liter).

Example: By dewatering, the water content of a sludge sample was reduced from 90 to 80%. The dry solids content therefore increased from 10 to 20%. If the original sludge volume was 50 m^3 then the total solids mass was $50 \times 10/100 =$

5 tons (1 m^3 of water or sludge weighs approximately 1 metric ton). The volume of the dewatered sludge is, however, $5 \times 20/100 = 25$ m^3, or one half of the original volume.

The organic and mineral content of the sludge can be determined by *ignition* of the dried sample in a laboratory oven at 600°C. The loss by ignition represents the organic content, whereas the residue (ash) is considered either mineral or inorganic. By digestion, about two-thirds of the organic content is decomposed, leaving about 40% organic and 60% mineral matter in the residue. Secondary sludges from trickling filters and activated sludge units are changed in a similar manner. It should be noted that separate sanitary sewerage systems yield sludges that are lower in mineral solids than sludges from combined sewerage systems.

Digestibility and expected gas production of anaerobic digestion are measured by placing a mixture of fresh and well-digested seed sludge into a bottle with an attached gas collector (Fig. 9.3) and noting the amount of gas produced during a given time by decomposition at a constant temperature (usually 20°C). The proportion of the fresh and digested sludge solids should be about 2 to 1 and the temperature in the container should be about the same as the temperature in the digester from which the seed sludge was withdrawn. Even relatively small deviations of about 5°C can make the results unreliable. The nitrogen content in the bottle should be at least 3.5 mg N per 1g of solids. If the nitrogen is present in insufficient amounts, supplemental nitrogen nutrients should be added. The gas volume produced can then be compared with a theoretical figure such as given in Figure 9.6. Secondary sludges yield gas volumes that are about 50% of those from primary or mixed sludges. Production of other components of the decomposition process (anaerobic fermentation)—such as volatile organics, changes in pH, organic and mineral content, gas composition (methane, carbon dioxide, hydrogen sulfide) and its fuel (calorific) value can also be observed. For anaerobically digested

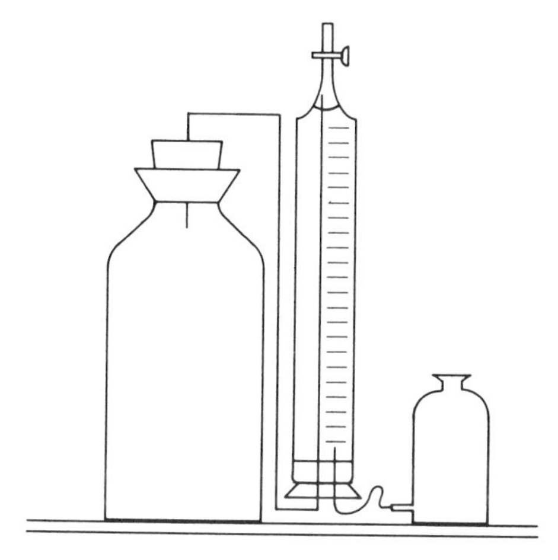

Figure 9.3 Measurement of sludge digestion activity.

sludges, the test involves measurement of the reduction of the volatile suspended solids of an aerated sludge sample and its oxygen uptake with time.

Dewaterability of the sludge is tested using the standard Büchner funnel procedure. A sample of sludge is placed on filter paper (or better, on a sand layer) in the funnel, and the amount of filtrate is measured without and with suction. This test is particularly useful if mechanical dewatering of thickened and/or digested sludge is considered.

For agricultural applications of the sludge, its *fertilizing value* (nitrogen, phosphorus, and humus contents) should be determined. Potassium content of municipal sludges is usually negligible. The metal content of the sludge should also be determined since lead, cadmium, and several other metals can be present in the sludge in quantities sufficient to preclude agricultural applications. Cadmium content is often the limiting factor since cadmium can be readily taken up by plants and incorporated into edible portions of crops.

If incineration is considered as a sludge disposal alternative, fuel (calorific) content of the sludge is measured in a calorimeter. The calorific content is inversely proportional to the moisture content.

For more information on the determination of sludge characteristics and sludge-handling parameters, readers are referred to Standard Methods for the Examination of Waters and Wastewaters[3] or to a publication by Adams, Ford, and Eckenfelder.[4]

9.2 SLUDGE VOLUMES

Sludge consists of mineral and organic pollutants that were removed from the wastewater in the treatment process. Its volume can be related to the number of inhabitants served by the sewage system. However, sludge quantity is also related to the degree of treatment and the type of treatment process. Plain sedimentation will remove most of the settleable suspended solids and biological treatment will remove a substantial portion of fine solids and dissolved and colloidal organics. However, the process of biological precipitation is complex and, as a result, it is not possible to identify exact characterization and sources of the sludge collected. Both decomposition and build up of suspended matter, as well as mineralization, dissolution, and volatilization, occur during biological treatment. Average quantities of sludge collected in different treatment processes are given in Table 9.1.

Sludge volumes are directly related to the moisture content of the sludge as documented by an example in Section 12.2. Appreciable departures from the values listed in Table 9.1 should be expected in practice if the sewage or the treatment process deviates from those typical for a conventional municipal wastewater. Allowances should be made if industrial wastes are present. Calculation of population equivalents can assist in estimates. The values given in Table 9.1 can be applied to separate sewerage systems or to combined sewer systems during prolonged dry periods. For combined sewers in general, the amounts of produced

TABLE 9.1 Weights and Volumes of Municipal Sludges

Treatment Process and Type of Sludge	A Solids Load (g/PE-day)	B Solids Content (%)	C Water Content (%)	D Sludge Volume (liters/PE-day)[b] $\left(\frac{A}{B} \times \frac{100}{1,000}\right)$
Plain Sedimentation[a]				
1. Raw primary sludge	45 (54)	2.5	97.5	1.8 (2.2)
2. Same but thickened	45 (54)	5.0	95.0	0.9 (1.1)
3. Thickened and digested sludge	30 (35)	10.0	90.0	0.3
4. Dewatered digested sludge	30 (35)	30.0	70.0	0.1
Trickling filtration				
5. Wet secondary sludge	25	4.0	96.0	0.63
6. Wet mixed primary and secondary sludge	70	4.7	95.3	1.5
7. Digested mixture	45	3.0	97.0	1.5
8. Dewatered digested sludge	45	28.0	72.0	0.16
Activated sludge process with anaerobic or aerobic stabilization				
9. Wet secondary sludge	35	0.7	99.3	5.0
10. Thickened mixed primary and secondary sludge	80	4.0	96.0	2.0
11. Wet digested mixed sludge	50	2.5	97.5	2.0
12. Dewatered digested mixed sludge	50	22.0	78.0	0.23
13. Aerobically stabilized thickened mixed sludge	50	2.5	97.5	2.0
14. Same as above, dewatered	50	20.0	80.0	0.25
Chemical Coagulation and Flocculation				
15. Sludge from primary flocculation, thickened	65	4.0	96.0	1.60
16. Digested and thickened primary sludge	45	5.0	95.0	0.90
17. Simultaneous precipitation (in activated sludge units) thickened primary and secondary sludge	90	4.0	96.0	2.25
18. Mixed sludge from simultaneous precipitation, thickned and digested	60	3.0	97.0	2.00
19. Tertiary precipitation, raw sludge, thickened	15	1.5	98.5	1.0

[a] The values in parentheses represent typical U.S. loads that include the effect of in-sink household refuse disposal.
[b] To convert from L/PE-day to ft^3/cap/day divide by 28.3.

sludge can fluctuate depending on the number and intensity of the overflows and/or on the volume of storage of overflows that will receive subsequent treatment. Also the number of homes using garbage grinders is another factor that may substantially alter the sludge volumes. In these cases, weekly average sludge production can increase up to 100% over the values given in Table 9.1.

9.3 THICKENING

Although sludge is thickened to some degree in the secondary clarifiers, the water content of the sludge from the clarifier underflow is still relatively high, around 98.5–99%. A separate thickening unit is usually needed to further reduce the moisture content of the sludge before stabilization and disposal.

A thickener can be either a continuous flow or a batch (intermittent) reactor. Thickening can be accomplished by gravity or by dissolved air flotation. In the latter process, air bubbles are attached to solid sludge particles that then will rise to the surface.[5,6] The most common method for thickening municipal sludges is by gravity.

Gravity Thickening

Figure 9.4 shows a continuous feed gravity thickener equipped with a rotating rake mechanism that breaks the bridge between the sludge particles. The rake mechanism should be heavy duty since it has to deal with a heavy, thick fluid. Because of higher solids concentrations in the reactor, settling is mostly of hindered character, in contrast to a secondary clarifier, in which settling is mostly flocculent. The incoming sludge is conveyed to the surface of the reactor and withdrawn from the conical bottom. The hydraulic loading should be less than 0.75 $m^3/(m^2\text{-hr})$ but is better to dimension the thickener using mass loading as guidance, which can be experimentally determined in a stirred laboratory cylinder test. Typical mass

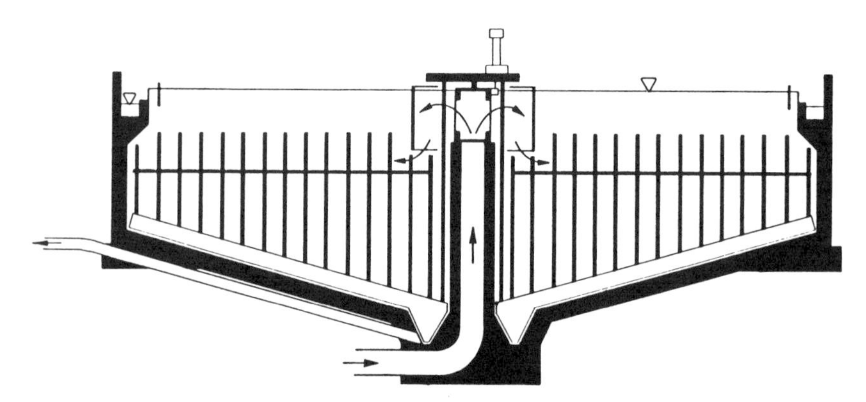

Figure 9.4 Gravity thickener with rake mechanism.

loadings for municipal sewage sludges should be expected to be around 20 kg/(m^2-day) (4 lb/ft^2/day) for waste activated sludge and around 100 kg/(m^2-day) (20 lb/ft^2/day) for primary sludges. For mixed sludge, a loading of 50 kg/(m^2-day) (10 lb/ft^2/day) is commonly used. The depth of the thickener is then related to the required detention time. Torpey[7] suggested computing the detention time from a ratio of the volume of the sludge thickener divided by the volumetric flow of the sludge. Most thickeners will have a minimum detention time of 6 hr. German practice recommends 1 day detention with a provision for 2 days detention during winter operations. The depth of thickeners range from 3 to 3.8 m (10–13 ft).

The thickened sludge moves toward the bottom of the reactor, whereas the supernatant liquid is collected at the surface. The BOD_5 of the sludge liquor (supernatant) is about 1500 mg/liter, therefore, the supernatant must be returned to the head of the plant for treatment. The thickening process is affected by temperature.

Chlorine or lime can be added to improve settleability of the sludge. These chemicals are primarily used to prevent fermentation in the thickener that would result in odor problems and gas development. The dosage of chlorine expressed per population equivalent is about 0.2–0.5 g/(PE-day). Lime dosage is about 500 g $Ca(OH)_2/m^3$ of sludge. By these means, the water content of the secondary sludge can be reduced to 97% (3% solids content) and that of the mixed sludge can be reduced even further to about 93% (7% solids). Joint thickening of primary and secondary sludges is not common in larger treatment plants in which it is not possible to handle the sludge in a fresh state. Partially fermented sludges are very difficult to dewater. The lime dosage to control fermentation is limited if the sludge is to be further dewatered by centrifuges or filters with the aid of organic polymers. In such cases it is better to thicken the primary and secondary sludges separately.

Batch thickeners are less common than continuous ones. The dimensions of the batch thickeners are based on the daily volume of the raw sludge. Since raw sludge,

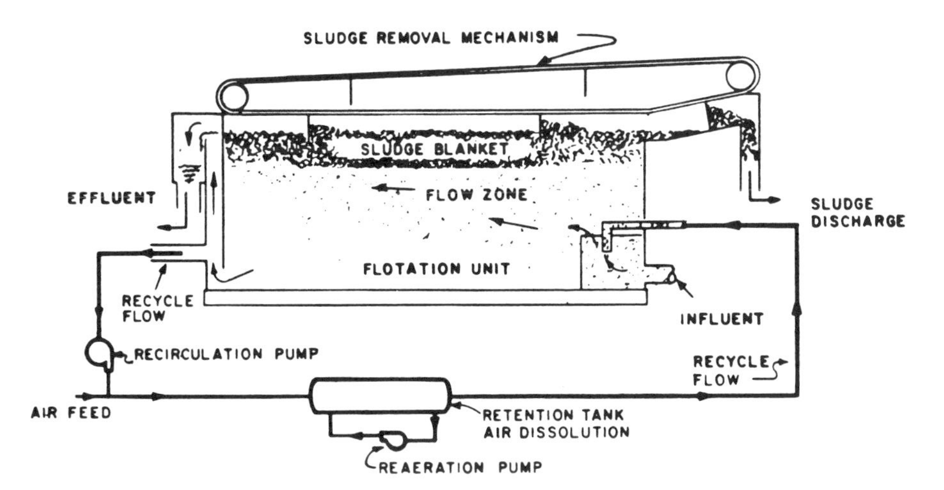

Figure 9.5 Schematics of a dissolved air flotation unit (Kompline–Sanderson).

thickened sludge, and sludge water must be handled in sequence, three separate thickeners should be chosen.

Flotation Thickening

Thickening by air flotation is a feasible alternative to gravity thickening. In the flotation units, a thickened sludge blanket is formed on the surface (Fig. 9.5). Typical sludge loading rates range from 10 to 20 kg/(m^2-hr) with an air to solids ratio of 0.03 to 0.05. Thickening performance is comparable to and often better than that of gravity thickeners.[8]

9.4 DIGESTION

Subsequent to the aerobic biological treatment, the raw sewage sludge still contains large quantities of decomposable organic materials and the decomposition process in the sludge will begin after a short time. If no oxygen is provided, the residual oxygen from the aeration units will be quickly exhausted, and anaerobic septic action of the decomposing bacteria will quickly begin. This process is called digestion or anaerobic fermentation. Traditionally, the term "digestion" has been used for anaerobic decomposition of sludges, but recently both aerobic and anaerobic digestion processes have been used in treatment of sludges from both municipal and industrial wastewater treatment plants.

Aerobic Stabilization

Aerobic stabilization of sludges occurs in extended aeration units if the F/M ratio is less than 0.05 g of BOD_5/(g MLSS-day). Under such low loadings, the sludge age becomes greater than 40 days, which results in an autostabilization of the sludge and wastewater in the same basin. The excess sludge from the secondary clarification can be also diverted after thickening to a separate aerated basin. By aerating the sludge for several days, a substantial portion of the decomposable solids is degraded and the sludge becomes amenable for land disposal. The detention time needed for decomposition of the sludge can be measured in a laboratory stirred and aerated reactor. In the experiment, the volatile suspended solids and the sludge oxygen uptake (measured by placing the aerated sludge sample in a BOD bottle and recording the dissolved oxygen loss by a DO probe with time) should be measured. The digestion process is completed when the volatile solids content of the sludge levels off and when the oxygen uptake rate drops to less than 0.1 g O_2/(g VSS-day). A BOD_5/COD ratio of the sludge below 0.15 would also indicate practical end of stabilization.

Aerobic stabilization is sometimes used in treatment plants that do not have separate primary settling units.

Typical design parameters for aerobic digestion units are[8]:

Hydraulic detention at 20°C*	
Activated sludge only	12–16 days
Activated sludge from plants without primary sedimentation	16–18 days
Mixed primary and secondary sludge	18–22 days
Solids loading	1.5–3 kg VSS/(m^3-day)
Oxygen requirement	0.12 kg O_2/(kg VSS-day)
	1.6 kg/(m^3-day)
Power requirement for mixing and aeration	0.05 kW/m^3

The energy consumption is, at the most, 18 kW-hr/m^3 of treated sludge, and on the average 12 kW-hr/m^3 of sludge. As can be seen, the process uses energy differently from anaerobic digestion, which is energy self-sufficient, and even produces energy in excess of the needs of the process. On the other hand, the aerobic digestion process is relatively odor free and does not require enclosed tanks. The construction cost of the tanks is similar to activated sludge units. Generally, the performance is reliable, the process is not corrosive to steel as is anaerobic digestion, and the supernatant is only moderately polluted. Worm eggs are not destroyed as they would be in the anaerobic digestion.

Experiments by Loll[9] indicated that thermophilic aerobic digestion is feasible at temperatures above 45°C, which will reduce the detention time to about 5–7 days. However, foam build up could be a problem.

Anaerobic Digestion[10–14]

Digestion Process. When the fresh fecal matter is allowed to decompose in water, the products of decomposition are hydrogen, carbon dioxide, and hydrogen sulfide. The sludge, which originally had a near neutral pH, becomes acidic during the first day of decomposition and the pH drops to 6 or even 5. This uncontrolled decay process, known as acid fermentation, is not desirable in wastewater treatment because it lasts a long time, the sludge volume is not significantly reduced, and the sludge is odorous and very difficult to dewater. It has a goldish-gray color, it is foamy, and tends to rise and form scums.

Fundamentally different from the acid fermentation process is the second kind of fermentation, which progresses under controlled conditions in a fully functioning digester and subsequent to acid fermentation. The products of this type of decomposition are methane and carbon dioxide. The pH remains near neutral and the odor of the digested sludge is pleasantly tarry and not objectionable. This anaerobic decomposition process is known as methane or alkaline fermentation. As

*Longer detention times are needed for colder temperatures.

previously described, anaerobic digestion progresses in two steps. In the first step, easily decomposable materials are broken down and converted to organic acids and carbon dioxide. Intensive acid formation drops the pH to about 6. The acid fermentation is followed by a period of acid regression, during which organic acids and nitrogen compounds are decomposed into ammonia compounds, carbonates, and methane. In a large digester at 15°C, the period of acid fermentation and acid regression lasts about 6 months.

Methane Fermentation. The periods of acid fermentation and regression constitute a breaking-in or ripening period through which sludge digesters must pass when they are started. They are followed by a period of alkaline or methane fermentation, in which more resistant materials, including proteins and organic acids, are decomposed. Large volumes of gas, primarily methane and some carbon dioxide, are released and the pH is near or above neutral.

Once methane fermentation takes over, the presence of well-buffered, enzyme-rich sludge and its microbial population make certain that the digestion of the incoming fresh solids will proceed with continuous evolution of the gases, which, in a well-functioning digester, will be 70% methane and 30% CO_2.

Although the methane fermentation sets in spontaneously, there may be conditions when acid fermentation cannot be overcome. As shown in Section 7.6 both acid and methane fermentations proceed simultaneously, however, in a broken-in digester the acids that are produced are rapidly decomposed. When the second reaction becomes slower, acid fermentation may prevail. This happens when the solids load is too large or when the temperature inside the digester drops. When during methane fermentation, the methane bacteria cannot keep pace with the produced acids, acids build up, the pH drops, the carbon dioxide content in the gas will increase, and the digester may fail. The remedy is to reduce the solids load and/or to raise the temperature. A well-functioning digester will have a fatty acid content expressed as acetic acid in the range from 100 to 1 000 mg/liter. An acid concentration of more than 2 000 mg/liter indicates a potential failure of the process. The addition of lime may be of assistance.

The digestion process works at an optimum temperature of 35°C. Lower and higher temperatures are also acceptable, however, the rate of digestion is substantially slower at lower temperatures. To achieve higher temperatures, overheating of the digester will result in a substantial increase of the energy use. However, the decomposition may be faster and the digester volume could be reduced.

A new digester can be broken in more rapidly by filling it with raw wastewater, warming the content to the optimal operating temperature, and adding the seed sludge from another well-functioning digester. To allow the methane bacteria to develop, the daily loading of the digester in the early stages should be low, not exceeding $\frac{1}{10}$ to $\frac{1}{8}$ of the solids present in the digester. If this procedure is followed, the methane fermentation will be in control of the acid fermentation. Rotting leaves can be sometimes used for seeding, also activated carbon can accelerate the maturing process. Without the seeding sludge, lime can be added to maintain the pH near 7 during the break-in period.

Methane fermentation is extraordinarily effective in destroying organic materials. There are only a few resistant substances, for example, hair. Also most plant seeds (excluding tomatoes) are decomposed, as well as worm eggs with the exception of *Ascavis* ova. With few exceptions (e.g., the bacteria causing tuberculosis), pathogenic bacteria will be killed to a large extent.

Reduction of sludge volumes in the digestion process is of great importance. This reduction is primarily due to the release of water from the sludge by destruction of waterbinding, more or less colloidal organic matter, and, to a lesser degree, the gasification and liquification of the sludge.

According to Table 9.1, comparison of items A2 and A3 demonstrates that the dry residue of a thickened primary sludge is on average reduced by digestion from 45 to 30 g/(PE-day), which is a 33% reduction. However, the water content drops from 95 to 90%, hence the volume of the primary digested sludge is 0.3 l/(PE-day), as compared to 0.9 l/(PE-day) of the raw thickened primary sludge (Item D3 compared with item D2 in Table 9.1). This represents a 66% reduction in sludge volume, which is significant in terms of transport.

In spite of its relatively low moisture content, digested sludge flows readily and can be pumped without difficulties. In contrast, raw sludge is viscous and bulky. Digested sludge can be quickly dewatered, dries easily, and does not give off offensive odors.

Rate of Digestion. The rate of digestion depends on temperature as shown in Figure 9.6, which shows the amount of gas produced in the digester in liters/kg of volatile suspended solids added to the digester. The values in the figure are based on original measurements of Fair and Moore[15] who used primary sludge. Newer statistical evaluation of gas production of mixed sludges from 23 studies published in the EPA manual on sludge treatment and disposal[8] showed that the rates of gas production measured on the United States mixed sludge are about 10% lower. Similar German data of gas production are about 35% lower than the original measurements by Fair and Moore. Figure 9.6 also relates the gas production, which is a measure of the sludge digestion rate, to the detention time of the sludge in the digestion tank.

The ultimate gas production rates are

Temperature, °C	10	15	20	25	30
Ultimate gas production, ℓ/kg of VSS added	459	530	610	710	760
Gas produced daily as percentage of the potential remaining gas production	4.3	6.0	8.4	11.4	16.4

It is important to keep the temperature in the digester relatively constant as even minor temperature fluctuations could result in reduction of the volume of produced gas.

The daily decomposition rate of sludges is much slower than the rate of aerobic BOD degradation. For example, the average daily BOD decomposition rate of

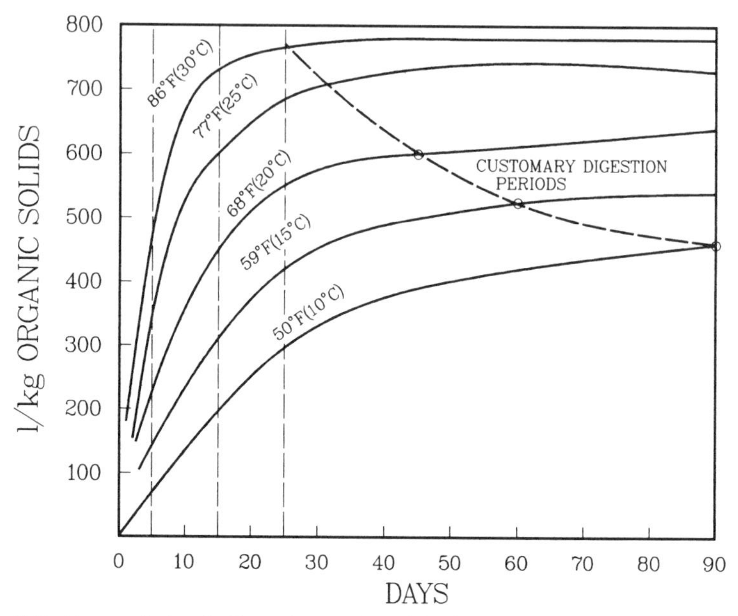

Figure 9.6 Gas evolution in liters from 1 kg of fresh primary organic sludge introduced into a conventional digester operated at different temperatures (after Fair and Moore[15]). Gas evolution from mixed primary and secondary sludges is about 90% of the values for the primary sludge.

typical municipal sewage at 20°C is about 20% of the BOD remaining compared to 8% for anaerobic decomposition of sludges.

The relation of the gas production to both temperature and digestion time is given in Figure 9.6. This enables one to compute the gas volume produced daily by the digested sludge. It should be noted that the connotation of *digestion time* implies that the volume of the sludge in the digester remains constant, that the same amount of sludge is added as is withdrawn, and that the sludge liquor is not separated from the sludge.

Example: A single stage digester serving a population of 50 000 has a volume of 500 m^3. The daily volumetric input of raw primary sludge (Table 9.1) is 0.9 ℓ/(PE-day) $\times$ 50 000 $\times$ 0.001 (m^3/liter) = 45 m^3. The sludge liquor is not separated. The digestion time is then 500/45 = 11.1 days. The digester is heated to 30°C. According to Figure 9.6, the volume of the produced gas is about 633 liters/kg of VSS.

The amount of solids in the raw primary sludge is 45 g/(PE-day), taken from Table 9.1. The sludge is about 70% volatile, hence 0.7 $\times$ 45 = 31.5 g of VSS/(PE-day). The gas volume is then 633 (liters/kg) $\times$ 31.5 [g/(PE-day)] $\times$ 0.001 (kg/g) = 19.9 liters/(PE-day) or 19.9 [liters/(PE-day)] $\times$ 50 000 $\times$ 0.001 (m^3/liter) = 995 m^3/day.

This computation is less reliable if the sludge liquor is separated in the digester, because of the uncertainty in determing the sludge moisture and, hence, the unreliable determination of the sludge volume and digestion time.

Using Figure 9.6, the gas volume for two-stage digestion can be determined separately for each stage.

Figure 9.7 shows that there are two optimum temperature ranges for the anaerobic digestion process: (1) a zone of high temperatures between 50° and 60°C in which heat preferring (thermophilic) bacteria digest the sludge, and (2) a zone of moderate temperatures between 30° and 37°C in which common (mesophilic) organisms are active. The gas production in both temperature zones is about the same, however, the digestion time as seen in Figure 9.7 for the thermophilic zone is about 50% of that for the mesophilic zone. Obviously, the heating requirement is much higher. Also development of the bacterial population in the thermophilic digesters is more difficult. Commonly, the use of thermophilic digesters is not justified on the basis of economy.

Maly and Fadrus[16] investigated the difference in gas production of digesters operating at 20°, 30°, and 50°C. They found no significant differences in methane production or gas composition per kg of volatile solids fed in the digester. However, the rate of optimum gas production and the digestion time differed. Approximately 80–100 days were required for complete digestion at 20°C, 33–50 days at 30°C, and only 20 days at 50°C. In the mesophilic range, more fats were degraded, whereas in the thermophilic range, more proteins and other complex nitrogen compounds were utilized.

The anaerobic bacteria responsible for digestion are sensitive to toxic metals and to a number of toxic organics. The inhibitory concentrations for cadmium and

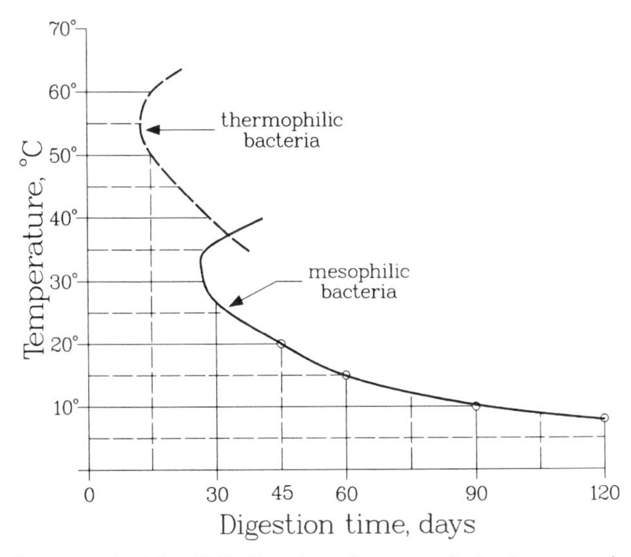

Figure 9.7 Time required for 90% digestion of sewage sludge in conventional digestion tanks.

zinc are 1 and 2%, respectively, expressed per weight of dry solids. When enough sulfides are available in the sludge, the metals can precipitate as insoluble sulfides. Only chromium remains in solution and must be removed by pretreatment of the wastewater. Adding sulfates can alleviate the toxicity of the metals.[17]

The pH in the digester should be between 6.8 and 7.4. This is normally automatically accomplished by cooperation of the two bacterial groups. Because of a number of factors, the acid formers can prevail and create problems. Once the operator notices a drop in pH, it might be too late to react. To have better warning, it is important to monitor the ratio between the alkalinity and volatile fatty acids contents. For safe operation, the ratio should be more than 4 to 1. The fatty acids content expressed as acetic acid is normally in the range of 50–300 mg/liter.

The alkalinity can be supplemented by adding lime, sodium carbonate, and bicarbonate (soda ash), or sodium hydroxide.

Type of Digesters

Combined Units for Separation and Digestion of Primary Sludges. According to the characteristics of the digestion unit and their relation to the sludge separation units the following types of sludge digestion units are recognized:

1. The earliest application of aerobic treatment in connection with sludge digestion was a simple *septic tank*, (Figure 11.4) that is still widely used for on-site household sewage disposal. The septic tank serves as a sedimentation basin, and provides anaerobic treatment of sewage and digestion of the deposited sludge.
2. The two-story *Imhoff or Emscher tanks* developed in the early 1900s combine primary treatment in a clarifier located on the top of the unit with digestion of the fallen sludge in the bottom (Fig. 9.8). In the Emscher tanks, the flowing sewage is introduced into the upper clarifier part and is separated from the digestion chamber. This differentiates them from historically older Travis tanks or recent upflow anaerobic sludge blanket systems in which sewage flows through the digestion volume.

In a situation in which primary treatment alone is sufficient, Emscher tanks provide a convenient method of clarification and sludge digestion in one unit. However, environmental regulations in most developed countries require secondary treatment of municipal wastewaters for which the two-stage Emscher tanks are not the best selection.

The top portion of the Emscher unit is basically a horizontal flow clarifier from which the sludge settles into the lower part through a 0.25-m-wide slot located in the bottom of the upper clarifying portion. Concrete slabs have a slope of at least 1.2–1.6 (vertical) to 1 horizontal toward the slots. The sludge can also be mechanically scraped into the digestion chamber. The sludge liquor is displaced from the

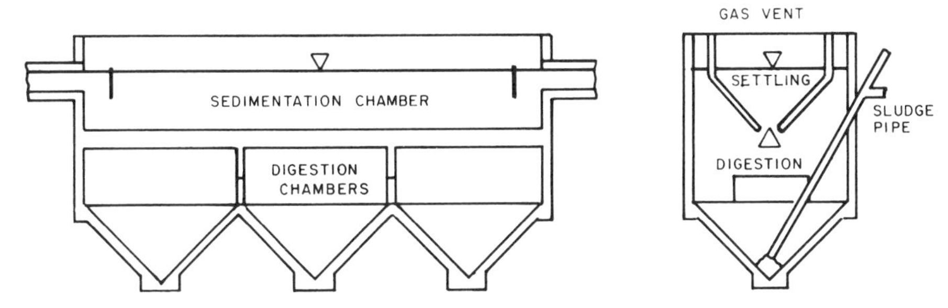

Figure 9.8 Imhoff (Emscher) tank.

digestion chamber in proportion to the volume of the entering settling solids. The lower story of Imhoff tanks is generally subdivided by cross walls into several compartments. In order to distribute the sewage solids uniformly into the digestion compartments, the flow direction should be reversed periodically after several weeks of operation.

The sludge compartment is kept at the temperature of the overflowing sewage, which means that the sludge is digested at temperatures well below the mesophilic optimum. However, no heating is needed.

A modification of the Emscher tank was built in 1929 in Rochester, N.Y. This type of tank is known as a "clarigester."

For a more detailed discussion on the construction or operation of the two-story Emscher tanks, readers are referred to the German version of this book[18] or to the previous U.S. edition of this book.[19]

3. *Separate digesters.* One of the disadvantages of the Emscher tanks is that the digestion chamber is located at greater depths below the clarifier chamber, which leads to increased construction costs. Another disadvantage is that they cannot be heated, hence, the temperature of digestion is not optimal. In modern treatment plants, the digester is a separate unit into which the sludge is pumped. The construction cost becomes smaller and the process can be optimized and better controlled. However, the operation is more complex because the units are commonly heated and mixed for optimum operation. Without heating, the temperature in the separate tanks would be low because of the heat losses through the walls and surface, which would have to be compensated for by increasing the volume. In comparison, the temperature in the unheated two-story tanks would be higher than that in unheated separate units as a result of the warming effect of the overflowing sewage above and minimum heat losses through the walls.

Sludge Lagoons. Lagoons are the simplest separate units in which sludge can be stabilized. They are essentially earth dug basins surrounded by dikes or concrete circular walls with bottom and sides lined with concrete or man-made plastic liners with a layer of clay below. The depth is between 3 and 5 m (10–16 ft). They should be preferably located high enough so that the sludge can be drawn by gravity onto the sludge-drying beds located in the vicinity of the lagoon. Raw sludge drawn into these basins must be mixed with older, well-stabilized sludge to ensure that acid fermentation does not prevail. A ratio of 2 to 2.5 parts of raw sludge to 1 part of digested sludge should be satisfactory. The incoming sludge should be introduced in rotation in several peripheral points and the outlet should have separate structures for withdrawing the sludge liquor and the digested sludge. Surface scum forms a natural cover for odor control and should not be removed or disturbed by added sludge. The lagoons must be completely emptied after several years of operation in order to remove the layers of accumulated sludge. Under favorable conditions in arid regions, the sludge liquor can evaporate from the lagoon, thus precluding the need for sludge liquor withdrawal. New lagoons must be constructed as the old ones are filled.

Sludge lagoons can also serve as a second stage of sludge digestion or as a temporary digester.

In the United States, emphasis on groundwater protection and public resentment against open sludge lagoons may present an insurmountable obstacle for this alternative of sludge digestion.

Enclosed Separate Tanks. Tanks made of reinforced or prestressed concrete or from steel or prefabricated modular pieces offer a more sound design than open sludge lagoons.[20] Three types of heated, separate sludge digestion tanks are common in U.S. sewage treatment plants. In order of their historical development they are (1) conventional or low-rate digesters, (2) two-stage digesters, and (3) high-rate or continuous digesters.

Structurally, all three types of digesters are similar. They are circular tanks with a roof or cover. The tanks include heating and sludge-mixing systems, a gas collection system, and a scum-breaking system. Typical designs are shown in Figures 9.9 and 9.10. For mixing the sludge, recirculating pumps are installed outside the

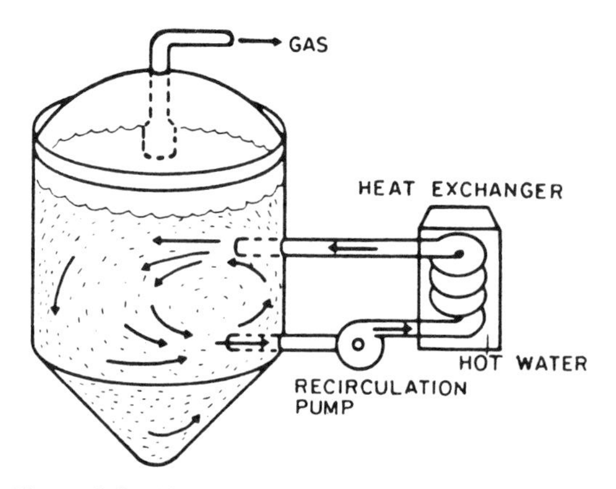

Figure 9.9 Single covered digester with recirculation.

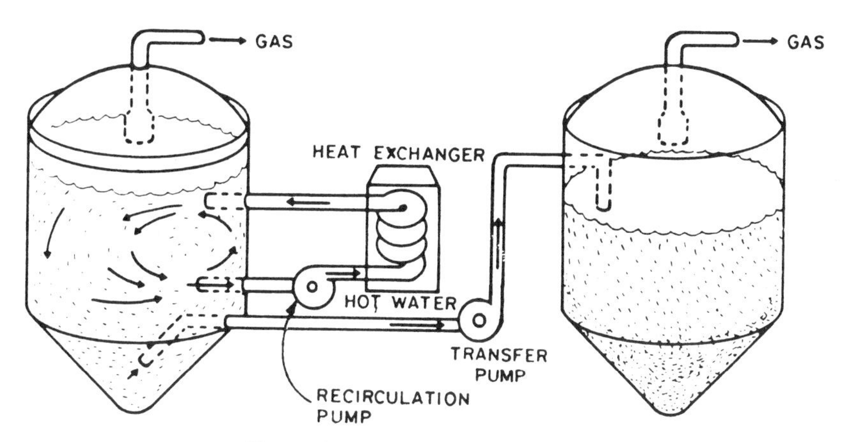

Figure 9.10 Two-stage digestion.

tank that take sludge from the bottom and return it to the top (Fig. 9.9). As an alternative, mixing by gas or an internal mixer can also be used.

Tanks are circular in plan with vertical walls and a conical or flat bottom. Tanks with flat or mildly sloped bottoms require a sludge-scraping mechanism. Single digesters have been built up to a size of 12 000 m^3 (3.2 mg) and 30 m in diameter, although more typical and economical sizes are up to 4 000 m^3 (1.1 mg). Large digesters are made of in-place poured reinforced concrete, whereas smaller digesters are more commonly modular or made of steel.

Since the digester gas contains methane it is highly explosive. Extreme care should be exercised to keep air from the system. To protect the gas collection system against the entrance of air and creation of explosive mixtures, the tanks are provided with tight covers that can be either

1. A floating cover (Fig. 9.11) that rises and falls with changing sludge and gas volumes and prevents air from entering the unit. The top can also be used as a floating gas holder.
2. A fixed cover (Fig. 9.12) with an outside pressurized gas holder that forces gas back into the tank when sludge or liquor is withdrawn, otherwise the vacuum that would be created by the withdrawals could damage the tank.

The scum that builds up on the surface of the enclosed digester should be kept at a minimum since it takes up the usable digester volume and interfaces with the heat input to the sludge. The floating scum should be destroyed continuously or at least once a day. The following methods can be used for breaking the scum layer:

1. Sprays using tap water, or better, using the sludge liquor, which supplies methane bacteria to the raw sludge and minimizes heat losses.
2. Spraying raw sludge over the top, which can be accomplished by means of a rapidly revolving, scattering disc (common in the United States).

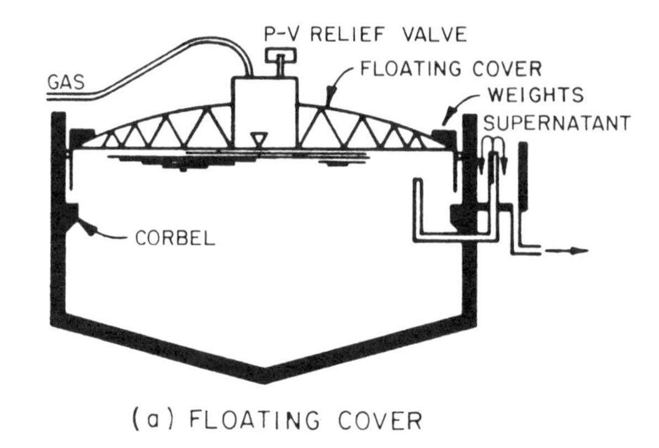

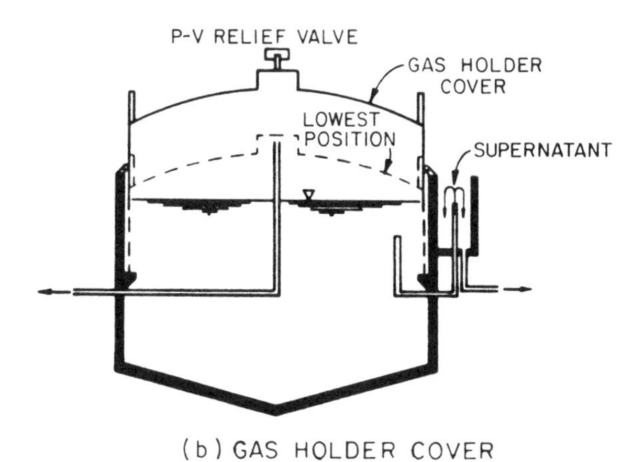

Figure 9.11 Floating cover digesters: (a) without gas storage, (b) with gas storage.

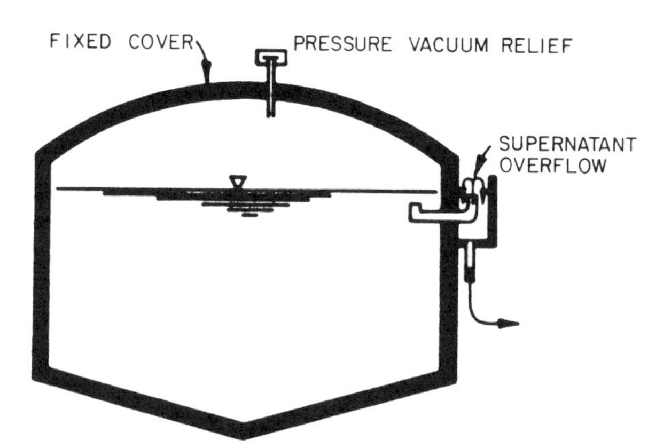

Figure 9.12 Fixed cover digester.

3. A rotating arm partially submerged in the sludge. This alternative is not favored because of the highly corrosive nature of the sludge inside the digester.
4. Good mixing of the sludge by outside recirculating pumps. It is important that the sludge or sludge liquor is directed onto or into the floating scum layer, which is then mechanically destroyed. When raw, cold sludge is pumped it should be heated to the inside temperature of the digester and mixed with older digested (seed) sludge.
5. Blowing the digester gas at rates of 1–1.5 m^3/min about 3 m below the scum layer.

The *supernatant* (sludge liquor) is a highly polluted liquid with high suspended solids, BOD_5, and ammonia. The strength of the liquor is 4 to 10 times that of raw sewage. The highest concentration of pollutants (about 10 times the strength of raw sewage) can be expected in the supernatant from digestion of activated sludge, and the lowest, but still 4 to 5 times the strength of raw sewage, is the liquor from the digestion of primary sludges.[21]

The amount of supernatant depends on the operating conditions. It can be calculated from values given in Table 9.1. Approximately, the amount of supernatant is 1 liter/(PE-day). As a rule, the supernatant is directed back to the influent of the plant. Sometimes, the supernatant concentration is so strong that it can upset the biological treatment unit. In such cases, the supernatant should be directed to a separate settling tank and receive at least primary treatment. Chemical treatment or sand filtration can also be used to reduce the concentration of the sludge liquor. In addition, pretreatment provides more even distribution of the supernatant discharge to the biological treatment plant.

Low-Rate Tanks are loaded with raw sludge in batches. There is no mechanical mixing, however, the tank content is mixed to some degree by the natural release of the gas and by the interchange of solids between the sludge and the scum zone. The tank content is divided into three layers: scum, supernatant, and sludge. Removal of the supernatant and transfer of scum to the sludge are done manually several times a day by the operator.

Two-Stage Digestion divides the needed digestion volume between two separate tanks. The two-stage process was originally used in conjunction with two-story Emscher tanks. In order to reduce the digester volume of the Emscher tanks in which digestion progresses at a lower, below optimum temperature, a separate digester was added in which stabilization of the partially digested sludge was completed. There was also the possibility of providing heating of the second digester. The second digester receives sludge that has less water than the original raw sludge, hence less supernatant is produced. In the Emscher tanks, the sludge liquor is returned directly to the clarification zone through the slot separating the two zones.

In newer treatment plants, two digesters in series are often used. The gas production in the first stage is so high that an adequate solid–liquid separation is not possible. If the first tank is, for example, heated to 30°C, two-thirds of the total gas production is achieved in 5 days and 90% in 14 days. No supernatant separation is installed in the first digester and the mixture of the sludge and sludge liquor is discharged into the second digester for solid–liquid separation. Structurally, the first reactor is similar to one-stage digesters (Fig. 9.10), which is equipped for a safe collection of the gas. Heating and scum separation are no longer needed. Because the gas production is greatly reduced the second tank does not have to be enclosed and the second digester can be built as a simpler unit, usually as an open tank made of reinforced concrete or from steel, or as an earth dug lagoon. The second unit can also provide storage of the sludge during the winter months or when drying beds are not functioning. However, the dimensions must be substantially increased to accommodate the sludge storage.

Elutriation (washing) of the sludge between the two stages removes salts from the sludge and improves its settling and thickening characteristics, which then results in a smaller size of the second unit and better dewatering of the sludge on drying beds.

High-Rate or Continuous Digestion is similar to the first stage of a two-stage process. Zonal differentiation, typical for a low-rate digester, is suppressed by (1) recycle pumps, and/or (2) higher gas production, and/or (3) compressed gas blown into the bottom of the tank. Sludge is added and withdrawn continuously to eliminate the formation of pockets of fresh sludge and also to ensure good seeding.

After 10–15 years of operation, the digester must be emptied for cleaning and layers of accumulated grit and hardened scum must be removed. The cleaning period usually lasts 4–6 weeks and more if repairs are needed. Therefore, it is advisable to have another digester or sludge storage lagoon available.

Sludge Pumping may become difficult if the sludge becomes too thick. Conventional, well-known centrifugal pumps will work well if the sludge is watery and without large solids. Centrifugal pumps can become clogged or be damaged if, for example, they have transported primary sludge-containing large materials such as pieces of wood, rags, and strings. For transporting raw primary and thick digested sludges, positive displacement pumps (plunger pumps) are more appropriate. A disadvantage of positive displacement pumps is that they produce a pulsating flow that may not be acceptable in some applications, such as feed to a centrifuge. Consequently, rotary screw pumps have become widely used in pumping sludges.[22,23]

Dimensioning of the Digester Volume[8,21,22,24–26]

The size of the digester can be determined from the digestion time provided that the temperature and sludge volume are known. Approximate sludge volumes can be obtained from Table 9.1 and the relation between the digestion time and temperature according to Fair and Moore is given in Figure 9.7.

It is simpler and more appropriate to estimate the digester volume from the

TABLE 9.2 Sizes of the Digester (m^3/PE)

	Digester Volume for		
Type of Sludge	Emscher Tank	Single Heated (30°C)[a]	Single Unheated
Primary[b]	0.05	0.02	0.15
Trickling filter	0.1	0.03	0.22
Activated sludge	0.15	0.04	0.32

[a] For thermophilic digestion at 55°C the volume can be reduced by about 30%.

[b] Increase the volume by 20% if solids from combined sewer overflow storage basins are present and by about 25–50% if the majority of households have in-sink garbage grinders.

loading expressed either as the number of the population (population equivalent) served by the plant or as volatile suspended solids. Using the population equivalent as a preliminary design parameter, the digester volume should be about 0.03 m^3/PE, or 35 PE/m^3 of the digester volume (1 ft^3/PE). This corresponds to an average load of volatile suspended solids (Table 5.2) of $39 \times 35/1\ 000 = 1.4$ kg VSS/(m^3-day), which is an approximate value on the border between high-rate and low-rate digestion. The loads to a conventional (low-rate) digester range between 0.5 and 1.5 kg VSS/(m^3-day) (0.03–0.1 lb/ft^3/day), whereas loads to a mixed high-rate digester would range between 1.5 and 6 kg VSS/(m^3-day) (0.1–0.40 lb/ft^3/day).

In Germany, digesters for industrial sludges are designed using a unit load of 1.5 kg of solids/(m^3-day).

In order to operate the high-rate digesters successfully at a rate of 3–6 kg VSS/(m^3-day), the sludge must be well thickened and well mixed with the digested sludge before entering the digester. Also the scum layer must be controlled and the temperature maintained at a constant optimal level. It is advisable to follow a high-load digester with a second stage for solid–liquid separation.

The digester volume depends more specifically on the type of the sludge, the type of digestion process, the operating temperature, and the degree of gasification of the sludge in the digester. The basic sizing parameters for preliminary design of single digesters that have withstood the test of practical experience are given in Table 9.2.

The recommended detention time depends on the operating temperature as shown in Table 9.2. If the quantity of the pumped primary sludge remains constant, as a rule the retention of the sludge in the digester should be at least 20 days.

9.5 GAS HANDLING AND SLUDGE HEATING

The digester gas from well-operating two-story tanks contains about 70–80% methane and 20–30% carbon dioxide and gas from separate sludge digesters is about 65–70% methane and 30–35% carbon dioxide. The difference between the two types of digestion is the result of the fact that the gas from the two-story units must pass through the overlying water, causing a part of the carbon dioxide to be absorbed into the water.

At typical temperatures in a conventional digestion process the volume of the gas produced by 1 kg of volatile sludge solids is about 720 liters/kg according to U.S. data (Fig. 9.6). With the organic solids load of 39 g/(cap-day) the gas production from primary sludges per population equivalent is 39 × 720/1 000 = 28 liters/(PE-day) (1.0 ft^3/cap). Mixed primary and secondary sludge from activated sludge units would produce about 36–41 liters of gas per capita per day (1.3–1.5 ft^3/cap-day). These values are typical for sludges from U.S. cities. German (European) data are about one-half of the U.S. values [13.5 liters/(PE-day) from primary sludges and 29 liters/(PE-day) for a mixture of primary and secondary sludges]. The difference is most likely the result of the presence of ground food residues from in-sink garbage disposal devices installed in most U.S. households, which are not common in Europe. The gas production values can increase if solids from retention basins of combined sewer overflows or industrial solids are added to the system.

Gas collection and storage are economically justified if the digestion tanks and the plant itself are sufficiently large. In two-story tanks, the floor of the settling compartment serves as a gas collection cover (Fig. 9.8), whereas in separate digesters, the cover is either fixed or floating. Floating covers adjust themselves to the fluctuating gas volume and the weight of the cover exerts pressure. In fixed roof tanks, the tank volume must be connected to a gas holder, from which the gas flows into the tank under pressure when the sludge is withdrawn from the tank. Keeping the gas inside the tank under pressure is an important safety precaution. Air should be prevented from entering the tank, otherwise an explosive methane–air mixture could be formed.

Sludge gas is generally odorless and, hence, very difficult to detect. It becomes violently explosive when it is mixed with air in a ratio of 1 part of methane to 5–15 parts of air.[27] For this reason, all enclosed spaces into which gas could leak must be ventilated, free of any flames, and spark proof. In addition, absolutely no smoking should be allowed in the vicinity.

Hydrogen sulfide is usually present in the gas. It is extremely toxic and its presence can be sensed at very low concentrations (0.001%), however, its odor may become less apparent at higher concentrations. Even brief exposures to concentrations of 0.1% or greater are fatal. Hydrogen sulfide is also corrosive. Therefore, it has to be removed prior to using the gas in mechanical devices, which can be accomplished either by passing the gas through iron filings or a caustic solution in which both carbon dioxide and hydrogen sulfide can be removed. The cost of chemicals may be significant, however.

Use of the Gas

The gas from anaerobic digestion can be used for several purposes:

1. For heating the treatment plant buildings and anaerobic digestion tanks and other units. The gas is burned in a heat exchanger. Excess gas is flared off and the gas is supplemented from purchased gas during shortages.

2. For gas supply. The gas can be upgraded to pipeline quality by removing the carbon dioxide and compressing it in order to enter it into a public gas distribution system for credit or sale.
3. For making electricity. The gas can be used to make electrical energy that can be sold to electric power companies if the plant has its own power station.

Steam Heating is the most economical and simplest utilization of sludge gas. There are basically three heating systems that are based on steam or hot water heating:

1. Internal Direct Heat. In this method, hot water or steam is added to the sludge. In a low-pressure boiler, about 20 kg/hr of steam can be produced over 1 m^2 of heated boiler area; 1 kg of steam can deliver about 0.7 kW-hr to the digester. Commonly, the steam or hot water is added to the bottom of the tank or to the raw sludge entering the digester. The temperature of the sludge is then elevated to the operating temperature of the digester. About 1.6 kg of low-pressure steam is needed to increase the temperature of 1 m^3 of sludge by 1°C. If hot water is used instead of steam, large amounts of water would be needed that could considerably dilute the sludge and reduce the digestion time.

Boiler water must be deionized, otherwise scale deposits will build up in the boiler and the pipelines.

2. Internal Indirect Heat. In this method, warm water is recirculated through a heat exchanger located in the digester. The temperature of water in the heating pipes or coils is held at 60°C to prevent sludge baking on the heating surface. At least 0.3 m^2 of coil surface should be provided for each 100 m^3 of the digester volume. To reduce heat losses, hot water pipes should be well insulated outside the digester. The heat transducer per unit surface of the coil can be computed as

$$\begin{aligned}&\text{Heat transfer } [\text{kW-hr}/(\text{m}^2\text{-hr})] = \\ &\quad \text{coefficient of heat conductance } [\text{kW-hr}/(\text{m}^2\text{-hr}/°\text{C})] \times \\ &\quad (\text{temperature of the heating surface} - \text{temperature of} \\ &\quad \text{the sludge})(°\text{C})\end{aligned}$$

The coefficient of heat conductance has an approximate value of 0.07 kW-hr/(m^2-hr-°C) (12 Btu/ft^2/°F). If the sludge is diluted then the value will be greater (up to 0.2 kW-hr/(m^2-hr/°C).

Assuming that the incoming water temperature in the coil is 60°C, the temperature of water leaving the digester is 40°C, and the sludge temperature is 35°C, the heat transfer per unit area of the coil becomes

$$0.07\,[\text{kW-hr}/(\text{m}^2\text{-hr-}°\text{C})] \times [(60 + 40)/2 - 35](°\text{C}) = 1.05\ \text{kW-hr}/(\text{m}^2\text{-hr})$$

3. External Indirect Heat. In large digesters, the sludge is pumped through an external heat exchanger in which the sludge temperature is maintained at the optimum level. The velocity in the recycle pipes should be at least 1–1.5 m/s to avoid sludge deposition. The diameter of the sludge recycle pipes should be at least 125 mm but 150 mm diameter (6 in.) provides more safety. Low-pressure steam or hot water flow in the opposite direction. The corresponding hot water heat transfer coefficient is about 1.05 kW-hr/(m^2-hr-°C). Then, assuming the incoming hot water temperature is 70°C, the temperature of water leaving the heat exchanger is 50°C and the sludge temperature is 35°C, the heat transferred per unit area is

$$1.05\,[\text{kW-hr}/(\text{m}^2\text{-hr-°C})] \times [(70 + 50)/2 - 35](\text{°C}) = 26.2\ \text{kW-hr}/(\text{m}^2\text{-hr})$$

When steam is used, condensation in the sludge pipes will result in a temperature of 100°C. Then the heat transfer per unit area becomes 1.05 × (100 − 35) = 68.3 kW-hr/(m^2-hr). The design of external heat exchangers should provide for easy cleaning of the sludge recycle pipes.

The heat supplied to digestion tanks must be sufficient (1) to raise the temperature of the incoming sludge to that in the tank, (2) to compensate for the heat losses through the walls, bottom, and cover, and (3) to compensate for the heat loss by evaporation of water from the sludge. Loss by evaporation is small and usually not considered.

The specific heat of sludge, which is the heat that is needed to raise the temperature of 1 m^3 of sludge by 1°C, is similar to that for water. In the SI unit system, it is 1.16 kW-hr/(m^3-°C). This is equivalent to a more commonly known physical unit of 1 cal/(cm^3-°C).

To select the level of heating, the operator has the following options:

1. Winter heating of the digester to a temperature of about 15°C, which will maintain the temperature more or less constant throughout the year. This type of strategy can be used for open sludge lagoons or earth-dug open digesters.
2. Year round heating to 35°C (mesophilic digestion) will result in significant reduction of the digester volume requirement when compared to alternative (1) (Fig. 5.66).
3. Overheating to 55°C (thermophilic digestion). The digester volume can be further reduced.

In the full-scale plant studies of Fisher and Greene,[28] overheating did not result in a better performance. The sludge was odorous and its dewatering characteristics were poor. On the other hand, by digesting the sludge at 55°C, Garber and coworkers[29] obtained better sludge dewatering characteristics and less pathogenic microorganisms. The change in gas production, methane content, and solids breakdown were found to be insignificant when compared to digestion in the mesophilic temperature range (35°C). A full-scale study in Chicago, Illinois by Rimkus et al.[30] showed about a 20% increase in gas production of a digester operated at 53°C when compared to a digester operated at 34°C. Most of the other parameters were similar at both temperatures. Also, in the Ruhr area of Germany, the thermophilic digesters were successfully employed.[31]

Heating by sludge gas produced by the digesters is usually self-sufficient to maintain a temperature of 35°C in the tank, provided that excess water is removed from the sludge prior to digestion. A large portion of the heat input is for warming the incoming sludge. Some heat is lost through the walls, cover, and in the pipelines. A typical heat loss in northern climatic conditions of the United States, southern Canada, and central Europe (heat loss coefficient) is about 1.2 W-hr/m^2 per 1°C of the temperature difference in 1 hr (= 0.21 Btu/ft^2/°F). For different materials the heat loss coefficient is

Sheet metal	11.6 W-hr/(m^2-hr-°C)*
Concrete, 30 cm	2.3 W-hr/(m^2-hr-°C)*
Concrete, 50 cm	1.7 W-hr/(m^2-hr-°C)*
Concrete in moist earth	1.5 W-hr/(m^2-hr-°C)*
Bricks	1.3 W-hr/(m^2-hr-°C)*
Insulated concrete cover	0.9 W-hr/(m^2-hr-°C)*
Concrete in dry earth	0.5 W-hr/(m^2-hr-°C)*

Example: Daily raw sludge input to a digester is 100 m^3. The sludge has a temperature of 10°C. The digester operating at 35°C is made of concrete. It has a volume of 2 000 m^3 and its height is 6 m. The outside temperature is 5°C. What is the size and capacity of the heat exchanger for warming the sludge?

1. For warming of the raw sludge from 10° to 35°C the needed heat input is

$$100\ (\text{m}^3\text{-day}) \times (35 - 10)(°\text{C}) \times 1.16\ [\text{kW-hr-}(\text{m}^3/°\text{C})] = 2\ 900\ \text{kW-hr/day}$$

2. The heat loss through the walls and cover if the digester is made of 30-cm-thick reinforced concrete. If the outside temperature is 5°C then

$$\text{Heat loss through the walls} = 3.14 \times 20.6\ (\text{m}) \times 6\ (\text{m}) \times (35 - 5)(°\text{C}) \times 2.3\ [\text{W-hr}/(\text{m}^2\text{-hr-}°\text{C})] \times 24\ (\text{hr}) \times 0.001\ (\text{kW-hr/W-hr}) = 643\ \text{kW-hr/day}$$

*To convert to Btu/ft^2/hr/°F multiply the values by 0.18.

$$\text{Heat loss through the cover (insulated)} = \frac{3.14}{4} \times (20.6^2)(\text{m}^2)$$

$$\times (35 - 5) \times 0.9 \times 24 \times 0.001 = 216.0 \text{ kW-hr/day}$$

Heat loss through the bottom (dry earth)

$$= \frac{3.14}{4} \times 20.6^2 \times (35 - 5) \times 0.5 \times 24 \times 0.001 = 120 \text{ kW-hr/day}$$

3. The total heat input and the heat capacity of the boiler must then be

$$2900 + 643 + 216 + 120 = 3\,879 \text{ kW-hr/day} = 162 \text{ kW-hr/hr}$$

4. If low-pressure steam heating is selected to deliver the heat directly to the incoming sludge, the amount of the needed steam is

$$162 \text{ (kW-hr/hr)}/0.7 \text{ (kW-hr/kg)} = 231 \text{ kg of steam}$$

5. An internal indirect heating system should have a coil surface area of $162/1.05 = 154$ m^2, an external indirect heat exchanger should have a coil area of $162/26.2 = 6.18$ m^2 if warm water is used to deliver the heat and $162/68.3 = 2.36$ m^2 if steam is used.

Gas Energy Value. The energy content of the gas can be compared to the following alternative energy sources:

Pure methane	10	kW-hr/m^3
Natural gas	9.5	kW-hr/m^3
Digester gas with 70% methane	7	kW-hr/m^3
Gasoline	8.5	kW-hr/liter
Heating oil	11	kW-hr/kg
Black coal	9.3	kW-hr/kg

It can be seen that 1 m^3 of digester gas is equivalent to about $7/8.5 = 0.82$ liters of gasoline (100 ft^3 of digester gas = 0.62 gal of gasoline). The use of methane gas as a substitute for liquid gasoline in cars was common during World War II, particularly in continental Europe. Presently, such substitution may not be economical in the United States and other industrialized countries.

9.6 SLUDGE DEWATERING[32]

During the digestion process, a significant portion of the organics is converted to gases. The residual sludge is not odorous, but it is still bulky, watery, and potentially unsafe because of the pathogenic microorganisms that may remain. The primary objective of the dewatering process is to remove as much water as possible and prepare the sludge for economic, hygienic, and safe final disposal.

Generally, most sewage sludges are not easily dewaterable without preparation and/or chemical treatment. Sludge dewaterability can be evaluated by general parameters that can be measured in the laboratory, such as the specific resistance to filtration, compressibility, and capillary suction time, and by the specific tests for each dewatering technique.[4,33]

Drying Beds

Sludge Air Drying Process. In the past, direct disposal of digested sludges by farmers on fields was common in small communities and still is practiced today under controlled conditions in some countries. In more recent facilities, the sludge is further dewatered and disposed of as filtered dewatered cake. Air drying of sludges on sand beds is the simplest way of dewatering the digested sludge.

When wet digested sludge is allowed to overflow on sand beds for dewatering, the gases that are dissolved or trapped in the sludge become significant. The sludge that is commonly withdrawn from the bottom of the digesters is under a hydrostatic pressure. Since the solubility of gases is proportional to the pressure, the dissolved and entrapped gases are released as bubbles when the sludge is discharged onto the beds, the sludge becomes foamy, and its specific weight is then less than that of water. As a result, the sludge floats on the water surface in a similar manner to flotation as discussed in Section 6.2. This effect can also be seen in laboratory test cylinders.[34]

The sludge therefore releases its water below, in contrast to degasified sludge that releases its water above. If the sludge is to be pumped onto the drying beds, pumps that work on the principle of creating a vacuum in the suction pipe and in the pump itself should not be used since the sludge might lose its gas content. In lieu of vacuum (centrifugal) pumping, air lift or compressed air ejector pumps are more appropriate. In the two-stage digestion process, the second unit can be located higher to permit gravity flow of the sludge onto the drying beds. In single, mixed digestion systems, a quiescent space should be provided near the digester outlet in which the sludge can rest, thicken, and produce a sufficient amount of gas.

Most of the water separation occurring on the drying beds is during the first day. The water layer below the floating sludge is drained by infiltration into the sand, which is then followed by 1–2 weeks of drying by evaporation. If gas was not present in the sludge, water would accumulate on the surface, which could be lost primarily by evaporation. Dewatering by evaporation only would take much longer. Digested sludge dewaters more readily than partially digested or raw undigested sludge.

The character of drying and its stage can be recognized by cracks that form on the sludge surface. A few small and thin cracks formed on the surface indicate a good quality, dewatered sludge, which floated on its own water released below and, as a result, did not lose its consistency when residual water was lost by evaporation. Numerous wide cracks on the sludge surface indicate that the sludge had poor dewatering characteristics caused either by its high content of poorly digested biological solids or by chemical solids, and, consequently, the sludge had lost its water primarily by evaporation. Few very wide cracks are typical for poorly digested viscous sludge that lost its water only by evaporation over a longer period of time.

In addition to its smell, poorly digested and/or raw sludge can also be recognized by the presence of sludge flies (*Eristalis tenax*) that are never found on well-digested sludges.

Design of Sludge Drying Beds.[35–38] The beds (Fig. 9.13) are generally about 0.25–0.4 m deep and consist of graded layers of gravel or crushed stone overlain by a 0.1- to 0.15-m layer of sand. It is recommended that the sand has an effective size of 0.3–0.5 mm and a uniformity coefficient of not greater than 4. Shallower depths and materials such as slug, cinders, and coke breeze have also been employed. Some sand is lost in stripping the dried sludge cake from the bed. Washed grit from grit chambers could provide a source of the replacement sand. Drying beds are underlaid with tile drains about 15 cm (6 in.) in diameter that have open joints between them. The width of the drying beds ranges from 5 to 20 m with no limits on length. In small plants with narrow-gauge railways and industrial dump cars, the width is kept below 6 m (20 ft). In large plants, sludge stripping machines moving on rails allow larger widths.

Individual beds are surrounded by concrete curbing. The digested sludge is applied in depths of about 0.2 m (8 in.). The area of the bed can be computed from an assumption that the beds are refilled about five to seven times per year, depending on climatic conditions. This corresponds to an annual total depth of the wet sludge of 1–1.4 m ($3\frac{1}{3}$ to $4\frac{2}{3}$ ft). Since the yield of the wet, digested sludge is about 0.3 liters/(cap-day) (Table 9.1), the required drying bed area is about $(0.3 \times 365)/(1\,000 \times 1.2) = 0.09$ m^2/capita or population equivalent (1 ft^2/cap)

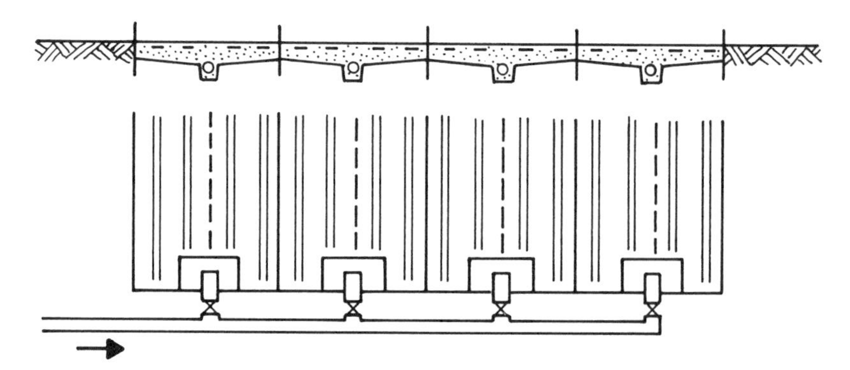

Figure 9.13 Open sludge drying beds.

TABLE 9.3 Typical Area Requirement for Drying Beds[8]

Sludge	Area[a] (m^2/cap)	Sludge Loading[b] (kg of Dry Solids/m^2-year)
Primary digested	0.09–0.14	120–200
Primary and humus digested	0.11–0.16	100–160
Primary and activated digested	0.16–0.275	60–100
Primary and chemically precipitated sludge	0.185–0.230	100–160

[a] To convert from m^2/cap to ft^2/cap multiply by 10.76.
[b] To convert from kg(m^2-year) to lb/ft^2/year multiply by 0.2.

or 11 population equivalents per square meter. This is a minimum value for primary digested sludges. Design loading parameters can also be expressed as solids mass per area per year. Recommended design area requirements for the northern United States, southern Canada, and central Europe are given in Table 9.3.

As reported by Swanwick[29], British researchers recommend drying bed areas that are about twice as much as those listed in Table 9.3. It is obvious that the presence of industrial sludges and solids from combined and storm sewer overflow detention basins would require an additional allowance in the sludge bed area. Area requirements must also be further adjusted to climatic conditions, operation procedures, and the size of the plant.

Glass-covered drying beds have been used in the United States in order to reduce the required drying bed area. When they are ventilated, the number of fillings can be increased by 50–100% and, consequently, the area can be reduced by the same ratio. This would bring the number of fillings and drying bed area requirements to those typical for southern United States.

The drying properties of poorly dewatering sludges can be improved by adding coagulant aids or by elutriation.

Bed-dried digested sludge contains air and has a black color. It can be easily raked and handled by a shovel, its moisture content being about 55–60% or solids content of 40–45%. Using the values for sludge volumes listed in Table 9.1, the volumes of the dried sludge cake can be estimated as

0.1–0.13 liters/(cap-day) for primary sludge
0.15–0.17 liters/(cap-day) for a mixture of primary and trickling filter sludges
0.17–0.27 liters/(cap-day) for a mixture of primary and activated sludges
(to convert to ft^3/cap/day divide by 28.3).

Sludge Lagoons

Lagoons are an inexpensive alternative for sludge dewatering and sometimes disposal. The lagoons are simple earthen basins into which the sludge is pumped in varying depths—0.75–1.2 m (2.5–4 ft). The lagoons can be equipped with manholes for observation of the water level. Usually there is no drainage installed in the lagoon and drying is primarily by evaporation. The area requirements for lagoons are about twice as much as that for conventional drying beds, hence 0.3–0.5 m^2/capita would be typical, depending on climatic conditions. At least two lagoons operated in parallel should be installed whereby one lagoon is being filled while the sludge in the other lagoon is drying. The sludge after drying can either be buried in the lagoon (a new lagoon is then needed) or the lagoon can be cleaned and the sludge hauled away for final disposal. The solids content of the dried sludge is about 20%. Because of groundwater contamination concerns, burial of sludges in a lagoon is not a recommended alternative in the United States, and in any case, a special permit subsequent to a detailed environmental impact analysis is required.

Sludge lagoons may become a feasible alternative if they are not located near residential areas, if they are environmentally safe and acceptable, and if the cost of sludge pumping is not prohibitive.[40] The soils must be reasonably porous and the highest groundwater table must be at least 1 m, preferably more, below the bottom of the lagoon.

Sludge Conditioning

Mixed primary and secondary sludge has a strong waterbinding capability and, consequently, without pretreatment it is very difficult to dewater in both the raw and stabilized states. For pretreatment or conditioning, the following methods can be used: heat conditioning or freezing, addition of coagulating chemicals and/or organic polymers, and addition of fly ash, and washing (elutriation) of the sludge with water.

Heat Treatment involves heating of thickened sludge in a heat exchanger (patented by Porteous and known as the Zimpro process) and subsequently pressurizing the sludge to about 10–25 atm. (150–400 psi) for about 30–45 min. The resulting sludge is both sterilized and conditioned and dewaters easily without adding chemicals. Heat treatment breaks down the organic matter of the sludge and the solid matter left behind is primarily made of mineral matter and cell residues.

As would happen in cooking a piece of meat, a significant portion of the organics may become dissolved, resulting in a highly concentrated supernatant that has a BOD_5 of around 10 000 to 15 000 mg/liter and an organic nitrogen content of around 1 000 mg/liter. Therefore, the return flow should be pretreated. Using anaerobic biological degradation, about 70% of the BOD_5 can be removed in 6 days at 37°C. The volume of the supernatant amounts to about 0.75–1% of the total wastewater flow. The return flow without a pretreatment will increase the load to the biological treatment units by about 25%. If the treatment system is not capable of handling organic nitrogen, the ammonia load to receiving water may increase significantly.[32,41]

Freezing is a process during which all solids are rejected from ice crystal lattice and, hence, ice crysals are made of relatively pure water. Therefore, water can be effectively separated and the sludge, after melting, has a better dewatering capability.

This procedure can provide reliable results in northern climates. Results with sludge freezing in Ontario, Canada and Hanover, New Hampshire showed that the solids content of anaerobically digested sludges (solids content 2–6%) can be increased by freezing to about 30%. Sludge supernatant after freezing contains about 1 000 mg/liter of BOD_5. A layer of sludge 100–150 cm thick ($3\frac{1}{3}$ to 5 ft) can be frozen during winter months in most of north-central and northeastern United States if the sludge is applied daily in layers that are maximum 8 cm thick. Sludge should not be applied under a deep snow layer that would act as an insulator and retard freezing. Thawing of previously frozen sludge is not a problem because the sludge solids will retain their transformed characteristics.

Warm weather storage of wastewater sludges may not be practical nor desirable. A cost-effective approach might be to combine winter freezing with off-winter polymer assisted dewatering on the same drying bed.[42]

Chemical Conditioning implies the addition of coagulating chemicals to the sludge. Ferric chloride, alum, ferrous sulfate, lime, and organic polymers are the most widely used chemicals. However, it must be recognized that typical dosages are relatively high, for example, the dosage of ferric chloride expressed per unit of dry sludge solids is about 2.5% for primary, digested, and raw sludges, and 7% for secondary activated sludges. In addition to ferric chloride, about 7% lime on a dry solids basis must be added to maintain the pH of the sludge near neutral.

Organic polymers can be used instead of inorganic coagulants. Their use becomes feasible when the required dosage is relatively low and the characteristics of the sludge are not adversely affected. Electrolytes provide an electrochemical charge needed for precipitation and adsorption and provide long molecular bridges between the sludge particles. The effective dosages are between 100 and 200 g/m^3 of sludge (= mg/liter). It is important to establish the exact dosage and optimal pH for flocculation by laboratory experiments (the jar test).

Bulking agents such as fly ash and diatomaceous earth improve mechanical dewatering of sludges. Fly ash must be added in large quantities, 1–2.5 kg of ash per 1 kg of sludge, however, this may lead to overloading of the mechanical dewatering equipment. Diatomaceous earth may be used as a filter precoat material by which the filter is covered in the filter cycle in order to produce a drier cake.

Lime stabilization of raw or digested sludges has been used as an emergency measure when, for example, digesters are out of service or during upgrading of existing facilities. Basically, lime increases the pH of the sludge until all biological life is terminated, which occurs when the pH reaches a value of about 12. Consequently, odor is eliminated or greatly reduced and pathogenic microorganisms are destroyed. Lime may also improve the dewatering characteristics of the sludge. Lime dosages are relatively high, about 12% of $Ca(OH)_2$ per weight of dry solids for primary sludges and up to 32% for waste-activated sludge.[8,43]

Mechanical Dewatering

Common devices used for mechanical dewatering of sludges include centrifuges, filter presses, rotary vacuum filters, and horizontal belt filters.[44]

Centrifuges[45,46] (Fig. 9.14). These devices are used more for thickening than dewatering. If they are to be used for sludge dewatering flocculant aids must be added.

The centrifuge uses centrifugal force to speed up the sedimentation of solid sludge particles. The principal elements of the solid bowl centrifuge as shown on Figure 9.14 are a conical rotating bowl and a screw-type conveyor of solids. The bowl has an adjustable overflow weir at its larger end for discharge of clarified effluent and solids discharge ports on the other conical end of the bowl. As the bowl rotates, the centrifugal force causes the sludge slurry to form a pool on the wall of the bowl, the depth of which can be adjusted by the weir. As the solids settle on the wall of the bowl, they are continuously moved toward the conical end by the conveyor scroll. At the same time, the supernatant liquid continuously overflows the effluent weir.

Solids contents of effluent sludges commonly range from 30 to 35% for primary, raw, or digested sludge, 20 to 30% for primary, raw, or digested sludge mixed with trickling filter humus, and 15 to 30% for mixed primary (raw or digested) and activated sludges.

Under a typical operation, the supernatant may contain significant quantities of sludge solids. Adding about 3 g of flocculant chemicals per 1 000 kg of dry sludge solids (0.3%) will significantly improve the solids recovery, resulting in a cleaner supernatant containing only about 2% or less of influent solids.

Another type of centrifuge used for solid–liquid separation is the basket centrifuge. This centrifuge works more or less on a batch feed–cake removal principle. Because of its batch feed concept, the centrifuge has a lower capacity than the continuous feed units, however, the solids recovery is higher.

Vacuum and Pressure Filtration. Both pressure and vacuum filters generate a pressure difference, thus forcing the fluid to pass from the compartment with a higher pressure into the one with a lower pressure, leaving the solids trapped on a

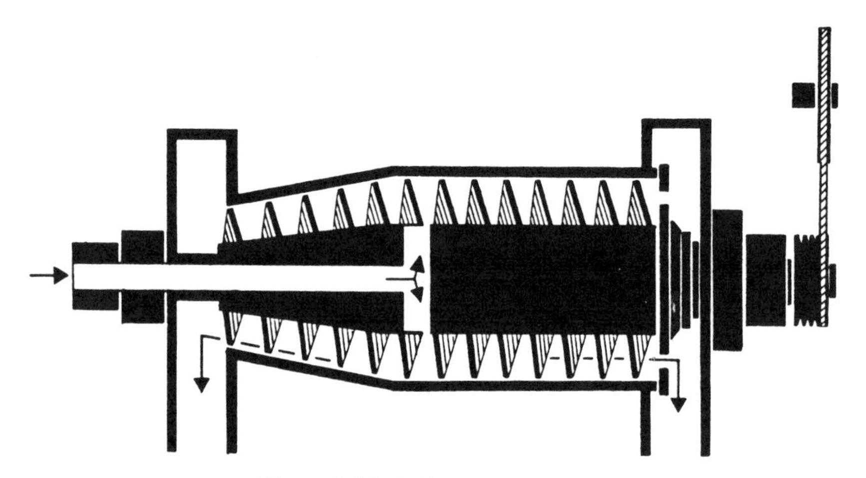

Figure 9.14 Solid bowl centrifuge.

filter media, usually a filter cloth. Selection of the filter media is important to facilitate an easy liquid–solid separation without clogging the media and/or without letting excessive amounts of solids pass into the filtrate. Municipal sludges usually must be conditioned to become amenable to filtration.

A Filter Press is a batch device that has been used in Europe and North America to process difficult to dewater sludges. The press consists of a series of vertical metal frames holding a stretched filter cloth. The frames are pressed together between the fixed and moving ends as illustrated in Figure 9.15.

Commonly, a municipal sludge is conditioned with the addition of about 4 kg of ferric chloride in a 40% solution and 6–10 kg of lime per 1 m^3 of the sludge. Heat conditioning is also possible and fly ash can be used for precoating. Sludge remains in the press for about 1–2 hr under an applied pressure of about 6–8 atm, during which time the sludge moisture content is reduced by filtration and screening from about 95% moisture to a moisture content as low as 50%. The filtrate, which is cleaner than that from centrifuges, is collected and returned for treatment, whereas solids remain in the press chamber.

At the end of the filtration cycle, which is recognizable by a filtrate flow of zero, the feed pump is turned off and the press is opened. The plates are then separated and the cake drops into a hopper or onto a belt conveyor. The filter cloth should be periodically washed by spraying to remove remaining sludge particles and to clean the openings in the cloth. The filter press produces a cake with a high solids content even from difficult to dewater sludges. However, batch operation and, hence, high labor cost are its main disadvantages.[8,47]

A Belt Filter Press (Fig. 9.16) has been manufactured in the United States since 1971. However, the system has been developed in Europe where it has been widely used. An endless filter belt runs over a drive and guides rollers at each end, similar to a conveyor belt. Above the filter bed is a press belt that runs in the same direction and at the same speed. The speed of the belts and the distance between them

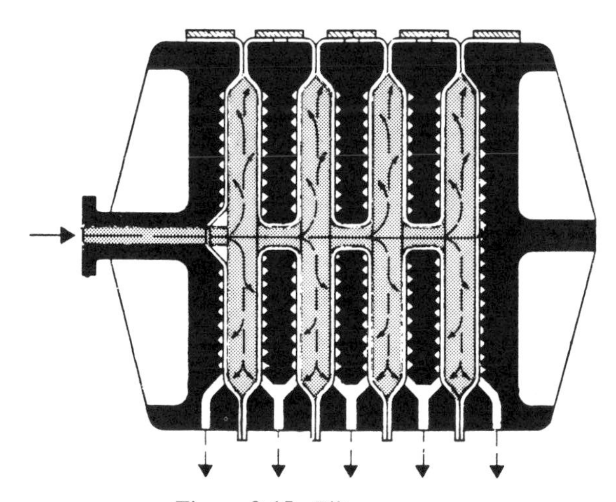

Figure 9.15 Filter press.

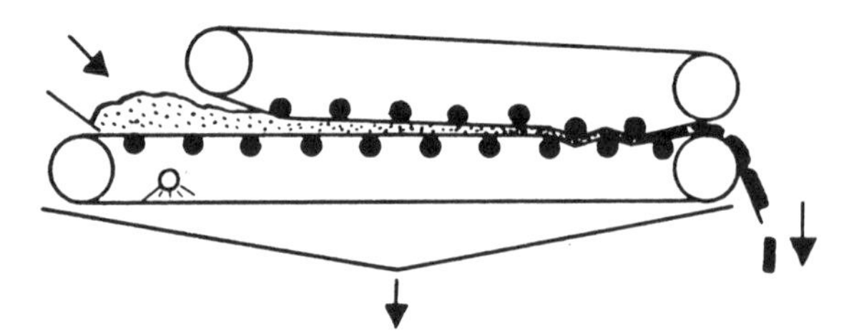

Figure 9.16 Belt press filter.

can be regulated. The sludge, conditioned by about 250 g of polyelectrolyte/m^3, is drawn between the belts, drained in the draining zone, pressurized, and dewatered. The dewatering capability of the press is enhanced by applying a shear force that is created in the third portion of the belt by adjusting the supporting and pressure rollers in such a way that the belts and the sludge between them follow an S-shape path. At the exit end, the sludge is removed by a scraper and the belt is sprayed with wash water. Imhoff[48] suggested a typical loading of the filter as being about 0.8 m^3 of sludge per m^2 of the filter area in 1 hr or 40 kg of dry sludge solids/(m^2-h) (8 lb/ft^2/hr).

Vacuum Filters have been very popular in the United States for a long period of time. The vacuum filter consists of a drum enveloped by a filter cloth. The drum is partially submerged in liquid sludge (Fig. 9.17), which is usually conditioned by chemicals. In a number of designs, the inside space of the drum is divided into sections running the length of the filter drum and a vacuum is generated in each individual cell. The drum revolves slowly ($\frac{1}{8}$ to 1 rpm), partly submerged (15–40%) in a sludge reservoir that can be equipped with an agitator. The pressure difference between the outside and inside of the drum forces the sludge to adhere to the filter cloth and form a 4- to 10-mm-thick continuous layer. During the drying time of the rotation cycle, the sludge liquor (filtrate) is drawn from the layer through the cloth into the sections inside the drum. At the end of the drying cycle just before the cloth is again submerged in the sludge reservoir, the dried sludge cake is scraped from the cloth by a blade and falls on a belt conveyor. If necessary, slight pressure is applied to the section reaching the blade, which lifts the cake from the cloth for easier removal. After the cake is removed, the cloth is washed and the cycle is repeated.

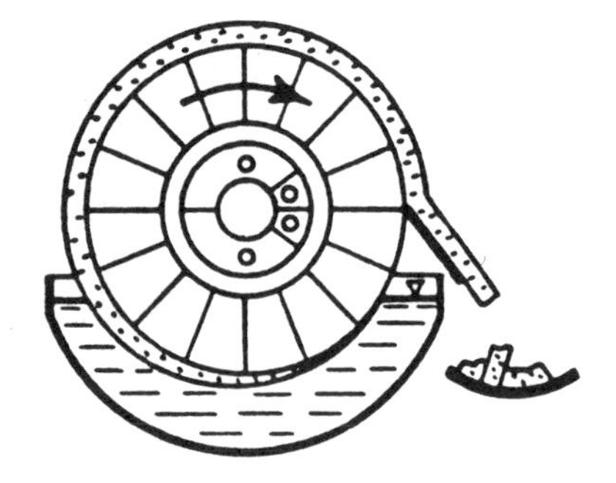

Figure 9.17 Vacuum filter.

The typical moisture content of the cake is 82% (18% solids) for activated sludge, and 70% for raw or digested primary sludge. Since the new environmental regulations for landfill disposal in the United States require that the sludge has a minimum of 20% solids, vacuum filters are often unsuitable and are being replaced by pressure filters at a number of locations. Other disadvantages of vacuum filters are (1) a large volume of wash water is required that subsequently becomes highly polluted and must be recycled for treatment, (2) the filtrate has relatively high BOD and solids content and must also be returned for treatment, and (3) the solids content of the cake is not very high. Continuous automatic operation is the main advantage of vacuum filters.

Filter loadings range from 10 to 30 kg of dry sludge solids per m^2/hr (2–6 lb/ft^2/hr) and the filtrate volume is about 200 liters/(m^2-hr). The optimum loading increases with the dry solids content of the incoming sludge and with the degree of digestion.[49,50] Filter loadings and other design parameters (drying time vs. submergence time, type of filter cloth, dosages of conditioning chemicals) are usually determined by laboratory experiment using Büchner funnel tests and a leaf test apparatus.[4,51] Ferric chloride and/or organic polyelectrolytes are the most common conditioning chemicals.

Heat Drying of sludges after vacuum and pressure filtration is commonly needed to reduce moisture levels prior to final disposal and to reduce odor problems with undigested sludges. The most common method of sludge drying is in a rotary dryer in which hot air flows countercurrent to the sludge movement. The temperature of the air entering the dryer is about 600°C and the exiting air has a temperature of about 300°C. The moisture content of the entering sludge should be about 50% in order to avoid operational difficulties. Inasmuch as such moisture levels are not easily achieved in typical mechanical dewatering devices, a portion of the dry effluent sludge should be recycled. The effluent sludge has a moisture content less than 10% and it can be sold as fertilizer (e.g., the Milorganite made from waste-activated sludge from Milwaukee) or incinerated.

It should be noted that the methane gas from digesters is a convenient source of heat for drying.

9.7 INCINERATION[52]

Types of Incineration

When the fertilizer value (nutrient content) of the sludge or the presence of toxic metals does not warrant agricultural use, sludge can be incinerated. Since the evaporation of water during incineration uses a substantial amount of heat, good dewatering is essential if the sludge is to be burned without additional fuel. Raw dewatered sludge has the highest calorific (heat) value, usually enough for burning after dewatering without supplemental fuel. Digested sludge loses about half of its calorific value during digestion and it may be very difficult to dewater it to the level needed for self-sustained burning. However, the volume of the digestion gas produced is usually more than adequate for fuel supplementation. From these considerations, it can be concluded that digestion is neither necessary nor economical if incineration is considered as a sludge disposal alternative.

A sludge incinerator requires a constant and stable supply of fuel. Therefore, sludge storage is needed, which could cause odor problems if raw dewatered sludge is burned. Digesters—if installed—and/or air drying beds provide the needed storage.

The following three factors can be used as guidance in determining whether the sludge can burn on its own and produce energy or whether an additional fuel supplement is needed: (1) the moisture content of the sludge, (2) the fuel (calorific) value of the dry sludge solids, and (3) the amount of excess air above that needed for incineration. High fuel content, low excess air, and low moisture content make incineration more energy efficient. Generally, heat drying precedes or is a part of the incineration process.

Three types of incineration devices are available: (1) rotary kiln furnaces, (2) fluidized bed incinerators, and (3) multiple hearth furnaces.

In the *rotary kiln furnace*, it is possible to burn large pieces of organic materials. The off gases must be scrubbed to reduce the odor, however.

Fluidized bed incinerators (Fig. 9.18) use a bed of fluidized sand as a heat reservoir inside the reactor to promote uniform and thorough combustion. The sludge is incinerated at a temperature of 815°C (1 500°F). The feeding is from the top of the incinerator and the hot air from the combustion flows upward and

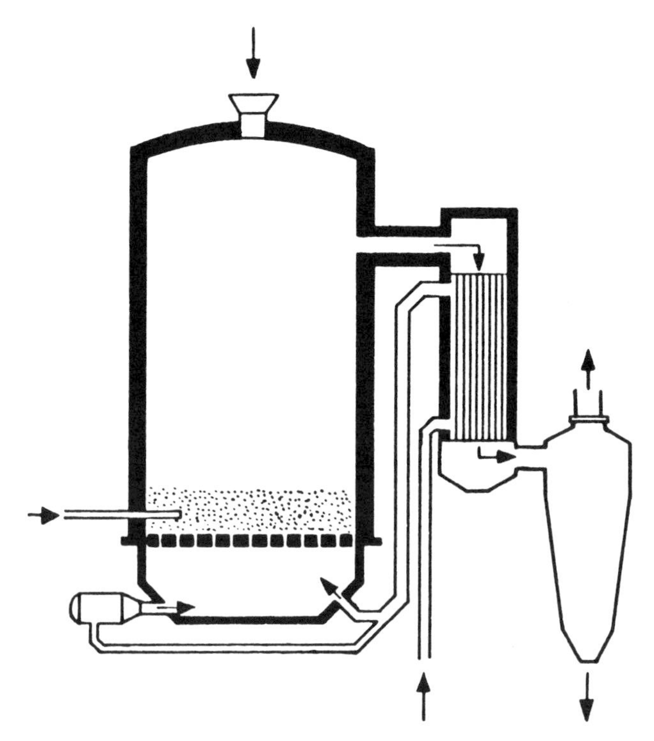

Figure 9.18 Fluidized bed incinerator.

fluidizes the mixture of the sand and sludge. Because of the uniform, high temperature, burning is more complete and odors are minimal. Most of the burned sludge residue is incorporated in the fly ash that is separated from the off gas by cyclones and by a wet scrubber using treated effluent. Electrostatic precipitators or fly ash lagoons have also been used for fly ash separation. The remaining heat of the exhaust gas is used to preheat the new air entering into the incineration process. Large pieces of organic materials are not suitable for burning in this type of incineration and must be pulverized.[53]

The most common method of sludge incineration is using the *multiple heart furnaces* (Fig. 9.19) that were originally introduced in the United States by Nichols decades ago. Dewatered sludge cake is fed into the uppermost hearth and moved (rabbled) from hearth to hearth by plows or teeth attached to hollow arms rotated by a central hollow shaft. The upper hearths serve for drying but, generally, the sludge dries on the first two hearts, then it ignites, burns, and cools on the remaining hearths. The temperature rises progressively to 925°C (1 700°F) and then it cools to about 315°C (600°F). Ash is removed from the lowest hearth. The exhaust gases from the furnace pass through a preheater or recuperator to the stack. Shaft and rabble arms are air cooled to keep them from burning. The air enters the shaft at the bottom and passes through the preheater in which it is heated before being blown into the lower hearth of the incinerator. Auxiliary fuel, if needed, is injected into the furnace on one side of the upper hearths. Air pollution control by scrubbing or electrostatic precipitators is required and the odor should be controlled by postburning the off gases.

A furnace 4 m in diameter and 5 m high can daily burn about 50 tons of dewatered sludge, however, capacities of furnaces range from 2.5 to 100 tons per

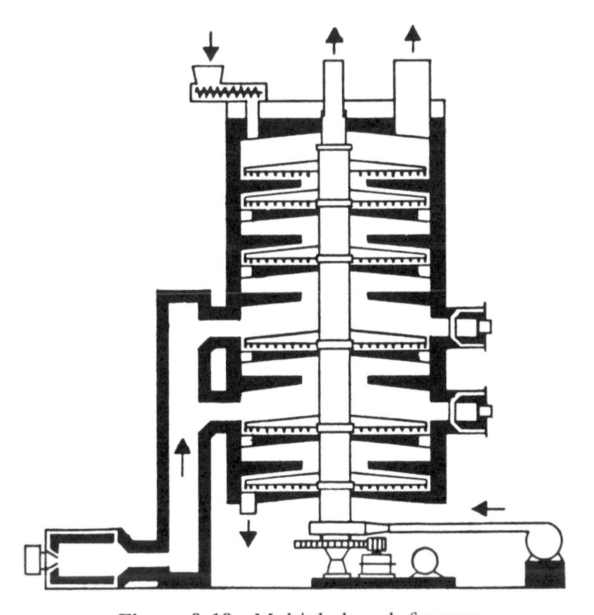

Figure 9.19 Multiple hearth furnace.

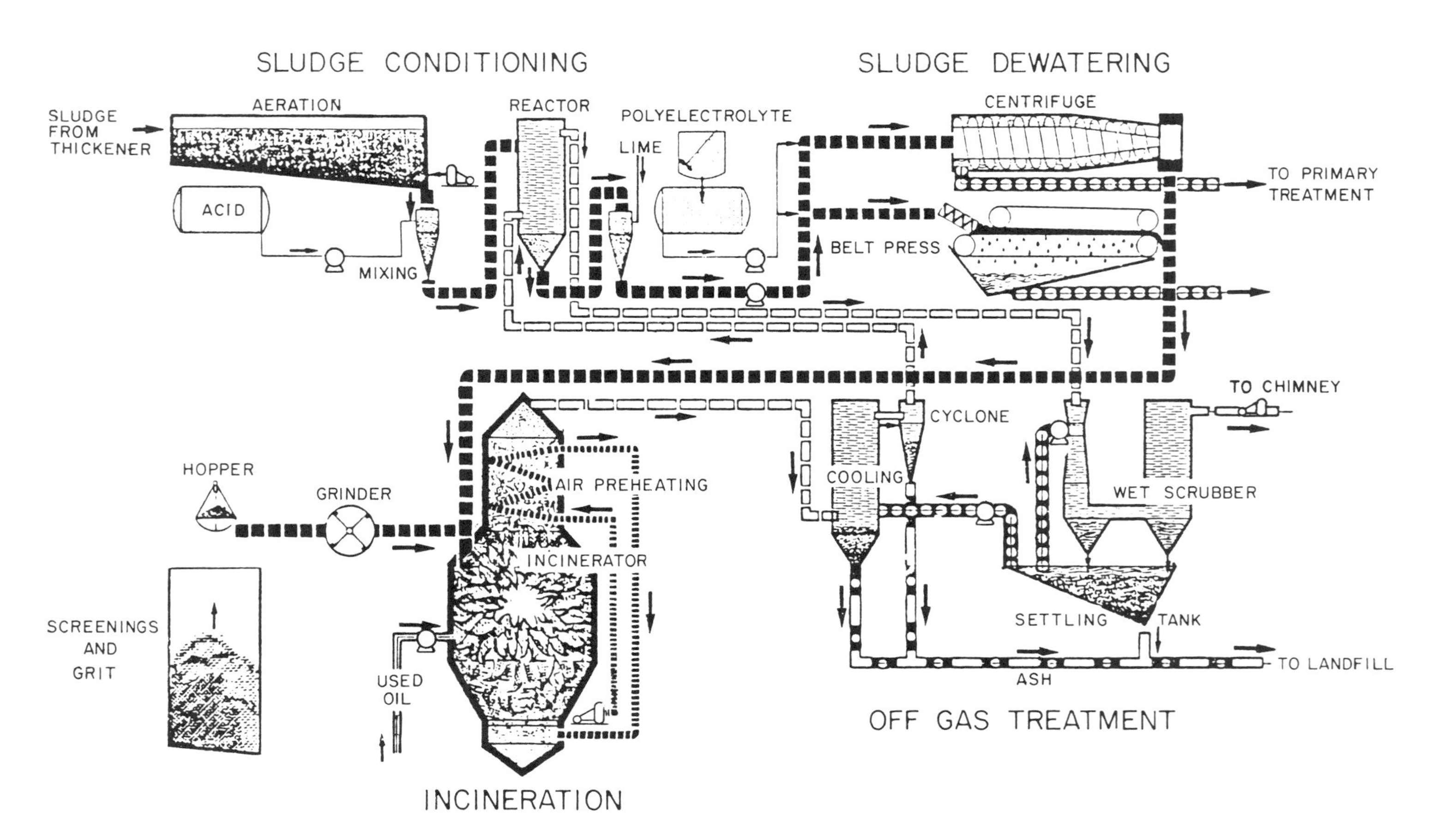

Figure 9.20 Flow schematics of sludge conditioning and incineration.

day of dry sludge solids.[47] The fuel requirement is about 2.3 liters of heating oil per ton of sludge and the mass of ash is about 15% of the original mass of sludge. The iron salts that were added during dewatering can be recovered from the ash.[54]

A block diagram of the unit processes employed in sludge dewatering and incineration is shown in Figure 9.20. The sludge is first stored and equalized in an aerated storage tank. In the next step, the sludge is heated by off gases and conditioning chemicals are also added. For dewatering, either a centrifuge or a belt press can be selected. The dewatered sludge is then mixed with solids from the grit chamber and the screens and the mixture is incinerated. Off gases from the incinerator are cooled, the fly ash is separated by cyclones, and the residues in the off gas are scrubbed.

Wet Oxidation of sludges is a process based on the principle that any substance can be oxidized in the presence of water provided that optimum temperature and pressure are applied. The wet oxidation (also called wet incineration) process operates at temperatures of 150–200°C (300–400°F) and pressures between 70 and 110 atm (100–1 700 psi). No prior dewatering is required and the temperature is much lower than that in the conventional incineration process, hence, fly ash, dust, sulfur dioxide, and nitric oxides (acid rain) emissions are not a problem. The major drawback of the process is its cost.[55,56]

Air Pollution Control

The designers and operators of a sludge incineration system must face the fact that all systems with the exception of wet oxidation create air pollution and odors. Wet scrubbers or electrostatic precipitators can minimize the problem, however, they cannot completely eliminate it. This may limit the applicability of sludge incineration in areas that already have an existing air pollution problem unless other air pollution sources reduce their emissions below the air pollution control limits. In the United States such plans are called the ''bubble'' air pollution control tradeoff.

Residual air pollution solids from incinerated sludges may contain heavy metals, however, it is believed that most heavy metals are incorporated in the fly ash and scrubbed from the effluent gases. The effluent (stack) gases contain sulfur dioxide, nitric oxides, and carbon monoxide and dioxide. Sulfur and nitric oxides are known causes of acid rain, however, the sulfur content of municipal sludges is generally low.

9.8 FINAL DISPOSAL OF SLUDGES

The ultimate disposal of sludges is a most pervasive problem because of the diminishing availability of suitable sites. Even incineration still produces fly ash residue in quantities that are significant (about 15% of the original volume of dewatered sludge). Eventually, land or sea disposal is the ultimate final step. It should be noted, however, that in the United States, ocean dumping is now possible only with a variance (a special permit based on an exception) and State regulations are making land disposal more difficult.

In comparison with commercially available fertilizers, the agricultural value of sludge is generally low. The nitrogen, phosphorus, and potassium contents of a typical stabilized municipal wastewater sludge are 3, 2.3, and 0.3% as compared to 5, 10, and 10% of these components in a typical commercial agricultural fertilizer.[57]

The agricultural worth of sludge, similar to manure, lies in its organic content and capability of forming soil humus. The heavy metal content of sludges is a major concern because metals can accumulate and concentrate in soil and crops. It has been documented that the removal of 1 mg/liter of a metal from a wastewater will result in about 10 000 mg/kg of the metal in sludge solids. Some metals can enter the food chain and several of them can bioaccumulate. A number of states in the United States have established strict standards on the disposal of metals on farm land. Similar guidelines are being enforced in Germany and many other countries.

In addition to metals, bacteria, viruses, and helminths in the sludge can also be a problem.[58]

Land Disposal of Wet or Dewatered Sludge

Guidelines and Limitations. If sludge is used as a fertilizer and/or soil conditioner on agricultural fields a decision should be made as to whether or not the sludge should be digested. In the digestion process, about 50–60% of the nitrogen is lost from sludge solids and become dissolved as ammonia in the sludge liquor. Therefore, the fertilizing value of raw sludge is greater than that of digested sludge. However, this does not preclude digestion. Milorganite from Milwaukee, Wisconsin is produced from dried and processed waste-activated sludge that retains most of its fertilizing nutrients. It has a granular texture and minimum odor problems. Production is energy consuming because fuel is needed for drying and heat treatment and no digestion gas is available. On the other hand, raw or only partially stabilized dewatered primary or mixed sludge would cause a number of problems in the soil. It forms greasy and sticky morsels, contains weed seeds, is odorous, and may contain pathogenic microorganisms. Generally, most land applications use digested (aerobically or anaerobically) sludge. If sludge is digested anaerobically, the methane production benefit is retained, however, the sludge contains less nitrogen.

The metallic content of sludges, primarily that of cadmium, lead, and zinc, somewhat limits the application rates and suitability of soils for sludge disposal. In a majority of cases, pretreatment of wastewater and removal of metals on the site of their origin should be considered. Nearly all sewage sludges contain cadmium, zinc, and lead in quantities that will increase the total concentrations of these metals in soils. The metals can then become available to crops and accumulate in the plant tissue. The concentrations of cadmium and zinc in plant tissue generally increase with a decrease of soil pH.[59]

In the United States, guidelines for land application of sludges are given by the Environmental Protection Agency, the Food and Drug Administration, and the Department of Agriculture.[60] The guidelines recommend the following:

1. Annual application rates of cadmium contained in the sludge should not exceed 0.5 kg/ha and there is a cumulative limit that should not be exceeded.
2. pH in the topsoil zone should not be less than 6.5 at time of application.
3. Sludges containing more than 10 mg/kg of polychlorinated biphenyls (PCBs) should not be applied.
4. Sludge should be treated by a pathogen reduction process before soil application. A waiting period of 12–18 months before a crop is grown may be required, depending on prior sludge processing and disinfection.
5. The sludge should not contain more than 25 mg/kg (dry weight basis) of cadmium and 1 000 mg/kg of lead.
6. Edible crops should not be planted within a period of 12–18 months after application of sludge.
7. Sludges should be applied directly to the soil and not to any human food crops.
8. The field must be regularly monitored.
9. Plants grown should be of the type that do not accumulate heavy metals.

German guidelines[61] require that the following limits on heavy metal concentrations should be observed:

	Heavy Metal Content (mg/kg) of Dry Solids	
	Sludge	Soil
Lead	1 200	100
Cadmium	20	3
Chromium	1 200	100
Copper	1 200	100
Nickel	200	50
Mercury	25	2
Zinc	5 000	300

A survey of environmental regulations dealing with sludge disposal on land in different countries has been compiled by Vesilind.[22] He concluded that most of the environmental concerns and regulations are surprisingly similar in many countries.

Sludge can be disposed of on land, either wet or dewatered, and plowed under immediately after application. It should not be applied during the growing season to root crops (carrots, beets, etc.) or low, leafy vegetables. Examples of modern controlled land disposal of digested wet sludges include the Maple Lodge treatment plant near London, England,[62] the Viersen treatment plant in the Niers river watershed in Germany,[63] and land disposal of sludges from the Metropolitan Sanitary District of Greater Chicago in Fulton County, Illinois.

Sludge Disinfection Methods. Disinfection for land disposal includes lime treatment, prepasteurization, or postpasteurization in combination with anaerobic stabilization and heat treatment.[26] Pasteurization is a process in which sludge is heated to about 70°C for a period of about 30 min, during which pathogenic organisms are effectively killed. In addition, weed seeds and worms will also be destroyed. This method has been used in Switzerland and other European countries. Gamma radiation has also been developed and used by the U.S. Department of Energy although the shielding requirements are a problem.

Composting. The composting of sludges has been utilized to increase the acceptability of the application of sludges onto agricultural lands. Composting has been popular in Europe and also in North America inasmuch as several municipalities have added composting facilities to their plants.

Composting is a process in which organic matter, such as dry sludge, undergoes aerobic oxidation and degradation to a stable, humus-like end product. A significant portion of sludge organics (20–30%) will be converted to carbon dioxide and water.

There are a number of composting methods. The best results have been obtained with cocomposting the dried sludge with a bulking material, either wood chips or municipal refuse. During composting, the sludge bulking agent mixture is placed in long rows (windrows), or piles, that are covered by a layer of screened compost for insulation and odor control. The piles are turned periodically to expose the organic materials to the ambient air.

The composting process lasts about 3–4 weeks. During this time, the mix is aerated and biological processes generate enough heat to raise the temperature to the thermophilic range. Therefore, the sludge becomes pasteurized and the pathogenic microorganisms are destroyed.

The oxygen needed for decomposition is supplied by (1) periodical mechanical turning over and mixing the piles, and (2) by forced aeration.

After composting, the material is usually cured for about 30 days. During this phase, further decomposition, stabilization, pathogens destruction, and degasification take place, which make the compost more marketable. The compost can then be used as a low-grade fertilizer–soil conditioner.

The nitrogen content of composted primary sludge is about 1.5% and that of phosphorus around 1%. The expected nitrogen and phosphorus contents of a composted digested municipal sludge are around 0.6 and 1%, respectively.[64]

If woodchips are used for bulking, they can be separated from the rest of the compost and recycled.

Municipal refuse, in addition to bulk, provides carbon and the mixture should have a carbon to nitrogen ratio of about 15 to 1. The feed sludge should have a moisture of about 70%.[64]

The heat treatment discussed in Section 9.6 provides a dry, disinfected sludge material that can be used as a fertilizer. The sludge temperature in this process is elevated to about 300°C, which provides a safe kill of pathogens. The sludge in its final form has a moisture content of about 10%, 3–6% of nitrogen, and 2–3% of phosphorus. Dried sludge is granular, has a slight fleeting odor, does not contain viable weed seeds, and blends readily with soil. Its appreciable nitrogen content and a slow release of nitrogen into soils make it an ideal fertilizer for golf courses and lawns. In large operations, these sludges can be bagged and commercially sold as fertilizer.

Landfill Disposal

Sanitary landfilling is an engineering operation requiring all necessary environmental precautions, including that of preventing groundwater contamination. This distinguishes sanitary landfill from uncontrolled dumping. The sludge is disposed either as a dewatered cake of stabilized sludge or as ash from incineration. Sludge should be covered daily by soil to prevent nuisance and odor problems. A codisposal with municipal refuse is common inasmuch as the addition of wastewater sludges to a sanitary refuse landfill has a beneficial effect on the anaerobic decomposition occurring therein.

Problems with landfilling include leachate and gas (methane) formation. Leachate is a heavily polluted liquid that can enter groundwater zones if the landfill is located and operated improperly. Gas from the landfills, primarily methane, can be collected and used as a source of energy, otherwise burners and relief pipes must be installed to prevent explosions and fires.

Presently, landfill disposal of dewatered sludge is still the most widely used method. However, with limitations on the availability of suitable landfill sites, this method is becoming less and less feasible in many areas.

Ocean Disposal is obviously available only to coastal cities or cities with a navigable waterway connecting them with the sea. However, this method has been subjected to substantial criticism in environmental and public circles and it is expected that it will be banned in the United States, although variances have been given.

The sludge is dumped in a liquid (thickened) form. Ocean currents and the probability of shoreline pollution are the major concerns.

Selecting a sludge disposal method is a complex process that may be tedious and requires considerable effort. It should also be noted that in addition to the economy of sludge disposal, other factors can play a major role. The most important decision factors are environmental considerations, availability of disposal sites and public acceptance. A detailed environmental impact statement must be prepared and accepted by regulatory agencies and the public.

The cost of sludge disposal is obviously related to the selected sludge dewatering, stabilization, and disposal processes. Usually about one-half of the total cost of wastewater treatment is attributed to sludge treatment and disposal. If one compares the cost of the three most typical sludge disposal alternatives, (1) disposal of liquid sludge on agricultural lands (LD), (2) dewatering and landfill disposal (DL), and (3) dewatering and incineration (DI), then the cost of the three alternatives would be in approximate proportions of LD : DL : DI = 1 : 2 : 4.

REFERENCES

1. W. W. Eckenfelder, Jr., *Water Quality Engineering for Practicing Engineers.* Barnes & Noble, New York, 1970.
2. K. R. Imhoff, *Münchner Beitrage. Abwasser-, Fischerei, Flußbiologie* **13,** 72 (1966).
3. *Standard Methods for the Examination of Waters and Wastewaters.* APHA, AWWA, WPCF, Washington, DC, 1985.
4. C. A. Adams, D. L. Ford, and W. W. Eckenfelder, Jr., *Development of Design and Operational Criteria for Waste-water Treatment.* CBI Publ., Boston, Mass., 1981.
5. W. J. Katz, *J. Water Pollut. Control Fed.* **36,** 407 (1964).
6. W. J. Katz, *J. Water Pollut. Control Fed.* **39,** 946 (1967).
7. W. N. Torpey, Trans. *Am. Soc. Civ. Eng,* **119,** 443 (1954).
8. *Sludge Treatment and Disposal,* Vols. 1 and 2, EPA Rep. No. 625/4-78-012. U.S. Environ. Prot. Agency, Cincinnati, OH, 1978.
9. U. Loll, *Korresp. Abwasser* **31,** 934 (1984).
10. A. M. Busswell, *Sewage Ind. Wastes* **29,** 717 (1957).
11. W. W. Eckenfelder, Jr., *Water Sewage Works* **114,** 207 (1967).
12. P. L. McCarty, *Public Works* **95,** No. 9, 109–112, No. 10, 123–112, No. 11, 91–95, No. 12, 95–99, (1964).
13. D. R. Rowe, *Water Sewage Works* **118** (3), 74–76 (1971).
14. R. E. Speece and P. L. McCarty, *Adv. Water Pollution Research,* Proc. 1st Int. Conf. IAWPR Vol. 2, 305–322 (1964).
15. G. M. Fair and E. W. Moore, *Sewage Works J.* **10,** 3 (1937).
16. J. Maly and H. Fadrus, *J. Water Pollut Control Fed.* **43,** 641 (1971).
17. J. W. Masseli, N. W. Masseli, and M. G. Burford, *J. Water Pollut. Control Fed.* **39,** 1369 (1967).
18. K. Imhoff, *Taschenbuch der Stadtentwässerung,* 26th ed. Oldenbourg Publ. Co., Munich, FRG, 1985.
19. K. Imhoff and G. M. Fair, *Sewage Treatment,* 2nd. ed. Wiley, New York, 1956.
20. G. P. Noone, C. E. Brade, and J. Whyley, in *Sewage Sludge Stabilization and Disinfection* (A. Bruce, ed.). Ellis Horwood Ltd., Chichester, England, 1984.
21. *Anaerobic Sludge Digestion-Operating Manual,* EPA Rep. No. 1430/9-76-001. U.S. Environ. Prot. Agency, Cincinnati, OH, 1976.
22. P. A. Vesilind, *Treatment and Disposal of Wastewater Sludges.* Ann Arbor Sci. Publ., Ann Arbor, MI, 1979.
23. A. P. Fisicheli, *J. Water Pollut. Control Fed.* **42,** 11, (1970).
24. *Sludge Stabilization-Manual of Practice, Facilities* Development-9. Water Pollution Control Federation, Washington, DC, 1985.
25. C. B. Townend, *Proc. Inst. Sewage Purif.,* Part 2, (1956).

26. A. M. Bruce (ed.), *Sewage Sludge Stabilization and Disinfection.* Ellis Horwood Ltd., Chichester, England, 1984.
27. S. G. Burgess et al., *Proc. Inst. Sew. Purif.* **24,** (1964).
28. A. J. Fisher and R. A. Greene, *Sewage Works J.* **17,** 718 (1945).
29. W. F. Garber, *J. Water Pollut. Control Fed.* **54,** 1170 (1982).
30. R. R. Rimkus, J. M. Ryan, and E. J. Cook, *J. Water Pollut. Control Fed.* **54,** 1447 (1982).
31. K. R. Imhoff, *GWF, das Gas-und Wasserfach: Wasser/Abwasser* **116,** 216 (1975).
32. *Dewatering Municipal Wastewater Sludges, Design Manual,* EPA Rep. No. 625/1-82-014. U.S. Environ. Prot. Agency, Cincinnati, OH, 1982.
33. L. Spinosa, *Waste Manag. Res.* **3,** 389–398 (1985).
34. K. Imhoff, *Schweiz. Z. Hydrol.* **22** (1), 475 (1960).
35. J. T. Calvert, *Civ. Eng.* (br.), p. 783 (1961).
36. C. B. Townend, *Proc. Inst. Sewage Purif.*, Part 4, p. 273 1961.
37. K. R. Imhoff, *GWF, das Gas-und Wasserfach* **105,** 710 (1964).
38. M. A. Kershaw, *J. Water Pollut Control Fed.* **37,** 674 (1965).
39. J. D. Swanwick, *Theoretical and Practical Aspects of Sludge Dewatering,* 2nd Eur. Sewage Refuse Symp. EAS, Munich, FRG, 1968.
40. N. S. Bubbis, *J. Water Pollut Control Fed.* **34,** 830 (1962).
41. R. B. Brooks, *Inst. Wat. Poll. Control* (br.), p. 92 (1970).
42. S. C. Reed, J. Bouzman, and W. Medding, *J. Water Pollut. Control Fed.* **58,** 911 (1986).
43. B. Paulsrud and A. S. Eikumn, in *Sewage Sludge Stabilization and Disinfection* (A. Bruce, ed.). Ellis Horwood Ltd., Chichester, England, 1984.
44. G. Mininni and L. Spinosa, *Phoenix-International,* **3**(3), 24–28 (1986).
45. J. R. Townsend, *Water Wastewater Eng.*, Nos. 11 and 12, (1966).
46. F. W. Keith and T. H. Little, *Chem. Eng. Prog.* **65** (11), 77–80 (1969).
47. *Process Design Manual for Sludge Treatment and Disposal,* EPA Rep. No. 625/1-74-006. U.S. Environ. Prot. Agency, Cincinnati, OH, 1974.
48. K. R. Imhoff, *Water Res.* **6,** 515, (1972).
49. A. L. Genter, *Sewage Ind. Wastes* **28,** 829 (1956).
50. P. L. McCarty, *J. Water Pollut. Control Fed.* **38,** 493 (1966).
51. R. C. Baskerville and R. S. Gale, *Water Pollut. Control* (*Maidstone eng.*), **68,** 233–241 (1969).
52. *Municipal Wastewater Sludge Combustion Technology,* EPA Rep. No. 625/4-85-015. U.S. Environ. Prot. Agency, Cincinnati, OH, 1985.
53. R. Dickens, B. Wallis, and J. Arundel, *J. Inst. Water Pollution Control* **79** (4), 431–441 (1980).
54. M. B. Owen, *J. Sanit. Eng. Div., Am. Soc. Civ. Eng.* **83,** 1172 (1957).
55. G. M. Teletzke, W. B. Gitchel, D. G. Diddams, and C. A. Hoffman, *J. Water Pollut. Control Fed.* **39,** 994 (1967).
56. E. Hurwitz and W. A. Dundas, *J. Water Pollut. Control Fed.* **32,** 918 (1960).
57. *Use and Disposal of Municipal Wastewater Sludges,* EPA Rep. No. 625/10-84-003. U.S. Environ. Prot. Agency, Cincinnati, OH, 1984.

58. *Health Effects of Land Application of Municipal Sludges,* EPA Rep. No. 600/1-85-015. U.S. Environ. Prot. Agency, Washington, DC, 1985.
59. Council for Agricultural Science and Technology, *Effects of Sewage Sludge on Cadmium and Zinc Content of Crops,* EPA Rep. No. 600/8-81-003. U.S. Environ. Prot. Agency, Washington, DC, 1981.
60. FDA and USDA, *Land Application of Municipal Sludge,* EPA Rep. No. 625/71-83-016. U.S. Environ. Prot. Agency, Washington, DC, 1983.
61. *Bundesgesetzblatt,* Part 1, p. 734 (1982).
62. R. C. Mertz, *Water Sewage Works,* p. 489 (1959).
63. K. Imhoff and K. R. Imhoff, *GWF, das Gas-und Wasserfach: Wasser/Abwasser,* **125,** 710 (1984).
64. C. A. Gordon, *J. Sanit. Eng. Div., Am. Soc. Civ. Eng.* **84**(SA6), 1852 (1958).

CHAPTER 10

Industrial Wastewater Treatment and Pretreatment

10.1 GENERAL CONSIDERATIONS[1-10]

Types of Industrial Wastewater Streams

Each industry produces several types of wastewater streams. In general, they can be categorized into (1) sanitary sewage from sanitary and washing facilities for the employees, (2) cooling water, (3) wastewater from washing raw materials, products, and production areas, (4) process (production line) wastewater, and (5) surface runoff from the premises.

Sanitary sewage is not very different from ordinary municipal sewage and can be disposed of directly into a nearby municipal sewer. Cooling waters represent the largest volume of industrial wastewaters and ordinarily they are relatively clean. However, they add heat to receiving waters. Most industries recycle their cooling waters, which requires incorporation of cooling towers in the system. Concentrated water from the cooling system, called ''blow-down,'' can be quite polluted by salts and anticorrosion additives and commonly requires physical–chemical pretreatment. Wastewater from washing operations are second in volume after cooling water. This type of wastewater can become highly polluted by suspended solids and other types of pollutants. Washwater can be recycled after treatment. Production line wastes are commonly polluted and require treatment. Surface runoff from many industries, including refineries, chemical industries, foundries, smelting operations, steel mills, and a number of others often contains as much pollution as the raw or partially treated wastewater and, consequently, surface water from areas outside the industry should not be allowed to enter the premises. Usually a retention basin with subsequent treatment is provided for surface runoff generated on the premises.

Generally speaking, water conservation and recovery of useful waste materials from wastewater are the governing principles of industrial waste control and the most effective method of reducing the waste load. Both call for engineering skills and a thorough knowledge of the manufacturing process.

Residual industrial wastes can be treated either by the industry itself in its own treatment plant, including effluent outfall into the receiving water body, or jointly with the municipal wastewater. In the former case, the discharge is subjected to the effluent standards that, in the United States, are dictated by the National Pollution Discharge Elimination Permit (NPDES) issued to the industry by a state regulatory agency (see also Section 4.1 for the discussion of the NPDES system). In the latter case, a municipality commonly has an ordinance governing the discharge of industrial wastewaters into its sewer system, which establishes maximum concentrations for certain toxic compounds, prohibits the discharge of dangerous compounds such as gasoline, and sets a rate schedule of payments (user charges) for discharge into municipal sewers of common pollutants, including suspended solids, BOD, and ammonia.

In the Ruhr area of Germany, the river basin agencies (Verbände) have a similar authority of imposing effluent limitations on industries, providing joint treatment, and setting user charges for disposal. The river basin agencies have full authority for both the wastewater collection and disposal from municipal and industrial sources, and for protection of receiving water bodies.

The brief discussion of types of industrial wastewater streams and appropriate treatment that follows is introductory. More detailed descriptions of each individual industrial wastewater and treatment technology can be found in the U.S. EPA monographs and in the references mentioned herein.

10.2 JOINT DISPOSAL AND TREATMENT WITH MUNICIPAL WASTEWATER

Limitations. The discharge of wastewater into public sewer systems relieves industry of the responsibility and technical problems of wastewater treatment and places the discharge under public control.

Most industrial wastewaters are amenable to treatment with domestic sewage, however, grease and oils, hot liquids ($>40°C$), gasoline and flammable solvents and other pollutants that create a fire or explosion hazard, and toxic metals and other toxic substances that can be damaging to the treatment process or treatment facilities must be excluded and removed by pretreatment at the site of origin. In addition, slug (shock) discharges of such magnitudes that could cause upsets in the treatment process and subsequent loss of treatment efficiency should be stored and equalized, or eliminated.[11]

Toxic industrial pollutants that can damage biological treatment or sludge stabilization include salts of copper, arsenic, cyanides, and chromium. Maximum concentrations of toxic substances, if not known, should be determined by toxicity bioassays. Other industrial wastewater components such as coal tar, wastepaper and cellulose, lanoline, phenols, and metallic salts should be removed and recovered from the wastewater by industries.

Secondary (biological) public treatment plants can effectively remove a number of toxic compounds provided that they are not present in concentrations that would be inhibitory to the treatment process (Table 10.1).

In addition to the pollutants listed in Table 10.1, a large number of organic chemicals have been designated as toxic and harmful to treatment. The list of

TABLE 10.1 Limiting Concentrations and Removals of Some Toxic Materials for a Biological Treatment Plant[a]

Pollutant	Activated Sludge or Trickling Filter (mg/liter)	Anaerobic Digestion (mg/liter)	Nitrification (mg/liter)	Removed (%)
Ammonia	500	1 500	NA[b]	10–80[c]
Arsenic	0.1	1.6	NA	NA
Cadmium	0.02	100	NA	50
Chromium(VI)	1–10	5–50	0.25	70
Copper	1.0	100	<0.5	80
Cyanide	1–5	4	0.3	60
Lead	0.1	NA	0.5	55
Mercury	1–5	1 365	NA	NA
Nickel	2.5	2	0.5	30
Sodium	NA	3 500	NA	0
Zinc	0.3	50	<0.5	75

[a] After U.S. EPA.[11]
[b] NA, not available or uncertain.
[c] Depending on the type of treatment and sludge disposal.

chemicals that must be excluded includes pesticides, solvents, and polychlorinated biphenyls.[11]

The type of joint treatment depends on the type of industrial discharge and its proportion of municipal (domestic) sewage. In industrialized areas in which industrial wastewater contributions are high, chemical pretreatment followed by biological treatment may be needed.

In the United States, two types of federal pretreatment standards are in place. The first group of standards defines prohibited discharges such as those listed in the preceding section and the second type of pretreatment regulations are "categorical standards," which are focused on each individual group of industries. This group of standards lists maximum effluent limitations on the concentrations or quantities of pollutants that can be discharged to public treatment works by new or existing industrial users.[12]

The type of treatment also depends on the type of pollutant and its character (mineral or organic, biodegradable or nonbiodegradable). The following compounds are of concern to a designer or operator of a joint treatment facility:

Soluble Inorganic Salts, such as those from potassium plants, mine water pumping, and steel mills. Mine water and wastewater from steel mills can also be acid. Such discharges can damage receiving water bodies, cause corrosion, and, in higher concentrations, adversely affect the biota of the treatment plant. Their concentrations are not affected by conventional treatment and the only feasible and economical way of control is by dilution. For this purpose, retention basins may be installed for storage of the wastewater during low flows (plus neutralization), from which the wastewater is released during high flows into a receiving water body. Storage and release depend on the concentration of the inorganic pollutants in the receiving water body upstream from the effluent discharge and on the in-stream standard for the given substance. Because conventional public treatment works cannot remove salts and because of the potential damage, wastewaters containing excessive amounts of salts should not be discharged into public sewers.

Acids can be discharged into streams or sewers only after neutralization. A crushed limestone filter bed is effective and economical for neutralization of acids containing hydrochloric or nitric acid. Neutralization of sulfuric acid can be accomplished by adding a lime slurry (5–10% of $Ca(OH)_2$ suspension in water). Limestone filters are not feasible for neutralization of wastes containing sulfuric acid because of gypsum ($CaSO_4$) formation during neutralization.

Inorganic Sludges, for example, from coal mines, steel mills, and chemical plants, or fly ash from power plant scrubbers, are removed in clarifiers or in lagoons. The settled sludge from clarifiers is then pumped onto drying beds, into sludge lagoons, or mechanically dewatered. Flowthrough sludge lagoons are designed to hold sludge quantities from several years of operation. After the lagoon is filled with sludge, a new lagoon must be built, and the filled lagoon is allowed to stay for several years, after which the land can be reclaimed. Effluent water from the lagoon can be reused. Organic biodegradable solids should not be discharged into sludge lagoons because they may cause odor nuisance and gas development.

Regulations on land disposal of industrial sludges in the United States and other developed countries has resulted in the phasing out of sludge lagoons. Industrial producers of inorganic sludges are now relying more on mechanical dewatering and off-site disposal of dewatered sludges into an approved landfill.

Excessive amounts of inorganic sludges may cause problems in sewers and increase the inorganic content of sludges, which, consequently, would reduce the sludge digestibility. Most of the municipal ordinances restrict suspended solids concentrations and assess a surcharge for concentrations exceeding those typical for municipal sewage (200–300 mg/liter).

Organic Industrial Pollutants in very high concentrations are also not acceptable for joint treatment and their concentrations must be reduced by on-site pretreatment. Examples of high-strength organic wastes include those from meat packing, dairies, and other food processing industries and leachate from landfills. If a joint treatment plant is specially designed to accept such wastes no problems should be encountered, provided that the discharge is fairly uniform and without shocks. Otherwise an equalization basin should be installed prior to treatment.

The organic chemical, steel, coal gasification, coal and wood distillation plants, oil refineries, coke plants, and a number of other industries discharge organic compounds that could be toxic to bacteria and could upset the biological treatment process. However, bacteria can adapt to fairly high concentrations of a number of such organics and thereafter decompose them successfully. Most problems with these organic compounds occur when they are discharged on an intermittent basis into an unacclimated biota in the biological units.

Phenolic Compounds

Origin and Classification. Phenolic compounds that arise from the distillation and gasification processes of coal and wood, from oil refineries, organic chemical plants, hydrolysis, chemical oxidation, and microbiological degradation of pesticides are examples of industrial organic compounds that can be treated successfully both by physical–chemical and biological processes.

Phenolic compounds include a variety of organic chemicals. Phenol itself is chemically represented by a hydrocarbon ring compound (benzene) with an attached hydroxyl group. It has a BOD_5 value of 1.7 mg of O_2 per 1 mg of phenol. Most of the processes for pretreatment and recovery of phenols from wastewaters fall into one of the following four groups: (1) extraction with solvents, such as benzene, pyridine, and tricresyl phosphate, (2) adsorption on activated carbon, (3) distillation, and (4) biological treatment, both aerobic (activated sludge and trickling filters) and anaerobic.

Treatment. Physical–chemical methods are used primarily for wastes with very high concentrations of phenolic compounds. Phenols that are recovered by physical–chemical methods can often be reused.

During biological treatment, phenols are destroyed by an acclimated microbiologic population. The first experiments with the removal of phenolics by biological treatment were conducted in Essen, Germany between 1910 and 1913. Since then numerous studies have proven that phenolic wastes with phenol concentration ranging up to 2 000 mg/liter can be successfully treated by acclimated bacteria, provided that enough nutrients (nitrogen and phosphorus) are present in or added to the influent. Activated sludge plants, trickling filters, soil bacteria, and anaerobic processes have been studied and shown to be effective, exhibiting removal rates exceeding 90% and yielding effluent phenol concentrations often below 1 mg/liter.[13–18]

The following loading design parameters for biological processes treating phenolic wastes are based on the recent German experience:

Process	Phenol Loading
Trickling filter with recycle	0.7 kg/(m^3-day)
Activated sludge plants	1.0 kg (m^3-day)

For treating phenolic wastewater from coke plants by the activated sludge processes, Noack[19] recommended daily loading of 3–4 kg of BOD_5 of the phenolic waste per m^3 of aeration volume with an addition of 25 mg/liter of potassium phosphate to the influent. The pH of the wastewater has to be neutralized with 450 mg/liter of caustic soda and the wastewater had to be cooled to a 20–25°C temperature range.

Phenolic compounds can adversely affect freshwater fish (1) by direct toxicity to both the fish and fishfood organisms, (2) by lowering the available oxygen because of their high oxygen demand, and (3) by the tainting of fish flesh.[20] In receiving water bodies at normal summer temperatures, phenols are oxidized and degraded in 3–4 days or faster if sufficient nutrients are available in the water.

The major problem associated with phenolic compounds is their organoleptic (taste and odor problem causing) properties in water or fish flesh. Chlorination of water containing phenols in water and wastewater treatment plants may form a persistent odor and bad taste causing compounds that are very difficult to remove by conventional water treatment techniques. McKee and Wolf,[21] following a review of world literature, concluded that phenol in concentration of 1 μg/liter would not interfere with fish and aquatic life, and 50 mg/liter would not interfere with irrigation. The U.S. receiving water quality standard for phenols is 1 μg/liter for waters designated for water supply.

In emergency situations that could occur during winter conditions in industrialized regions, phenols found in the intakes of water supply works can be removed by carbon adsorption. Activated powdered carbon can be added to water before filtration or filtered water can be treated on carbon adsorption columns.[22] Superchlorination followed by dechlorination or ozonization can also be used as an emergency measure for controlling problems associated with the presence of phenolics in raw potable water.

Biological Treatment of Industrial Wastewater

Activated carbon added to biological treatment units or incorporated in separate columns is also employed for the removal of a number of synthetic organics from industrial wastewaters. Adams[23] has reviewed the use of powdered activated carbon in activated sludge basins. Powdered carbon is added to the basin and recycled with the sludge. The benefits of powdered carbon, in addition to removal of toxic organics, also included an improvement in the overall performance of the units.

Biological treatment processes are also amenable to joint or separate treatment of wastewaters from food processing and producing industries such as dairies, distilleries, breweries, sugar refineries, and yeast and starch processing industries. They are also applicable, often with precautions, to wastewaters from industries that process wood and animal raw materials such as pulp and paper mills, flax retting industries, textile and wool processing, and tanneries. With on-site pretreatment such wastes can be jointly treated with municipal sewage. When these wastes are treated separately, the treatment processes are not much different from those used for municipal wastewater.

Very often, nutrients (nitrogen and phosphorus) must be added into aerobic and anaerobic treatment units to offset the nutrient deficiency of some industrial wastes. Needed nutrients can be derived either from added municipal sewage or from man-made nitrogen and phosphorus chemical additives. According to Helmers et al.[24] about 3–4% of nitrogen and 0.5–0.7% of phosphorus are needed for each mass unit of BOD_5 removed in the plant. On the average, the ratio of BOD_5:N:P should be about 100 : 5 : 1.

Wastewaters and sludges from food processing and meat-packing industries tend to quickly turn anaerobic and undergo acid fermentation. However, anaerobic treatment at elevated or nonelevated temperatures is often used as a pretreatment step for such wastes in order to reduce their high BOD concentrations. The effluents from anaerobic units can be limed (if needed) and treated aerobically in the next step, jointly with the municipal wastewater or separately. However, it is better to chlorinate the effluent from anaerobic units or the raw wastewater if primary sedimentation is used in order to suppress the acid fermentation process during settling time in open tanks.

A compendium of process modifications and reuse/recycle methods for several industries is included in a publication by Overcash[9]; publications by Tavlarides[25] and by Noll et al.[26] are also available.

10.3 WASTEWATER LOADS

The waste loading from industries can be either estimated for raw wastewater (flow and pollutant loads) or for treated effluents. It should be noted that wastewater loads vary significantly among industries, even in the same category group. The implementation of water conservation measures, product recovery, and enforcement of environmental regulations will usually result in substantial load reduction in the production process. Therefore, the loadings are best measured than estimated.

A number of U.S. EPA publications contain information on wastewater loads, industry by industry. In Europe, the raw wastewater load is often expressed in population equivalents related to a unit of production or some other parameter. If one assumes that one population equivalent unit (PE) represents 60 g of BOD_5 per day the following loading values can be used for guidance:

Raw Wastewater Loads	
Dairy without cheese production	25–70 PE/1 000 liters of milk
Dairy with cheese production	45–230 PE/1 000 liters of milk
Brewery	150–350 PE/1 000 liters of beer
Slaughterhouse	130–400 PE/1 ton of live weight
Beet sugar refinery	45–70 PE/1 ton of beets
Paper pulp–sulfide	3 500–5 500 PE/1 ton of kraft
Paper making	200–900 PE/1 ton of paper
Sulfur dye	2 000–3 000 PE/1 ton of product
Wool washing	2 000–4 500 PE/1 ton of wool
Tannery	1 000–3 000 PE/1 ton of hides
Laundry	300–900 PE/1 ton of clothes
Spilled oil	11 000 PE/1 ton of oil
Landfill leachate	45 PE/1 ha of area

The above loading figures are only approximate and do not reflect recirculation or product recovery measures. Also the values of the population equivalents express only the carbonaceous BOD. In many industries, significant quantities of ammonia appear in wastewater and must be dealt with. In addition, suspended solids, toxic components, acids, and salts present in the wastewater and their effect on the treatment process efficiency must be evaluated.

In the United States, all separately treated industrial effluents from municipal wastewaters must comply with the effluent standards that are based on the U.S. EPA guidelines. The standards are enforced by the National Pollution Discharge Elimination System permits issued by the state pollution control agencies. The effluent standards are based, industry by industry, on production units and capacity of the plant, similar to the raw waste loads elucidated in the preceding paragraph. The standards for existing and new industries are based on the Best Available Treatment which is Economically Achievable (BATEA). The effluent standards specify how much of the BOD_5, suspended solids, and some other pollutants are allowed in the effluent in terms of maxima for 1 day and for 30 consecutive days. It is the industry's responsibility to achieve and maintain these standards.

10.4 INDIVIDUAL INDUSTRIAL WASTEWATERS AND THEIR TREATMENT

In the following section, characteristics of wastewater streams from most common industries and their treatment will be briefly discussed.

Dairy Wastes. Wastewater from dairies consists principally of various dilutions of milk (e.g., from washing milk containers), separated milk, buttermilk, and whey from accidental spills. Consequently, much can be done in the plant to reduce the volume of wastewater. The strength and volume of dairy wastewater depend on the type of product, for example, whether cheese or butter is produced or not. Milk solids ferment quickly to lactic acid and strong odors are generated if milk wastes are allowed to decompose anaerobically.

Dairy wastewaters are generally high in BOD. Undiluted milk has a BOD_5 value of about 80 g/liter and typical BOD_5 values of dairy wastewaters range from 1000 to 4000 mg/liter in the majority of the plants. The pH of the wastewater depends on the type of product, with cheese plant wastes being decisively acidic because of the presence of whey, which is a high BOD_5 content liquid ($>$ 10 000 mg/liter). Rather than discharging this waste into a sewer, it is more economical to recover the organics in the whey and use them as feed for cattle. Nitrogen and phosphorus are generally present in amounts higher than those found in municipal sewage but not in sufficient amounts to provide adequate nutrient supply in a biological treatment process.

Dairy wastewater is suitable for joint treatment with municipal wastewater and can be discharged into sewers, usually without pretreatment. In order to reduce the user charge, industrial plant operators may choose to reduce the BOD load, commonly by installing conventional aerobic or anaerobic biological treatment.

In a separate treatment process, both aerobic and anaerobic systems are feasible.

Because the flow and organic load from dairy plants are highly variable, equalization must be provided, however, the odor caused by acidic fermentation in equalization and settling tanks must be controlled by liming or chlorination or by aeration.[27] Anaerobic treatment units can be used with methane fermentation that can be achieved after seeding the units with well-digested municipal sludge. Another treatment alternative is using trickling filters with a high recycle.[28,29] Recently, rotating biological contactors have also been tested.[30]

The excess activated sludge is completely liquified or gasified during anaerobic digestion, hence, its volume is quite small.

Dairy effluents disposed of on land must be chlorinated to control odor and for disinfection of possible pathogenic microorganisms such as the bacteria causing tuberculosis.

Meat Packing Wastes.[31–33] The meat packing industry is divided into the following three categories:

1. slaughterhouses (abatoirs) in which killing and dressing of animals is carried out,
2. packinghouses in which, in addition to killing the animals and dressing, the meat is also processed by curing, cooking, smoking, and pickling; packinghouses also include the manufacture of sausages, the canning of meat, the rendering of fats, and a number of other meat and byproduct processing operations,
3. meat processing plants in which meat from a slaughterhouse is processed.

The wastewater originates from different stages of the production process and cleaning of the carcases or products. The wastewater from slaughterhouses and packinghouses usually has very high BOD and organic contents, contains suspended and floating solids and grease, and, furthermore, has an elevated temperature. Larger pieces of flesh, fat, manure, dirt, and other materials must be removed from the wastewater on screens or shredded by comminuter. The waste contains 4–8 kg of BOD_5, 3–17 kg of suspended solids, and 1.5–12 kg of fat per 1 ton of animal weight killed or processed. The flow volume is 5–20 m^3 per 1 ton of live animal weight.

In the majority of plants, most of the waste materials are recovered in the form of inedible grease and tankage (dried residues used as feed or fertilizer). Separate paunch manure handling and an efficient rendering operation also will reduce waste loads. Blood, which has a very high BOD value (about 160–210 g/liter) and fat are recovered and sent back onto rendering for sale. Remaining fat should be collected in grease traps. The recovered fatty substances are used as raw material for manufacturing soap, wax, glue, or cosmetics. Various methods of recycling and material recovery from slaughterhouse wastes have been described by Onufrio and Fierro.[34]

Common treatment methods used for separate treatment of meat packing wastes are similar to conventional sewage treatment, however, grease traps must be included. High load trickling filters with a high recycle or activated sludge with a high sludge load are most common. In a joint treatment, large solids must be removed from the wastewater by screens or shredded by comminuters prior to discharge into sewers. Further pretreatment may be required by local ordinances.

Wastewater from sanitary slaughterhouses (facilities for disposing of sick animals) will contain numerous pathogenic microorganisms that can be best eliminated by pasteurization of the waste at 110°C for about 15 min.[35]

Vegetable and Fruit Processing.[36] The wastes from plants processing and canning vegetables and fruits consist of (1) the liquid wastes from preparation of the processed materials containing the fruits and vegetable residues or blanching the stock, and (2) cooling water from cooling cans after sterilization.

The wastewater load is highly seasonal and conventional biological treatment may not be feasible. The screenings can be used as cattle feed or as a raw material for production of ethanol. The screened wastewater can be discharged into sewers and jointly treated.

The wastewater itself and the sludge tend to undergo acidic fermentation, which results in foul odors. Anaerobic treatment is a feasible alternative if the wastewater cannot be discharged into public sewers.[37,38] Chemical precipitation produces excellent results but is expensive. Facultative lagoons are also popular. The odor emanating from the lagoon can be checked by addition of sodium nitrate or by adequate aeration.

The steep water from *cornstarch plants* is evaporated and the wastewater that escapes is amenable to a conventional, joint, or separate biological treatment.

Land disposal by spray irrigation may cause problems with odor. In addition, safe hydraulic loadings are very difficult to establish.[39,40]

Wastewaters from *sugar production and refineries* are seasonal, imposing heavy loads on receiving waters during late fall and winter periods. Sugar is made either from sugar beets (common in the northern United States and throughout Europe) or from sugar cane (southern United States, Hawaii, and most of Latin America).

The sugar beet processing ''campaign'' usually lasts 2–3 months, beginning in October. The process of extracting sugar from beets produces four principal wastes: (1) flume and wash water derived from transportation and washing the beets, (2) process water, consisting of washings from the extraction or diffusion vessels and pulp presses, (3) lime wastes produced by the purification of the extract, and (4) Steffen's waste, which is the residual liquor after the final extraction of sugar from molasses by the Steffen's process.

The flume and wash water is high in suspended solids. These can be removed by screening and/or settling and, subsequently, the clarified water can be reused. Chlorination of recycled water can prevent it from turning septic. The sludge settling is granular. German data and experience suggest a typical hydraulic overflow rate of settlers as 0.9 m/hr (530 gpd/ft^2).

The wastewater from slicers and diffusers and pulp presses (process waste) is high in BOD and is recycled in some modern plants. For treatment, a two-step anaerobic process can be used. In the first step, the waste undergoes acidic fermentation that is interrupted by liming, after which methane fermentation commences. Stoppok and Buchholz[41] recommended loading rates for the acidification reactor in the range of 5–15 kg/(m^3-day) (0.3–0.9 lb/ft^3/day) with a hydraulic detention time of 10–30 hr. The corresponding values for the methane reactor varied from 2 to 6 hr.

Smith and Hayden[42] reported excellent results in using sugar beet processing wastewater for hay field irrigation at a rate of approximately 2.5 cm/week (1 in/week). They concluded that with good management and proper loading, sugar beet processing wastewater can be used for irrigation.

Lagooning is the most common method of treatment and disposal. Lagoons should be large enough to store most or all of the wastewater and provide sufficient time for settling and biological degradation. The lagoons can be followed by a conventional aerated biological process, however, in most cases, conventional activated sludge or trickling filters are impractical due to the seasonal nature of the waste production. The principal objection against lagooning is the strong odor that will emanate if inadequate aeration is implemented.

The odor and taste problems of receiving waters caused by discharges of sugar beet processing wastes are very difficult to control and the receiving stream can become unsuitable for water supply. In addition to biological treatment, Nemerow[8] also listed several physical–chemical treatment processes, including adsorption of pollutants from sugar beet processing wastes on sawdust or coke and sparging the alkaline wastewater with CO_2.

Typical sugar beet wastewater reuse steps are as follows: (1) condenser water can be reused as make up water for both beet fluming and washing, (2) flume and wash water can be recirculated after settling with or without coagulation and/or screening, (3) pulp screen and pulp press waters can be recirculated to the diffuser, (4) lime cake water can be burned and the lime reused, and (5) in plants utilizing Steffen's process, the Steffen's waste can be evaporated (e.g., by spraying into stack gases) and the residue used for livestock feed.[7]

Sugar cane refinery wastewaters can have a very high organic content and, often, they may be very difficult to treat. The acids that are formed during the first stage acidic fermentation would normally require a relatively high dosage of lime for neutralization and conversion to methane fermentation.

Wastes from Fermentation and Related Industries. These include those from breweries and distilleries, wineries, yeast production, fermentation for the production of higher alcohols, and manufacture of antibiotics.

The principal waste from breweries and distilleries is the spent grain that is collected on screens. It has a very high food value and, consequently, its recovery as feed for cattle is profitable. Nutrients must be added for subsequent biological treatment or pretreatment.

Wastewater from fermentation industries can be treated jointly with municipal sewage. In situations in which joint treatment is not feasible, separate biological treatment plants are designed. In general, the BOD of high-strength organic wastes, such as those from breweries, yeast production, and distilleries, is first reduced by anaerobic fermentation followed by a high recycle trickling filter or activated sludge process. A two-stage anaerobic process can also be used.

The production of antibiotics releases wastes that are similar to brewery wastes and can be treated using the same methods. The waste materials contain spent yeast that can be recovered as feed for cattle.

The residues from wineries include pomace (largely grape pulp, skins, and seeds), lees (sludge left in fermenting and aging tanks that is high in tartrates), and argols or wine stone (crystallized tartrate). The solids are separated and directly conveyed to anaerobic digestion. The pretreated wastewater is best handled by joint treatment with municipal wastewater. Recovery of tartrates may be profitable in large establishments.

Tannery Wastewater. The tanning of leather produces wastes that are very strong in organic and inorganic substances. They contain biodegradable organics, mainly from spent solutions, inorganic toxic compounds, including chromium from chrome tanning waters, soda ash, and sulfuric acid and their salts (sulfates) from bleaching and rinsing operations. Chrome tanning is used mainly for light leather whereas vegetable tanning is typical for heavy leather manufacturing. The composite waste usually has an olive green color and contains grease, particles of flesh, hair, and free and combined lime. It has a very high content of salts, primarily sodium chloride.

Tannery wastes cannot be discharged into public sewers without pretreatment, which should include screening and sedimentation supported by chemical coagulants. Because of the intermittent nature of tannery waste discharges, equalization is needed to prevent shock loads. The wastewater can be jointly treated only if large volumes of municipal sewage are available for dilution.[43] In a separate treatment process, equalization, coagulation, flocculation, and sedimentation unit processes can be followed by a two-stage biological filtration and clarification. The sludge can be mechanically dewatered and burned or stabilized by thermophilic (high temperature) digestion.

Sulfide concentrations can be very high and pose a problem when the wastewater is discharged into sewers in which it can cause the formation of toxic hydrogen sulfide gases. The major problem associated with the hydrogen sulfide formation is corrosion and degradation of sewer materials. Oxidation of sulfides can be achieved by aeration with a catalyst such as manganese.

Textile Wastes.[44] Textile wastewaters vary significantly with the type of the processed fiber (wool, silk, hair, linen, cotton, semisynthetics, and true synthetics) and with the number and extent of the manufacturing operations.

The wastewater from *raw wool scouring and preparation* contains dirt, manure, soaps, alkali or detergents, and large amounts of wool fibers and grease. As a result, it has a high BOD content (about 100 mg BOD_5/liter) and significant quantities of suspended and colloidal matter. The settleable solids are first removed in settling tanks. Wool grease can be recovered, for example, by high-speed centrifuges.[45] Also a common practice is to acidify the waste to a pH of 3–4 and to agitate the mixture with compressed air. The grease is then released from the emulsion and separated as scum.[46,47] After the grease is removed, the wastewater can be biologically treated in aerobic or anaerobic reactors.[48,49]

Silk-Washing Wastes contain soapy liquor and rinse waters. Chemical precipitation of the wastes requires large quantities of coagulants and produces a bulky, slow-drying sludge. Treatment with sulfuric acid yields an oil that can be reused in making soap for silk mill.

Cotton is cleaned and dewaxed after it is woven into fabric. This is often followed by bleaching. The BOD_5 of wastewater from cotton mills is in the range of 200–600 mg/liter, and it is strongly alkaline (pH 10–12). Consequently, biological treatment is not possible until the wastewater is neutralized, which can be achieved by absorption of carbon dioxide from the atmosphere while the wastewater is stored in a lagoon, or by treating the wastewater with flue gas. The alkaline wastes can also be neutralized by acidic wastewater from different parts of the mill, which also contributes to color removal.

Most Dye Wastes contain color that may be very difficult, if not impossible, to remove. A good equalization and a settling basin are the minimum treatment that should be provided. By mixing different wastewater streams, color can be neutralized. In this sense, waste bleaching liquids are effective and, similarly, chlorination can also reduce or destroy some dyes. Color is also removed by chemical precipitation with iron salts or alum or by the carbon adsorption process.[50]

Sulfur and vat dyes have a very high chemical oxygen demand and a sulfide content that will cause emanation of hydrogen sulfide at a lower pH. Acidification will precipitate the dyes and it is possible to recover indigo from the dye baths by oxidation and acidification. Biological treatment is possible if the waste is pretreated by chemical precipitation, neutralization, and if enough municipal sewage is available for dilution.

Wastewaters from *manufacturing of synthetic fabrics* are normally amenable to biological treatment after neutralization.

Generally, chemical precipitation and biological treatment are the primary methods for treatment of textile wastewaters. As most textile wastes are released in batches, equalization basins should be an integral part of the treatment process. Trickling filtration is an attractive method of biological treatment because of its flexibility; however, activated sludge processes have also been used. After pretreatment (equalization, neutralization, and chemical precipitation) textile wastewaters can be jointly treated with municipal sewage. Good housekeeping and in-plant water-saving measures are very effective. Reference 44 can be consulted for further details.

Pulp and Paper Wastes. Paper making includes two operations: preparation of pulp and actual manufacture of paper. Four varieties of wood pulp are produced, groundwood, sulfite, soda, and sulfate. Pulp is also made from straw, hemp, jute, and rags.

Groundpulp is made by mechanically grinding debarked logs. The wastewater contains mostly wood particles and fibers and seldom causes a serious water pollution problem. A complete recirculation of water is feasible.

Sulfite pulp is produced by cooking wood chips at high temperatures and pressure in a solution containing calcium bisulfite. After cooking the pulp is washed and bleached. The cooker liquor is one of the most difficult to handle industrial wastes. About 10 m^3 of the liquor is produced per each ton of pulp. The liquor contains about 12% or 1.2 tons of solids per ton of pulp from which about 90% or 1.1 tons is organic. The sulfur compounds in the liquor impose an immediate oxygen demand, which amounts to about 11% of the total BOD of the waste.[8] The pollution load from one sulfite pulp mill and the burden it imposes on the dissolved oxygen regime of receiving waters can be compared to that from untreated sewage of a large city.

Anaerobic treatment of sulfite liquor is feasible after nutrient addition and the BOD removals range from 70 to 90%.[51] The sulfite liquor can also be separated from the waste stream, its water content reduced by evaporation and subsequently incinerated.[52] The solids after evaporation can also be used as an adhesive for making briquettes or foundry molds.

The residual lignosulfonic acid liquor can be chemically precipitated at a pH of about 12 and converted to sludge. After neutralization, joint treatment with municipal wastewater is feasible.[53] According to experiments conducted in Germany, waste solids from sulfite pulp wastewater can be absorbed by aluminum oxides that can subsequently be thermally regenerated.[54]

In the *soda process*, the wood is cooked in a 10% solution of caustic soda. Normally, there is little waste since the liquors are evaporated, charred, and leached, and the leached sodium carbonate is causticized for soda recovery. Lime sludge from the causticization is disposed on landfills, reburnt to quicklime, or processed to pure calcium carbonate.

In the *sulfate process*, the wood is digested in a solution of caustic soda and sodium sulfate. The liquor is evaporated and calcined for recovery of the chemicals and the waste consists of lime sludge and pulp washings. In efficient processes, the wastewater should not have a very high pollution content. Sulfate pulps do not bleach readily and are used mostly for brown or colored paper.

Wastes from *manufacturing of paper* contain so-called "white waters" that contain the fiber and clay filler used in the manufacturing process. Sedimentation with or without chemical precipitation is the most common treatment method.

Recovery of the materials present in the waste and recycling of water are the most important components of modern pollution abatement in the pulp and paper industry. Chlorination of the recycled water is necessary for control of slime growths in the system. Several possibilities for recycling were presented by Wyvill et al.[55]

The color of pulp and paper wastewater is dark and causes an objectionable colorization of the receiving waters. Color can be removed by precipitation with lime or iron salts or by acidification of the effluent.[56] Biologically, color can be removed by fungi.[57] It should be noted that color removal is difficult and expensive.

The National Council of the Paper Industry for Air and Stream Improvement, Inc., New York, which is a research arm of the pulp and paper industry, publishes annual literature reviews on the treatment and disposal of pulp and paper wastewater.

Chemical Industry Wastes. The chemical industry can be divided into several categories, each with its own wastewater characterization, effluent limitations, and feasible treatment methods. Chemical industries can be generally divided according to their production line into:

Inorganic Chemicals	Organic Chemicals
Acids and alkali	Soaps and detergents
Powder and explosives	Plastics and resins
Silicones	Pesticides
Inorganic fertilizers	Solvents

In addition, chemical plants also produce a large number of raw chemicals that are then used for manufacturing other products.

In the past, many waste products from chemical industries were not disposed of in an environmentally safe way, thus creating a number of so-called "Hazardous Waste Sites." Cleanup of these unsafe and environmentally as well as health threatening chemical dump sites and contaminated groundwater is one of the current challenges facing environmental engineers.

In the plant, the waste streams should be kept separated into strong and weak waste streams. Waste materials should be recovered whenever possible. The best and most economical recovery can be accomplished when the waste materials are concentrated. A summary of water conservation and reuse technologies suitable for use in the chemical processing industry was presented by Holiday[58] and by Overcash.[9]

Most wastewaters from chemical industries must be treated first by physical–chemical processes to remove toxic materials. The available processes include (1) incineration (e.g., of PCBs or toxic organic wastes), (2) evaporation, (3) adsorption, (4) reverse osmosis and ultrafiltration, (5) extraction, (6) precipitation, and (7) ozonization.

The effluent from chemical plants should be equalized and neutralized. Flotation will remove insoluble floating materials. With the addition of nutrients, the biodegradable organics can be biologically removed by activated sludge or trickling filters processes. Foam formation in activated sludge units caused by detergents can be abated by spraying the surface. If volatile solvents or odor emanate during treatment, the units may have to be covered and the exhaust gases scrubbed.

Some landfills can accept liquid or solid waste chemicals, however, disposal of hazardous chemicals on land, which is now strictly regulated, should be the last alternative.

Landfill Leachate is an extremely polluted liquid that collects at the bottom of municipal and industrial landfills for the disposal of solid wastes. Leachate must be pumped out and safely disposed of, otherwise, it could enter and contaminate groundwater. Discharge into municipal sewers should not be allowed without pretreatment. Table 4.3 shows typical composition of landfill leachate.

Leachate can be recycled back onto the landfill and, in this case, the landfill itself can act as an anaerobic filter and stabilize the leachate.[59] However, in humid and colder climatic conditions such as those prevalent in Europe and the northern portion of the United States and southern Canada, the practice of leachate recycling may not be feasible.

Treatment alternatives for leachate also include physical–chemical treatment followed by biological activated sludge unit[60–62] or anaerobic treatment.[63] Phosphate must be supplemented in the biological units and very long detention times are required if nitrification is needed. BOD and COD concentrations can be reduced by 90%.

Coal Mine Drainage and Coal Processing Wastewaters (Colliery wastewaters). Coal mine drainage may contain large amounts of sulfuric acid and iron that result from oxidation of iron pyrite (FeS_2). In most cases, neutralization by lime may not be feasible because of the large amounts of lime that may be needed. In all cases, surface water inputs into mines must be minimized by diversion.[64] A review of possible acid drainage control technologies was published by Heunish.[65]

Wastewater produced from coal washing and transport contains waste solids (clay, sand, and fine coal). The removal of these solids is by granular settling (an example was presented in Section 6.3). The effluent after settling is recycled back into the washing process. A portion of the treated recycle should be wasted to control salt buildup and increase in acidity. Flocculation of colloids is induced by lime and alum, which should then be followed by clarification and/or filtration.

Foundries. The wastewater from foundries originates from scrubbing the off gases from the Cupola, which is a blast furnace in which coke and iron are mixed. The water contains significant amounts of phenols, sulfides, and cyanides. It can be recycled after treatment and the blowdown (wasted portion of the recycle to control pollution buildup) can be jointly treated with municipal wastewater.

The Iron and Steel Industry produces a variety of wastewaters, most of which are inorganic and are treated on the site. Because of the large volume and inorganic nature of the pollutants, the effluent may be discharged after treatment into the receiving waters that must provide adequate dilution. Blowdown from blast furnace recycle systems and from coke areas contains phenols and cyanides and needs pretreatment before it can be discharged into sewers, or full treatment before a discharge into receiving waters. The phenol and cyanide concentrations in the blast furnace blowdown are not high, however, coke wastes contain very high concentrations of ammonia (50–200 mg/liter), phenols (500–1 000 mg/liter), and cyanides (5–20 mg/liter). These wastes can be biologically treated when nutrients are added, however, the bacteria must be acclimated to the wastewater. Most steel companies have their own wastewater treatment plants.

In *the pickling process,* dirt and oxides are removed from steel by submerging it in a weak sulfuric acid (carbon steel) or in a sequence of baths made of sulfuric and nitric/fluoric acids (stainless steel). The steel is then rinsed and the rinse water is neutralized, which is followed by clarification to remove the precipitated metals. The spent liquor from pickling baths can either be reclaimed or neutralized. It should be noted that the amount of sludge from neutralization and precipitation is quite large.

Small quantities of waste pickle liquor can be used in municipal treatment plants as an additive for phosphorus removal. The addition of chlorine to the liquor produces ferric chloride, which can be used as a coagulant in water or wastewater treatment and for sludge conditioning.

Wastewaters Containing Copper are produced during the pickling and washing of copper and copper alloys. Copper is toxic in relatively small amounts to aquatic life, to biota in biological treatment units, and to bacteria in sludge digesters. If possible, copper should be reclaimed from wastewater by one of the following methods: (1) evaporation, (2) electrolysis of the pickling liquor, (3) crystallization of copper sulfate after neutralization, and (4) neutralization by lime followed by sedimentation.

Wastes from Galvanizing and Electroplating. These wastes originate from electroplating baths that contain chromates and cyanides and, as a result, wastes are extremely toxic. Depending on the manufacturing process, the wastewater may contain 10–100 mg/liter of metals (cadmium, nickel, copper, lead, hexavalent chromium) and cyanides.

Both hexavalent chromium and cyanides in concentrations above 1 mg/liter will interfere with biological treatment. In the plant, cyanide wastes must be separated from acid wastes because lethal hydrogen cyanide gases (HCN) can be volatilized by acidification. Raising the pH of cyanide containing wastewater to about 10.5 by the addition of caustic soda and subsequent chlorination by sodium hypochlorite or by chlorine gas will convert the toxic cyanides to less toxic cyanates such as sodium cyanate (NaOCN). Further chlorination at neutral pH results in the decomposition of cyanates to harmless carbon dioxide and nitrogen.

Hexavalent chromium, Cr^{VI}, can be reduced to trivalent chromium at a low pH, which is accomplished by adding sulfuric acid and reducing the pH to about 2.5 and by subsequent addition of sulfur dioxide or sodium sulfite.

The combined wastewater is then neutralized by lime and the pH is raised to about 8.5. The metallic ions present in the wastewater will then precipitate as hydroxide sludges that can be removed by sedimentation.[66]

The removal of metals from wastewater can also be accomplished by ion-exchange processes, however, economics is a major factor.

Table 10.2 contains the pretreatment standards established by the U.S. Environmental Protection Agency.

TABLE 10.2 Pretreatment and Effluent Guidelines for Metals[a]

	Established Standard (mg/liter)	
	Daily Maximum	30-Day Average
Cadmium		
Existing sources	1.29	0.27
New sources	0.064	0.018
Chromium—total	2.87	0.80
Copper	3.72	1.09
Nickel	3.51	1.26
Lead	0.67	0.23
Silver	0.44	0.13
Zinc	2.64	0.80
Cyanide, total	1.30	0.28
Total toxic organics	0.58	—
Total suspended solids	61.0	22.9

[a]After U.S. EPA.[67]

Radioactive Wastes. Radioactive wastes are produced by the nuclear industry (uranium ore processing, fuel preparation and processing), by hospitals and research institutions, and by laundries servicing these institutions. Also nuclear detonations and malfunctions of nuclear power plants (such as the Chernobyl plant in the Soviet Union in 1986) and nuclear weapon manufacturing can contaminate surface and groundwater with large doses of radioactivity. With time, radioactivity is reduced by its own decay so safe storage is a major consideration. Low-level radioactive wastes can be stored for a period of time to allow the radioactivity to decay and then they can be discharged into receiving waters provided that enough dilution is available. Burial of low-level wastes is another acceptable method.

Many radioactive materials require a long time to decay and the half-time (the time in which the radioactivity is reduced to $\frac{1}{2}$ of its original level) may exceed thousands of years. In such cases concentration and inactivation methods are used. These include (1) chemical precipitation, (2) ion exchange, (3) adsorption, and (4) evaporation. The concentrated, high-level radioactive wastes can then be solidified in concrete and ceramic blocks and buried. It should be noted that the burial of high level wastes in the United States is still a subject of much controversy and the ultimate disposal method has not yet been decided. Suitable burial sites of solidified high-level radioactive wastes have included abandoned rock salt mines, sea disposal, and tuff. The reader is referred to the many publications by the U.S. Department of Energy for further details.

In the United States, handling and disposal of radioactive wastes are regulated by the Nuclear Waste Policy Act of 1982 and the Low-level Radioactive Policy Act of 1980. The current status of repository programs in the German Federal Republic was summarized by Malting et al.[68] The French government has developed an integrated waste management system that is mandated both as a management tool to optimize low-level waste processing, transportation, and disposal, and for selection of acceptable waste forms and disposal sites in accordance with safety standards.[69]

Oil-Field Brines and Oil Refinery Wastes. In natural geological formations, petroleum is generally associated with salt water and much of this brine resurfaces with the oil. Usually the quantities of brine water exceed the quantity of pumped oil. The brine contains drilling muds and is high in sodium, sulfates, and chlorides. Its salinity may be 10 times more than that of seawater. Diversion of this liquid into surface waters must be controlled to prevent salinity shocks. Entrained oil is removed by an extended period of storage and sedimentation.

Wastes from *petroleum refineries* originate from a number of processes. The volume of wastewater is very large but about 80–90% of it is cooling water that generally is not polluted by conventional pollutants, however, heat addition to receiving waters is considered as pollution and temperature standards have been established. Wastes polluted by conventional pollutants (BOD, suspended solids, toxics) are derived mainly from washing and chemical treatment of the oils during the refining. The wastewater contains oil emulsions and particles of coke from stills. It also includes hydrogen sulfide, mercaptans, phenols, and other compounds. Oil separation processes include gravity separators, coalescers, filters, various oil absorbers, flotation units, centrifuges, and membranes. Oil skimmers installed in sedimentation basins or flotation units with polymer addition are the most frequently used processes.

After oil separation, refinery wastes can be treated biologically by the activated sludge process provided that missing nutrients in the wastewater (nitrogen and phosphorus) are supplemented. Extensive review of wastewater treatment in the petroleum industry was published by Vernick et al.[70] Loehr et al.[71] have shown that land disposal of oily wastes is a feasible method, however, a fraction of the oil will accumulate in the soils. Chirers[72] and Borup and Middlebrooks[73] provided comprehensive overviews of all aspects of *petrochemical pollution control,* including air, wastewater, and hazardous wastes. Precautions necessary for development of petrochemical plants in developing countries were also discussed.[73]

Oil spills into receiving waters can be contained by floating plastic skimmers that surround the spilled oil and concentrate it in the smallest possible area from which the oil can be skimmed and pumped out. The chemistry and fate of oil spills are covered in a publication by Philips and Payne.[74]

Cooling Water and Thermal Pollution. A modern, coal fired power plant that uses once-through cooling needs a flow of 4 m^3/s of cooling water per each 100 MW capacity. The need is even greater for nuclear power plants. Other industries with heavy requirements for cooling include most of the heavy industry (steel mills, foundries, machinery and automobile manufacturing), paper industry, oil refineries, and a number of chemical industries.

Most inland streams cannot provide the flow volumes during low flow conditions that are needed for once-through cooling of power plants, therefore, this type of cooling is used only when the power plant can be situated near a very large stream, lake, or at a seaside. The amount of heat that is given to cooling water and subsequently dissipated into the environment depends on the type and the size of the plant. Because of their lower efficiency, nuclear power plants waste more heat than conventional power plants that use fossil fuel (coal or oil) as a source of energy. Typically, 56% (fossil fuel plants) to 64% (nuclear power plants) of the heat contained in the fuel must be disposed as waste heat to the environment.

Cooling water after it passes the cooling system of the plant is warmed up about 8–10°C. Then if ΔT is the temperature rise in the cooling system and *Ef* is the overall efficiency of the power plant, the amount of cooling water needed for once-through cooling can be estimated as (assuming 10% energy loss in furnace and turbine)

$$Q(\mathrm{m}^3/\mathrm{hr}) = \frac{860 \times \left(\dfrac{90}{Ef(\%)} - 1\right) \times \text{power capacity (MW)}}{\Delta T\ (^\circ\mathrm{C})}$$

If enough dilution by the receiving water flow is not available the heat discharge may be deleterious to aquatic life. Thermal tolerance of fish and other aquatic life depends on the species, acclimation, period of exposure, presence of toxic materials, oxygen levels, etc. Temperature also affects spawning. Logically, cold water game fish (trout, salmon) are more sensitive to increased temperatures than warm water fish.

The U.S. thermal pollution standards are based on the time of exposure to elevated temperatures for short exposures and on weekly average temperatures for longer exposures, type of the season (spawning or not spawning), acclimation, and type of fish.[21]

The waste-assimilative capacity of receiving water bodies for oxygen-demanding compounds such as BOD is reduced if the temperature is increased.[75]

In water-short areas, once-through cooling is not feasible and recycling of cooling water is required whereby the waste heat is dissipated into the atmosphere by evaporation (note that about 580 cal is dissipated by 1 g of evaporated cooling water or 673 kw-hr/m^3). Cooling towers or cooling ponds are the most common heat dissipators. The area requirement for a cooling pond is very large, about 0.5–1 ha (1–2 acres) of pond surface area per 1 MW of power plant capacity. The area requirement can be somewhat reduced if sprays are installed in the pond. About one-half of the heat loss from the pond is by evaporation; the remainder is by back radiation and conduction of the heat to the atmosphere. Water loss from the pond amounts to about 1% of the total recycle flow.

Cooling towers are relatively expensive and not without environmental ramifications. The heat loss is amost solely by evaporation and the water loss amounts to about 1–2% of the total recycle flow. In order to prevent a significant increase in salinity in the recycled water, more water must be added in the cycle than is lost by evaporation. This excess water, called blow-down, is highly polluted and requires physical–chemical treatment.

In designing cooling water intakes, the impingement and entrainment of organisms caused by the large quantities of water pumped through intake screens and through the system, must be considered.

For a detailed discussion of the thermal pollution problem the reader is referred to publications by Krenkel and Parker.[76,77]

REFERENCES

1. E. W. Moore in *Sewage Treatment* (K. Imhoff and G. M. Fair). Wiley, New York, 1940 and 1956.
2. W. W. Eckenfelder, Jr., *Industrial Water Pollution Control.* McGraw-Hill, New York, 1988.
3. K. Imhoff, *Wasser, Luft Betr.* **2,** 150 (1958).
4. F. Meinck, H. Stoof, and H. Kohlschütter, *Industrie-Abwasser* (*Industrial Wastewater*). Fischer, Stuttgart, 1968.
5. C. V. Gibbs and R. H. Bothel, *J. Water Pollut. Control Fed.* **37,** 1417 (1965).
6. Industrial Wastewater, *Prog. Water Technol.* Nos. 2/3, (1976).
7. H. F. Lund (ed.), *Industrial Pollution Control Handbook.* McGraw-Hill, New York, 1971.
8. N. L. Nemerow, *Industrial Water Pollution.* Addison-Wesley, Reading, Mass., 1978.
9. M. R. Overcash, *Techniques for Industrial Pollution Prevention.* Lewis Publ., Chelsea, MI, 1986.
10. J. T. Patterson (ed.), *Industrial Waste Management,* 3 vols. Lewis Publ., Chelsea, MI, 1985.
11. *Guidance Manual for POTW Pretreatment Program Development.* U.S. Environ. Prot. Agency, Washington, DC, 1983.
12. U.S. Environ. Prot. Agency, *General Pretreatment Regulations for Existing and New Sources.* Federal Register, Washington, DC, 1984.
13. N. T. Putila, *Mikrobiologia* **38,** 703 (1969).
14. I. P. S. Rao, P. M. Rao, D. S. Rao, G. K. Seth, *Environmental Health* (India) **11,** 23–31 (1969).
15. J. V. Dust and W. S. Thompson, *For. Prod. J.* **23** (9), 59 (1973).
16. J. J. Ganczarczyk, *Water Res.* **13,** 337–342 (1979).
17. R. J. Luthy and L. D. Jones, *J. Environ. Eng. Div.,* (*Am. Soc. Civ. Eng.*) **106,** 847 (1980).
18. H. Mercer, S. Nutt, I. Marvan, and P. Sutton, *J. Water Pollut. Control Fed.* **56,** 192 (1984).
19. W. Noack, *Wasser, Luft u. Betrieb Wasser, Luft u. Betrieb* **13,** 319 (1969).
20. European Inland Fisheries Advisory Commission, *Water Res.* 7, 929 (1973).
21. *Quality Criteria for Water,* U.S. Environ. Prot. Agency, Washington, DC, 1976.
22. K. Imhoff and F. Sierp, *Water Sewage Works J.,* p. 197 (1958).
23. A. D. Adams, *Water Wastes Eng.* **11,** B8 (1974).

24. E. N. Helmers, J. D. Frame, J. D. Greenstieg, and C. N. Sawyer, *Sewage Ind. Wastes* **24,** 496 (1952).
25. L. L. Tavlarides, *Process Modification for Industrial Pollution Source Reduction.* Lewis Publ., Chelsea, MI, 1985.
26. K. E. Noll, C. N. Haas, C. Schmidt, and P. Kodokula, *Recovery, Recycle and Reuse of Industrial Wastes.* Lewis Publ., Chelsea, MI, 1985.
27. M. Strell, *Staedtereinigung* **32,** 85 (1940).
28. K. Imhoff, *Gesund. Ing.* **64,** 367 (1941).
29. J. Garison, W. Bough, D. Landers, and G. Reeds, *Ind. Wastes* (*Chicago*) **29** (3), 12–14 (1983).
30. I. C. Agarwaland and P. S. Suria Pandian, *Indian J. Environ. Health* **23**(1), 27 (1981).
31. *Upgrading Meat Processing Facilities to Reduce Pollution,* Vols. 1 and 2, EPA Rep. No. 625/3-74-003. U.S. Environ. Prot. Agency, Washington, DC, 1974.
32. M. A. Bull, R. M. Sterrit, and J. N. Lester, *Environ. Technol. Lett.* **3,** 117, (1982).
33. S. E. Hrudey, in *Surv. Ind. Wastewater Treatment* (D. Barnes, C. F. Forster, and S. E. Hrudey, eds.) Vol. 1, p. 128. Pitman, Boston, MA, 1984.
34. G. Onufrio and F. Fierro, *Phoenix Int.* **3** (5), 26 (1986).
35. D. Scharfe, *Wasserwirtsch.-Wassertech. WWT* **36,** 366 (1986).
36. *Pollution Abatement in the Fruit and Vegetable Industry,* Vols. 1–3, EPA Rep. No. 625/3-77-0007. U.S. Environ. Prot. Agency, Cincinnati, OH, 1977.
37. R. C. Landane et al., *Agric. Wastes,* **5** (13), 111 (1983).
38. K. J. Kennedy and L. van der Berg, *Proc. Ind. Waste Conf.,* (*Purdue Univ.*) **37,** 71–76 (1983).
39. C. W. Scheffield and M. D. Sims, *Proc. Ind. Waste Conf.,* (*Purdue Univ.*) **37,** 61–69 (1983).
40. J. A. Gerick, *J. Water Pollut. Control Fed.* **56,** 287 (1984).
41. E. Stoppok and K. Buchholz, *Biotechnol Lett.* **6,** 119 (1984).
42. A. E. Smith and B. Hayden, *Agric. Wastes* **5** (8), 83 (1983).
43. T. R. Haseltine, *Sewage Ind. Wastes* **30,** 65, (1958).
44. S. G. Cooper, *The Textile Industry Environmental Control and Energy Conservation.* Noyes Data Corp., Park Ridge, NJ, 1978.
45. *Upgrading Textile Operations to Reduce Pollution,* Vols. 1 and 2, EPA Rep. No. 625/3-74-004, U.S. Environ. Prot. Agency, Cincinnati, OH, 1974.
46. J. J. Warner, *Text. Res. J.* **55** (2), 133 (1985).
47. W. H. Hiller, *Surv. Munic. Cty. Eng.,* p. 769, (1946).
48. M. T. Singleton, *Sewage Works J.* **21,** 286 (1949).
49. F. Wilson and P. H. King, *Proc. Ind. Waste Conf.,* (*Purdue Univ.*) **38,** 193–200 (1984).
50. G. McKay, *Chem. Eng. J.* (*Lausanne*) **21,** 95 (1984).
51. E. J. Donovan, TAPPI Environ. Conf. [Proc.], p. 209 (1984).
52. P. Sander, *Wasser Abwasser* **39,** 57 (1941).
53. O. Klee, *Wasserwirtschaft* **73,** 183 (1983).
54. H. Ulrich, *Prog. Wat. Technol.* **9,** 89 (1978).
55. C. Wyvill, J. C. Adams, T. C. Shelnutt, and G. E. Valentine, *TAPPI Env. Conf.* [*Proc.*], p. 111 (1984).

56. D. C. Eaton, *TAPPI* **65**(3), 83 (1982).
57. G. Royer, M. Desroches, L. Jurasek, D. Rouleau and R. C. Mayer, *J. Chem. Tech. Biotechnol.* **35** (1), 14 (1985).
58. A. D. Holiday, *Chem. Eng.* (*N.Y.*) **89** (8), 118 (1982).
59. M. Tittlebaum, *J. Water Pollut. Control Fed.* **54,** 428 (1982).
60. J. D. Keenan, R. L. Steiner, and A. A. Fungaroli, *J. Water Pollut. Control Fed.* **56,** 27 (1984).
61. H. D. Robinson and P. J. Maris, *Water Res.* **17,** 1537 (1983).
62. H. D. Robinson and P. J. Maris, *J. Water Pollut. Control Fed.* **57,** 30 (1985).
63. P. E. Schafer, G. C. Woelfel, and J. L. Carter, *Proc. Ind. Waste Conf.*, (*Purdue Univ.*) **41,** 383–389 (1987).
64. B. F. Ferguson, *Water Resour. Bull.* **21,** 253 (1985).
65. G. W. Heunish, *J. Coal Qual.* **5,** 18 (1985).
66. E. A. Ramirez and O. F. D'Alessio, *Met. Finish.* **5**(11), 15 (1984).
67. *Environmental Regulations and Technology: The Electroplating Industry,* EPA Rep. No. 625/10-80-001. U.S. Environ. Prot. Agency, Cincinnati, OH, 1980.
68. P. Malting et al., *Nucl. Eng. Int.* **30,** 366 (1985).
69. C. A. Hutchinson and B. Vigreux, *Trans. Am. Nucl. Soc.* **36,** 107 (1985).
70. A. S. Vernick, B. S. Langer, P. D. Lanik, and S. E. Hrudey, in *Surv. Ind. Wastewater Treatment* (D. Barnes, C. F. Forster, and S. E. Hrudey, eds.) Vol. 2, p. 1. Pitman, Boston, MA, 1984.
71. R. C. Loehr, J. H. Martin, Jr., and E. F. Neuhauser, *Proc. Ind. Waste Conf,* (*Purdue Univ.*) **38,** 1–12 (1984).
72. G. E. Chivers, in *Surv. Ind. Wastewater Treatment* (D. Barnes, C. F. Forster, and S. E. Hrudey, eds.) Vol. 2, p. 130. Pitman, Boston MA, 1984.
73. B. Borup and E. J. Middlebrooks, *Pollution Control for the Petrochemical Industry.* Lewis Publ., Chelsea, MI, 1986.
74. C. Phillips and J. Paynes, *Petroleum Spills in the Marine Environment.* Lewis Publ., Chelsea, MI, 1985.
75. P. A. Krenkel and V. Novotny, *Water Quality Management.* Academic Press, New York, 1980.
76. P. A. Krenkel and F. L. Parker, *Thermal Pollution: Biological Aspects.* Vanderbilt Univ. Press, Nashville, TN, 1969.
77. F. L. Parker and P. A. Krenkel, *Thermal Pollution: Engineering Aspects.* Vanderbilt Univ. Press, Nashville, TN, 1969.

CHAPTER 11

Disposal of Household Wastes from Unsewered Areas

11.1 ON-SITE DISPOSAL[1-6]

Individual, rather than collective sewage disposal is considered when public sewerage cannot be provided at a reasonable cost. It should be noted that sewers and sewage conveyance to a municipal treatment plant are possible only when household water use is relatively high. In many developing countries with water uses less than 50 liters/(cap-day) installation of municipal sewers is not feasible.

Pit privies, pour-flush toilets, composting toilets, aquaprivies, and septic tanks are the major types of individual household sanitation systems. The distinguishing feature of these systems compared to the sewer-based community systems is that they require low investment and very little supervision and maintenance.

Privies

There are many types of privies that have been used in the United States and throughout the world. The pit privy was the most common (Fig. 11.1). The pit is a leaching vault in which the fecal solids are stored and decomposed and from which urine seeps into the soil. In its most elementary form, the pit privy has three components: the pit, the seat and riser, and the structure.

The amount of solids stored in the pit is about 45 liters (1.5 ft^3) per capita per year. The pit is about 1.5 m deep. The odor is controlled by ventilation of the pit space and periodically adding lime. For maximum odor control, the vent pipe should be about 150 mm in diameter, painted black, and located on the sunny side of the privy so that the air in the pipe will heat up and create an updraft. The pit should be cleaned periodically, at least once per year, and the solids can be buried. An alternative is to move the structure with the seat to another location, excavate a new pit, and cover the old pit with soil.

A tight vault may be built into the pit for sanitary reasons, usually for the protection of groundwater and nearby wells. The volume of the pit should be large enough to also store urine. The combined volume of urine and fecal solids is about 1.5 liters/(cap-day) = 45 liters/(cap-month) for an adult with about half that amount for a child.

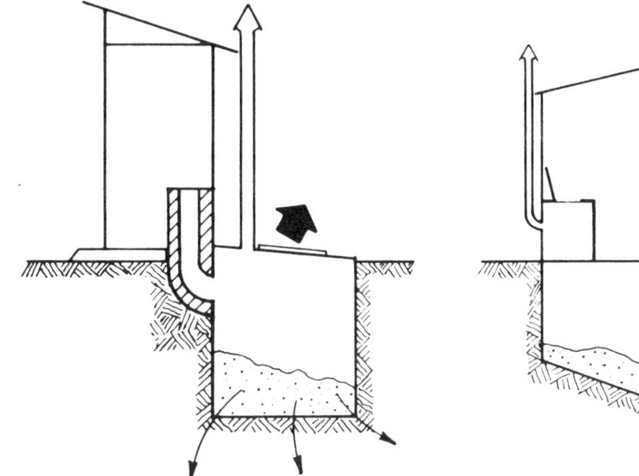

Figure 11.1 Pit privy.

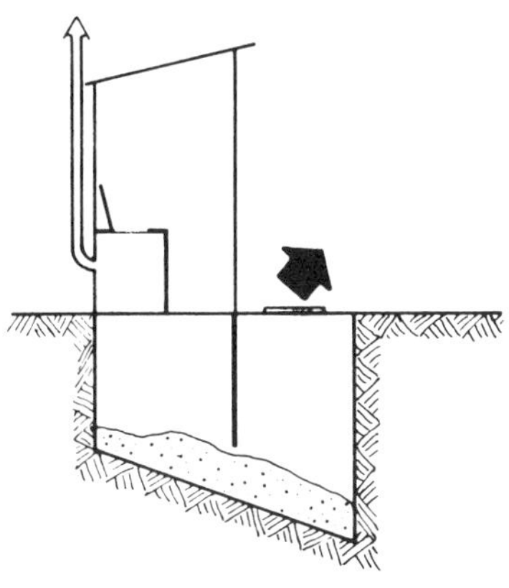

Figure 11.2 Composting toilet.

Composting Toilets. There are two basic types of composting toilets: continuous and batch (Fig. 11.2). The continuous composters were developed from a Swedish design known as "multrums." The composting pit, which is immediately below the squatting plate, has a sloping floor with inverted U- or V-shaped channels suspended above it to promote aerobic decomposition. Grass, ash, or household refuse are added to the pit to attain the necessary carbon–nitrogen ratios for promoting the composting process.

The functioning of composting toilets is very sensitive to the degree of user care, otherwise the humus may be only partially decomposed and could contain appreciable quantities of fresh excreta.

Aquaprivies. A conventional aquaprivy consists of a squatting plate above a small septic tank that discharges into an adjacent soakway (Fig. 11.3). The squat-

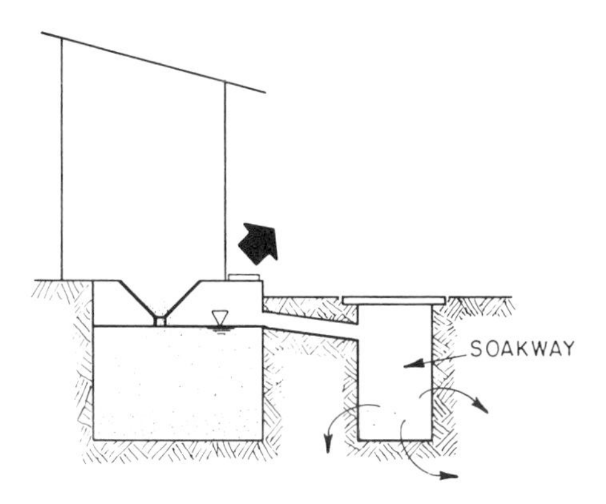

Figure 11.3 Aquaprivy.

ting plate has an integral drop pipe that is submerged into the water in the tank to form a simple water seal. As long as the water level in the tank is properly maintained, odor and fly nuisance should be minimal. In order to maintain the water level, the vault must be watertight and the user must flush sufficient water into the tank to replace any losses by evaporation. The tank requires sludge removal when it is about two-thirds full, usually every 2–3 years.

The sanitation systems mentioned above should not receive any waste other than fecal matter or urine. *Sullage or gray wastewater,* which is the remaining wastewater containing, for example, laundry, kitchen, and bath water, is a polluted wastewater that must be safely disposed of. In developing countries with relatively low water use, sullage has the same organic pollution potential as raw sewage in North America. Although its environmental hazard can be great, it is generally less than that of raw sewage. Therefore, when choosing sullage disposal facilities, the important factor becomes how much the community is willing to spend on environmental protection.

There are basically four kinds of sullage disposal systems[1-3,5]: casual tipping in the yard or garden, on-site disposal in seepage pits, disposal in open drains (usually stormwater drains), or disposal in covered drains or sewers. The first one is adequate where water use is low and the soil and climatic conditions are such that the yard remains dry. A seepage pit requires suitable soils. Disposal into open or covered storm drainage conveys the sullage to the nearest receiving water body where it can cause the same problems (depletion of dissolved oxygen, bacterial contamination) as sewage.

11.2 SEPTIC SYSTEMS AND SOIL DISPOSAL

Septic Tanks[6-9] are individual household disposal systems that provide the luxury of house plumbing fixtures common to municipal sewered communities at lower expense and with little or no attention. Water carriage toilets, shower and bathtubs, washing and dishwashing machines, and even garbage grinders can be used in country homes as if the wastes were going into a collector municipal sewer. Septic tanks are simple anaerobic pretreatment units that accomplish (1) separation of the solids from liquid sewage, (2) provide for some anaerobic digestion of organic matter, and (3) store the accumulated decomposing solids. The septic tanks (Fig. 11.4) are always used with soil adsorption systems in which the liquid effluent is disposed of by percolation through the soil.

The septic tank shown in Figure 11.4 is usually a single-story rectangular or circular tank divided into two or more compartments by a submerged overflow wall or a wall pierced by 10-cm holes located 30–45 cm below the surface of the liquid. In the United States, the size of the tank is based on the size of the house or the population it serves. For a typical two bedroom house (four people) the size of the tank is about 2.8 m^3 (750 gal) and increases in size by about 0.6 m^3 (150 gal) for each additional bedroom or 0.3 m^3 (75 gal) for each person connected. The minimum detention time of sewage in the tank should be at least 12 hr. For

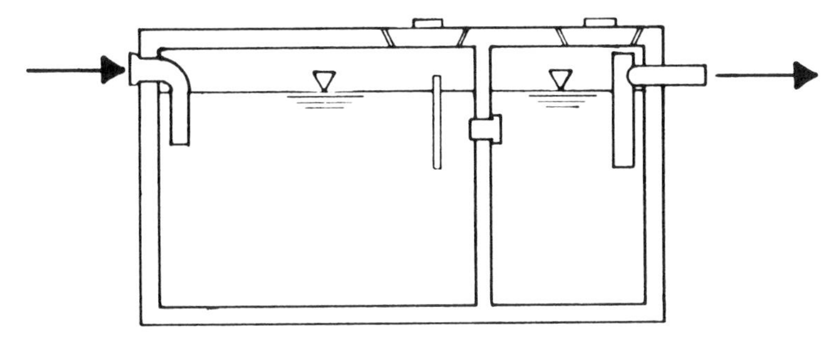

Figure 11.4 Typical two-compartment septic tank.

commercial/institutional applications, the minimum effective tank volume can be calculated as

$$V = 4.25 + 0.75Q$$

where V is the tank volume in m^3 and Q is the wastewater flow in m^3/day.

The treatment efficiency of the septic tank alone is very low and most of the BOD and bacteria removal occurs in the soil. The sludge accumulated in the tank must be pumped out at least once per year and the tank should be professionally cleaned. The solids and liquid pumped out from the septic tank during cleaning are called *septage* and must be disposed of safely either in the nearest municipal treatment plant or by safe and controlled land disposal.

In Germany and throughout Europe, small *Imhoff (Emscher) tanks* (Fig. 11.5)

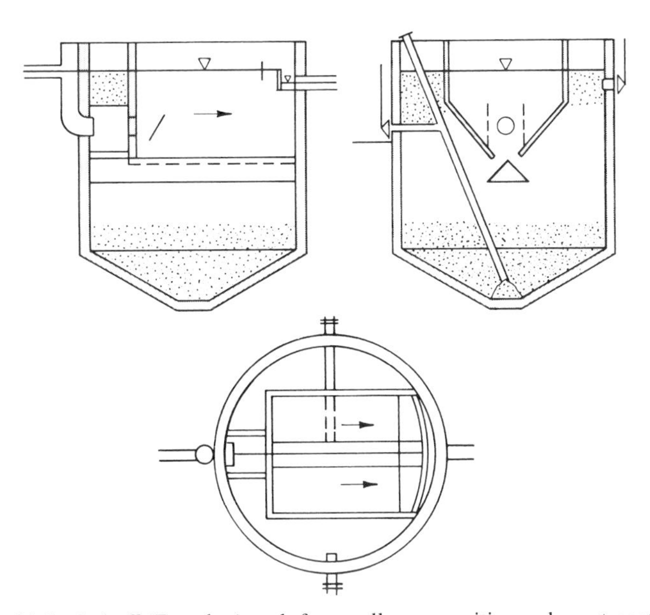

Figure 11.5 Imhoff (Emscher) tank for small communities and apartment houses.

are often used instead of septic tanks that are more popular in the United States. The small Imhoff, two-story tanks are sized as follows: settling compartment volume 0.03 m^3/cap, minimum volume 1.5 m^3, digestion volume 0.06 m^3/cap, minimum 3 m^3, and the upper volume for floating solids 0.03 m^3/cap, minimum 1.5 m^3.

The effluent from a septic tank or Imhoff tank enters a 100-mm-diameter pipe and flows to the soil absorption area. Here, the effluent trickles through a 100-mm perforated pipe into a gravel bed and into the subsoil for further treatment. If the seepage area is properly designed and located in permeable soils, aerobic degradation by soil bacteria will accomplish rapid degradation of BOD and pathogenic bacteria and nitrification of ammonia to nitrate.

For further discussion, the term septic tank will refer to the rectangular or circular horizontal flow tanks used primarily in the United States, whereas the term septic system will refer to systems using solids separation either in the conventional septic tank or Imhoff tank and effluent disposal by soil absorption or filtration.

Since the treatment of the septic system effluent relies on soil adsorption and degradation in a leach bed, soils must be suitable for such disposal. Unfortunately, more than half of the soils in the United States are unsuitable for adequate purification of wastewater by conventional septic systems, therefore, failures of septic tank systems built in past on marginal soils are quite common. More then one-half of such systems built in the United States 15 or more years ago have failed. The suitability of soils is determined by soil boring analysis and by a standard percolation test. In the test, a test boring is filled with water and the decrease of water level is observed with time. The measure is then the time it takes the water surface to drop 1 cm (1 in.).

Conventional septic systems are unsuited for sites with slow permeable, clayey soils, high or perched water tables, shallow soil depth to bedrock, excessive slope, or seasonal flooding. Rainwater from roof drains or groundwater from sump pumps should never be discharged into the septic tank drainage. The required area for leach beds is related to the type of soil and percolation rate measured during the on-site testing. The values are given in Table 11.1. Soils with percolation times

TABLE 11.1 Recommended Rates of Wastewater Application for Soil Leach Areas[7]

Soil	Percolation Rate (min/cm)[a]	Application Rate [liter/(m^2-day)][b]
Gravel, coarse sand	<0.4	Not suitable[c]
Coarse to medium sand	0.4–2	50
Fine sand, loamy sand	2.5–6	32
Sandy loam, loam	6.5–12	25
Loam, porous silt loam	12.5–24	18

[a] To convert from min/cm to min/in. multiply by 2.5.
[b] To convert from liters/(m^2-day) to gpd/ft^2 divide by 40.7.
[c] Soils with percolation rates of less than 0.4 min/cm (<1 min/in.) can be used if the soil is replaced with a 0.6-m-thick layer of loamy sand or sand.

of more than 24 min/cm (60 min/in.) are usually considered unsuitable for septic systems.

Excessive groundwater contamination by nitrates has been detected in densely populated suburban areas that use septic systems for sewage disposal.

Mound Systems.[10–12] A more expensive mound system can be used in lieu of a conventional soil absorption field (Fig. 11.6) in areas in which soils are tight or areas with a shallow groundwater table. The treatment of sewage occurs in a suitable fill material (sand) incorporated in the mound rather than in the underlying soil. Wisconsin regulations allow mound systems on soils that have percolation times up to 48 min/cm (120 min/in.).

Other Systems. *The intermittent sand filtration* process is similar to the rapid infiltration process described in Section 7.2. The effluent from a septic tank or from an Emscher two-story unit passes through a 0.7- to 1-m thick sand bed and the perched, cleaned wastewater is collected by underdrains that are made of perforated plastic or open-joint 100-mm-diameter pipes that are surrounded by gravel. The distribution lines above the bed are of similar design.

The filters can be designed as open or buried. Similarly to rapid infiltration systems, wastewater is applied intermittently in order to maintain aerobic conditions. Dosing frequency is about two to four times per day and the hydraulic loading of the filter should be between 40 and 80 liters/(m^2-day) or about 4–8 m^2 per population equivalent. The hydraulic loading can be increased if the effluent is recirculated with a pump.

In arid or semiarid areas, *evapotranspiration* (ET) *systems* can be used. An ET system normally consists of a sand bed with an impermeable liner and wastewater distributor piping.

Where no other on-site treatment alternative is available household sewage can be diverted to a *holding tank,* usually made of steel. The holding tank has no other purpose than to store the sewage that must be then regularly pumped out and trucked to the nearest communal sewage disposal site or a special sewer inlet in a nearby

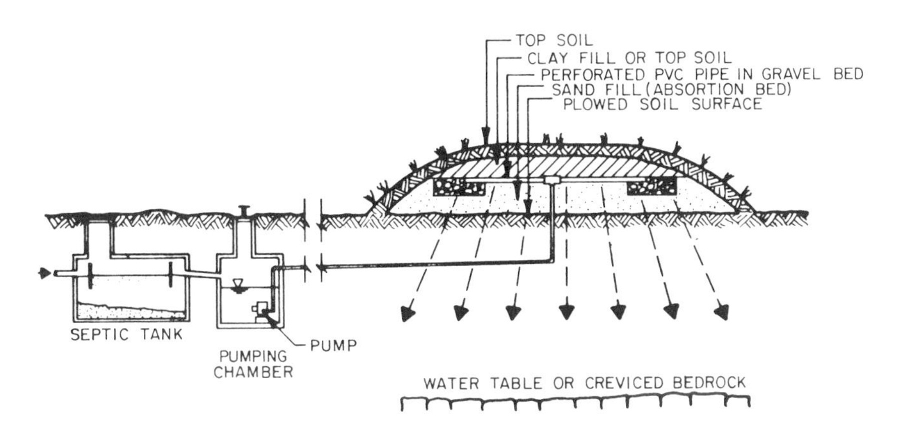

Figure 11.6 Mound system.

sewered community. An alarm must be installed indicating when the tank is about two-thirds full and requiring pumping. Using holding tanks and trucking the sewage is very costly. For a typical two bedroom house, the minimum size of the holding tank should be about 3 m^3, with 1.5 m^3 volume storage added for each additional bedroom. Depending on household water use, this will provide about 2 weeks of storage.

Most of the small systems mentioned above can also be used for wastewater disposal from public facilities such as rural and suburban schools, restaurants, overnight camps, highway rests, and rural hospitals. Recommended population equivalents for these facilities are 1 PE = 10 school children or guests in restaurants = 2 guests in summer cottages = 0.5 beds in a hospital. Wastewaters from hospitals must be disinfected by lime, chlorinated lime, or chlorine gas.

11.3 SMALL TREATMENT PLANTS[13,14]

In densely populated areas, individual on-site disposal systems may not be feasible. The installation of septic systems in the United States is usually limited to housing developments (single or multiple family) in low-density areas. The maximum population density (minimum land area) requirement is specified by local ordinances and varies from community to community. Soil and groundwater conditions should be the guiding factors for planners.

In developing countries, privies and other simple on-site sanitation may not be possible when the population density is greater than about 200 people/ha.[1]

When individual on-site systems are not feasible, a small cluster treatment plant can be the solution. In developing countries, waste stabilization ponds may be the best and most economical treatment. Other more advanced systems may fail because of common operational problems and lack of maintenance.

In industrialized countries, there is more variety to the selection of a suitable treatment and disposal system. The primary requirement is simplicity of operation and reliability.

Wastewater flows from small communities will generally have greatly accentuated peaks and minima. Small sources also sometimes exhibit a much stronger waste in terms of suspended solids, organic matter, nutrients, and grease, particularly if septage is added to the wastewater being treated. Therefore, sizing of small treatment works is based on design factors that are about 1.5 times greater than the corresponding factors for large municipal plants.

Prefabricated "package" plants are available, even in large sizes, but their most common application is for plants with a flow capacity of less than 200 m^3/day, which in the United States would represent 400–500 connected people. Most of these plants are biological facilities using the extended aeration activated sludge process. Many of the smaller package plants can serve as emergency or temporary treatment. Advantages of package plants over standard construction include smaller land area requirement, smaller hydraulic head loss, fast and low-cost field installation, reduced excess activated sludge, generally little or no odor, and low capital cost. Disadvantages include relatively higher power and operation costs.

The packaged plants usually do not employ primary clarification. Typical design parameters for extended aeration package treatment plants are[14]

Design flow	0.004 liters/(cap-sec)
Volumetric loading	$\leq$250 g $BOD_5/(m^3$-day)
Mixed liquor suspended solids	3–5 g/liter
F/M ratio	0.05–0.15 g BOD_5/(g MLVSS-day)
Sludge excess	0.3–0.5 kg solids/kg BOD_5 removed
	20–50 g/(cap-day)
Power input (for mechanical aeration)	20–25 W/m^3
Air supply (diffused aeration)	1.8 $m^3/(m^3$-hr)

Package plants are also available that use contact stabilization units or rotating biological contactors (RBCs). The EPA publication mentioned herein[14] also contains a detailed list of manufacturers of package plants and the types and design capacities of the available products.

Other reliable and simple systems include oxidation ditches or Imhoff (Emscher) tanks followed by a simple trickling filter or even by a man-made wetland.[15] The influent to the plant with a small Imhoff two-story tank should be directed into the lower digestion compartment to achieve separation of scum and floating solids. A typical trickling filter sizing parameter for small treatment plants should be about 0.3 kg of BOD_5 per m^3 of filter volume per day (about 5 persons/m^3), which corresponds to low-rate filters (Table 7.5). The depth of the filter should be about 1.5 m (6 ft).

11.4 DESIGN EXAMPLES OF SMALL WASTEWATER DISPOSAL SYSTEMS

1. Campground with central bathhouses—150 sites. Selected design—septic tank with soil adsorption field.

   Sewage flow = 200 liters/(cap-day)

   $200 \times 150/1000 = 30\ m^3$/day; most of the flow occurs during a 16-hr period

   Size of the tank based on 12-hr detention

   $$30\ m^3 \times 12/16 = 22.5\ m^3$$

   Volume of accumulated solids—3 months storage

   $$45\ \text{liters/(cap-year)} \times 150\ \text{people} \times 3/12 = 1{,}687\ \text{liters} = 1.7\ m^3$$

Total volume of the tank = 22.5 + 1.7 m^3 = 24.2 m^3
Select a tank with dimensions of 2.5 × 1.5 × 6.5 m (width × depth × length)

Size of the seepage bed: The percolation test resulted in a percolation rate of 10 min/cm. From Table 11.1, the hydraulic loading should be 25 liters/(m^2-day) = 0.025 m^3/(m^2-day), hence the seepage bed area is

$$30(\text{m}^3/\text{day})/0.025\,[\text{m}^3/(\text{m}^2\text{-day})] = 1200\ \text{m}^2$$

Design the tile field 30 m wide and 40 m long. The effluent from the tank is diverted to distribution boxes by a 20-cm-diameter pipe. The distribution tiles will be located at a depth of 0.6 m below the surface. Each tile will be made from 100-mm-diameter perforated pipes that will be laid onto a gravel bed.

2. An army camp with 220 persons and flushing toilets. Selected design—two-story Imhoff tank with a trickling filter.

Sewage production: 150 liters/(person-day)
220 × 150/1000 = 33 m^3/day
Peak daily loading 33/10 = 3.3 m^3/hr = 11 ℓ/s

Size of the Imhoff tank (Fig. 11.4):
Settling volume (0.03 m^3/person) = 0.03 × 220 = 6.6 m^3
Detention time = 6.6/3.3 = 2 hr
Sludge digestion volume

$$0.06\ \text{m}^3/\text{cap} = 0.06 \times 220 = 13.2\ \text{m}^3$$

Trickling filter:
Volumetric loading (low rate) 5 persons/m^3
Volume = 220/5 = 44 m^3, selected depth = 1.5 m
Surface area of the filter 44/1.5 = 29.3 m^2, diameter of the filter = 6.1 m
The filter will be equipped with a perforated bottom made of concrete with a slope of 1/20 toward the center. Filter media—crushed rock, filter constructed from welded or ready made steel rings. Recycle pump for recycle of flow if the sewage flow drops below 1 m^3/m^2-day) = 1 × 29.2/24 = 1.2 m^3/hr.

For a trickling filter, use of vitrified clay underdrain blocks will provide adequate support and ventilation.

Alternative: A waste stabilization pond—aerobic, preceded by a sedimentation pond (suitable for developing countries).

Settling pond: 0.5 m^3/cap = 0.5 × 220 = 110 m^3, selected depth 1.5 m, surface area 110/1.5 = 73.3 m^2, select dimensions: width 5 m, length 15 m.

Aerobic pond:

Area requirement (depending on the geographical location)—5 m^2/person = 5 × 220 = 1 100 m^2

Depth—1 m

Select two alternating ponds each with a surface area of 550 m^2.

In northern colder climates the area requirement for the aerobic pond is about twice as much (based on a unit area requirement of about 10 m^2/cap). To reduce the solids load in the effluent from the aerobic pond, a final polishing pond should be added.

3. Oxidation ditch for 600 people.

Volumetric loading: 3 persons/m^3, volume = 600/3 = 200 m^3, depth of the ditch 1 m side bank slope 1–1.5 (height to length), bottom width = 0.65 m. Length of the Kessener brush = 1.5 m. No primary treatment. The incoming wastewater will be aerated by the rotating action of the brush, which will keep the wastewater in a forward recirculating motion with a velocity of 0.3 m/s. The activated sludge will be separated in a circular clarifier and returned back in the ditch by a screw pump. Waste sludge can be disposed of on a field and plowed in.

REFERENCES

1. J. M. Kalbermatten, D. S. Julius, and C. G. Gunnerson, *Appropriate Technology for Water Supply and Sanitation.* World Bank, Washington, DC, 1980.
2. D. A. Okun and G. Ponghis, *Community Wastewater Collection and Disposal.* World Health Organization (WHO), Geneva, 1975.
3. A. Pacey (ed.), *Sanitation in Developing Countries.* Wiley, New York, 1978.
4. E. C. Wagner and J. N. Lanoix, *Excreta Disposal for Rural Areas and Small Communities,* Monog. No. 39. World Health Organization, Geneva, 1958.
5. J. H. T. Winnenberg, *Manual of Greywater Treatment Practice.* Ann Arbor Sci., Ann Arbor, MI, 1974.
6. J. A. Salvato, *Environmental Engineering and Sanitation.* Wiley (Interscience), New York, 1972.
7. *Manual of Septic Tank Practice,* U.S. Publ. Health Serv. Publ. No. 526. U.S. Govt. Printing Office, Washington, DC, 1972.
8. R. H. Laak, K. A. Healy, and D. M. Hardisty, *Ground Water* **12,** 348–352 (1974).
9. L. W. Canter and R. C. Knox, *Septic Tank System Effect on Groundwater Quality.* Lewis Publ., Chelsea, MI, 1985.

10. J. J. Bouma, J. C. Converse, R. J. Otis, W. J. Walker, and W. A. Ziebel, *J. Environ. Qual.* **2,** 382–388 (1973).
11. J. C. Converse, *Design and Construction Manual for Wisconsin Mounds.* University of Wisconsin, Madison, 1978.
12. J. M. Harkin, C. J. Fitzgerald, C. P. Duffy, and D. G. Kroll, *Evaluation of Mound Systems for Purification of Septic Tank Effluents,* Tech. Rep. No. WIS WRC 79-05. University of Wisconsin Water Res. Cent., Madison, 1979.
13. *Small Scale Waste Management Projects,* EPA Rep. No. 600/2-78-173. U.S. Environ. Prot. Agency, Cincinnati, OH, 1978.
14. *Process Design Manual: Wastewater Treatment Facilities for Sewered Small Communities,* EPA Rep. No.-625/1-77-009. U.S. Environ. Prot. Agency, Cincinnati, OH, 1977.
15. Design Manual. *Constructed Wetlands and Aquatic Plant Systems for Wastewater Treatment*, EPA/625/1-88/022. U.S. Environ. Prot. Agency, Washington, DC, 1988.

CHAPTER 12

Planning and Design of Treatment Plants

12.1 SYSTEM FEATURES AND LAYOUT

The most important task is the site selection for the treatment plant. The economy of scale makes one large treatment plant more efficient than several smaller plants even though the cost of transporting sewage for larger distances may be considerable. The decision as to whether one large plant or several smaller plants phased-in in stages also depends on the discount rate at which the capital and operation maintenance costs are amortized.

The selection of treatment efficiency generally depends on the waste assimilative capacity of the receiving water body. In the United States, secondary treatment of municipal wastewaters is mandated by the Clean Water Act and quantity and quality of industrial wastewater effluents is regulated by the NPDES permit system.

From German experience, approximate treatment plant area requirements suitable for preparation of preliminary plans are:

Population	1000 to 10 000	1.3 m^2/capita
	10 000 to 50 000	0.65 m^2/capita
	50 000 to 100 000	0.50 m^2/capita

Referring to these values, the building site must have sufficient area with adequate ground and foundation conditions and must be accessible. The effects on the environment and on the surrounding residental areas must be considered. In the United States, all public works receiving federal funding and grants must prepare a detailed Environmental Impact Statement as mandated by the National Environmental Policy Act enacted by the Congress in 1969.

It is advantageous to locate the treatment plant at the lowest point, near the natural watershed outlet and close to the receiving water body into which the effluent will be discharged. This takes advantage of gravity flow. However, in low-lying ground areas foundations may be poor and the groundwater table high. In flat regions, lift stations are needed. If the works are to be protected from flooding, they should be situated outside the flood plain, otherwise, the plant and the access roads should be protected by dikes.

In the next step, it is necessary to find an optimum flow sheet for various treatment and sludge-handling units. This depends on the magnitudes of the wastewater flow, environmental constraints, and foundation conditions. Various alternatives should be considered and the size of the units should be determined along with a preliminary design of all units. The units should be spatially arranged and a longitudinal hydraulic profile should be prepared, considering several alternatives. A simple solids balance flow sheet is also helpful. The longitudinal (vertical) profile of the units is prepared according to foundation conditions. Pumps should be used to raise the flow line of the treatment works to an elevation that provides for the most economical construction and operation. Subsequently, costs of excavation, building, pumping, and water and sludge storage should be optimized. The hydraulic profile of water levels in the units should be prepared for the characteristic design flows (Q_{min}), and design dry weather (Q_d) and wet weather (Q_{max}) flows. Provision should be made to drain or empty all tanks or enclosures by gravity or pumping.

Reinforced or prestressed concrete is the principal material for construction of municipal treatment plants, however, tanks made from welded steel plates or rings are also common. Circular tanks are cheaper than rectangular units, however, the choice is also governed by the performance of the units. The depth of the tanks is an important economical consideration. An assumption that a flat shallow unit is cheaper than a deep unit of the same volume is usually not correct. Most of the treatment plants are constructed in low-lying areas, usually with a high groundwater table. Shallow basins must then have thick, heavy concrete walls and the bottom of the structure must be anchored by driven piles or the bottom must have openings for the passage of groundwater. The use of building materials is therefore smaller for deeper basins. The inflow of groundwater into foundation pits during construction of deeper units is not a major problem since the cost of pumping and controlling groundwater inflow represents only a small fraction of the construction cost.

For flexibility in operation and convenience of repair, treatment structures should be subdivided into two or more units. The size of individual units is dictated by hydraulic considerations and by the dimensions of available equipment.

Means of flow and quality measurements throughout the plant must be considered. Flow measuring devices include Parshall flumes, Palmer–Bowlus flumes, and weirs in open channels equipped with water level measuring and recording instruments. Dissolved oxygen and temperature measuring probes should be located in the aeration tanks whereas pH probes should be installed in chemical treatment units.

Modern treatment plants are often partially automated and equipped with a number of telemetric sensors and measuring devices. The signals reflecting the operating conditions of various units are transmitted to a centrally located control room. Computerized real time or off-line control and supervision of modern treatment plants are feasible and advantageous.

The central building should also house locker and sanitation rooms for the employees, laboratories for wastewater analyses, instrumentations, and switches (in the control room). Large treatment plants should also have a maintenance shop,

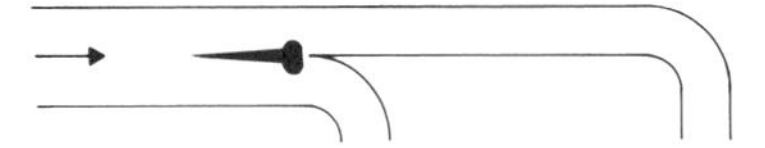

Figure 12.1 Flow distributor.

central heating (using digestion gas if possible), and offices for the engineers. Small plants with one or two operators usually require only one or two control rooms that can be located in a pumping station, blower house, or a similar building.

Plant Conduits. Plant conduits should be so arranged that incoming gravity sewers are not placed under pressure. Open, rectangular channels are used for conveyance and distribution of the sewage to the various treatment units. The channels can be designed using the Nomographs in section 2.2 or in the Appendix. The minimum velocity in the channels should be above 0.4 m/s in order to prevent deposition of solids, and 0.6 m/s in the channels ahead of the grit chamber. Channels that branch off into individual tanks should be equipped with adjustable guide vanes, or they should be so subdivided that fluctuating rates of sewage flow will be evenly distributed between the units (Fig. 12.1). The circular distributor shown in Figure 12.2 is the most reliable. In large plants with many subdivisions, it may be more economical to employ small velocities of flow and prevent deposition of sludge by blowing diffused air into all channels that carry sewage with solids. Uniform load distribution can then be secured with the benefit of preaeration.

Large solids should be kept from the sludge by screening or the solids should be reduced in size by comminution. Almost any kind of sludge can then be pumped, however, when the sludge is to be pumped large distances it is better to digest the sludge first. Centrally located sludge-handling units provide the most optimal solution.

Pumping. For pumping sewage, centrifugal pumps are most common and the same type of pumps can also be used for recirculation of secondary sludges. To prevent clogging during pumping primary sludges, the pumps should have a large-diameter impeller, which means that the capacity should be larger. Therefore, pumping of sludges is intermittent and the pumping time is shorter. Vents should be provided for the escape of sludge gas from sludge lines and pumps.

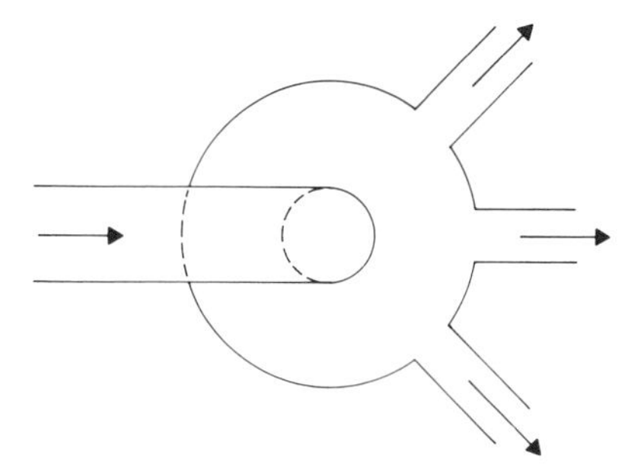

Figure 12.2 Circular flow distributor.

Sludge Conduits. Hydraulic computation of sludge conduits is similar to that for wastewater, however, the resistance factor and, hence, the headloss are twice as much as those for water. This occurs in the optimum velocity range of 0.5–1 m/s. If the velocity is small, the headloss is more than twice that of water. Even larger heads are needed for sludge-withdrawal pipes from digestion and sludge-holding tanks (such as sludge thickeners), particularly when the tanks have hopper bottoms in which the sludge is compacted or when sand or other heavy grit adds to the weight of the sludge. Provisions should then be made to stir or hydraulically or mechanically break up the sludge. The velocity in sludge lines should be at least 1 m/s to prevent deposition of grit.

In developing countries, it is common to use manual labor for cleaning and servicing the coarse screens, grit chambers, and sludge-drying beds. In developed countries, the trend is toward further mechanization and automation. The designer should know enough about the performance characteristics of available equipment to incorporate it in the plant, but it is not necessary to be an expert in design of the machinery itself. That is the responsibility of the engineering staff of equipment manufacturers.

Architectural Features of Plants. The architecture of plant structures should express their functions. At the same time, the architecture must fit into the surroundings. Landscaping adds much to the attractiveness and little to the cost and maintenance.

Plants should be fenced in, and all open channels and tanks within them should be surrounded by railings for the safety of operators and visitors. Lighting should be adequate.

Surface runoff should be kept away from the plant area and from the units.

There should be an adequate supply of water for washing and cleaning. Washing is usually done by spraying from hoses attached to the water supply. Cross-connections of water supply with wastewater are not allowed and should be guarded against both in plant operation and in the design.

12.2. DESIGN EXAMPLES OF WASTEWATER TREATMENT FACILITIES

Wastewater treatment plants are typically designed for average organic loads and not for a maximum load. Hydraulically, the system should be capable of handling the maximum flow but the size of the treatment units is based on average loading. For sizing of the units, U.S. practice uses mostly the 24-hr average values whereas European practice recommends sizing of the units based on an hourly load that is a $\frac{1}{18}$ fraction of the average daily load. When comparing the design parameters between European and North American specifications one must consider the hourly load on which they are based.

The design examples presented herein cover the following most common treatment processes:

1. Primary treatment with separate sludge digestion.
2. Low-rate trickling filter.
3. High-rate trickling filter.
4. Activated sludge.
5. Rapid Infiltration.

Example 1. *Primary clarifier with separate sludge digestion.*
50 000 inhabitants, sewage discharge 300 liters/(cap-day), combined sewer system with no storm water retention, very little industrial wastewater contributions.

$$\text{Dry weather flow (DWF)} = 300 \times 50\,000/1\,000 = 15\,000 \text{ m}^3/\text{day}$$

$$\text{Average daily flow} = 15\,000/24 = 625 \text{ m}^3/\text{hr}$$

$$= 615 \times 1\,000/(60 \times 60) = 173.6 \text{ liters/s}$$

The plant will have automatically cleaned screens and grit chamber.

Clarifier Design according to Figure 6.10 or 6.12. Select the detention time of 2 hr. The useful volume of the clarifier is then

$$2 \times 625 = 1\,250 \text{ m}^3$$

The dimensions: circular clarifier—diameter 29 m and the depth 2 m; as an alternative select two rectangular clarifiers 6.25 wide, 50 m long, and 2 m deep. During a peak hydraulic load of five times the dry weather flow the clarifier will still provide a detention time of 24 min.

Sludge Pump capacity should be about 30 liters/s to allow for adequate velocity in the minimum diameter pipelines. The sludge volume is 0.9 liter/(cap-day) (according to Table 9.1). Hence, $0.9 \times 50\,000/1\,000 = 45$ m^3/day of the sludge must be periodically pumped throughout the day to the digester.

The Sludge Digester will be heated to maintain the temperature of sludge at 37°C. According to Table 9.2, the digester volume is 0.02 m^3/cap, therefore the required volume is $0.02 \times 50\,000 = 1\,000$ m^3. The volume of gas produced for sludge from U.S. cities is about 20 liters/(cap-day) or $20 \times 50\,000/1\,000 = 1\,000$ m^3/day. For European cities, the gas production would be lower, about 13.5 liters/(cap-day). A part of this gas can be used for heating the digester while the remaining portion is used for other purposes.

The Drying Beds need an area of 50 000 × 0.09 m^2/cap = 4 500 m^2. Select five mechanically cleaned beds, each 9 m wide and 100 m long.

Example 2. *Low-rate trickling filter with primary clarifier, digester, and final clarifier.*

50 000 inhabitants, 300 liters/(cap-day), combined sewers, no stormwater retention facilities, little industrial wastewater.

$$\text{DWF} = 300 \times 50\,000/1\,000 = 15\,000 \text{ m}^3/\text{day}$$

Average hourly flow is the same as in the previous example:

$$625 \text{ m}^3/\text{hr} = 173.6 \text{ liters/s}$$

Mechanically cleaned screens and grit chamber.

Primary Clarifier with a separate, heated digester according to Example 1. Hence, the clarifier volume is 1 250 m^3.

The Digester Volume according to Table 9.2 is 0.03 m^3/cap or 50 000 × 0.03 = 1500 m^3. Gas volume is 25% higher than that estimated in Example 1, or 1 250 m^3/day.

Trickling Filter. At a volumetric loading of 5 PE/m^3 (Table 7.5), the total required volume is 50 000/5 = 10 000 m^3. Select three trickling filters, each with a diameter of 36 m and 3.6 m average depth.

Final Clarifier design according to Figure 6.12. Detention time 3 hr, hence the useful volume is 3 (hr) × 625 (m^3/hr) = 1 875 m^3. With a selected average depth of 2.4 m, the surface area of the clarifier should be 1 875/2.4 = 781.25 m^2. Select a circular clarifier with a diameter of 32 m, 2.4 m deep, or two rectangular clarifiers each 40 m long and 10 m wide with the surface area of 800 m^2. The clarifier overflow rate is then

$$625 \ (\text{m}^3/\text{hr})/800 \ (\text{m}^2) = 0.78 \text{ m}^3/\text{m}^2\text{-hr})$$

which is acceptable.

Assuming that the acceptable maximum overflow rate during wet weather flow is 2.5 m^3/(m^2-hr) [about 1.5 m^3/(m^2-hr) for German conditions], the maximum wet weather flow that can be safely treated is 2.5/0.78 × DWF = 4 × DWF.

The sludge is pumped continuously by a 10 liters/s capacity pump to the inflow of the primary settling tank.

Sludge Lagoon. According to Table 9.1, the volume of the digested sludge with 3% solids is 1.5 liters/(cap-day), or 1.5 × 50 000/1 000 = 75 m^3/day or 75 ×

365 = 27 375 m^3/year. After about 5 years of detention in the lagoon the solids content of the deposited sludge will increase to about 20%. The annual sludge volume generation then becomes 27 275 × 3/20 = 4 106 m^3. The required area is 0.3 (m^2/cap) × 50 000 = 15 000 m^2. Select two lagoons, each with 7 500 m^2 surface area and 3 m depth, that will provide 45 000 m^3 or 45 000/4 106 = 11 years of sludge storage.

Example 3. *Conventional (normal-rate) trickling filter with a primary clarifier, sludge digester, and final clarifier (for the flow sheet of the plant see Fig. 12.3).*

50 000 inhabitants, combined sewers with no stormwater retention, little industrial wastewater contribution.

Pretreatment, primary clarifier, digester, and sludge lagoons per Examples 1 and 2.

Trickling Filter loading (Table 7.5) is 10 PE/m^3, therefore the volume is 50 000/10 = 5 000 m^3. Select the depth at 3 m and the surface 5 000/3 = 1 666 m^2. Select two trickling filters each with a diameter of 33 m and a total surface area of 1 710 m^2.

Assuming a one to one recycle ratio, the hydraulic surface loading of the filter is 2 × 625 (m^3/hr)/1 710 (m^2) = 0.73 m^3/(m^2-hr) × 24 (hr) = 17.5 m^3/(m^2-day). The distributors should be designed for a flow of 2 × 625 = 1 710 m^3/hr.

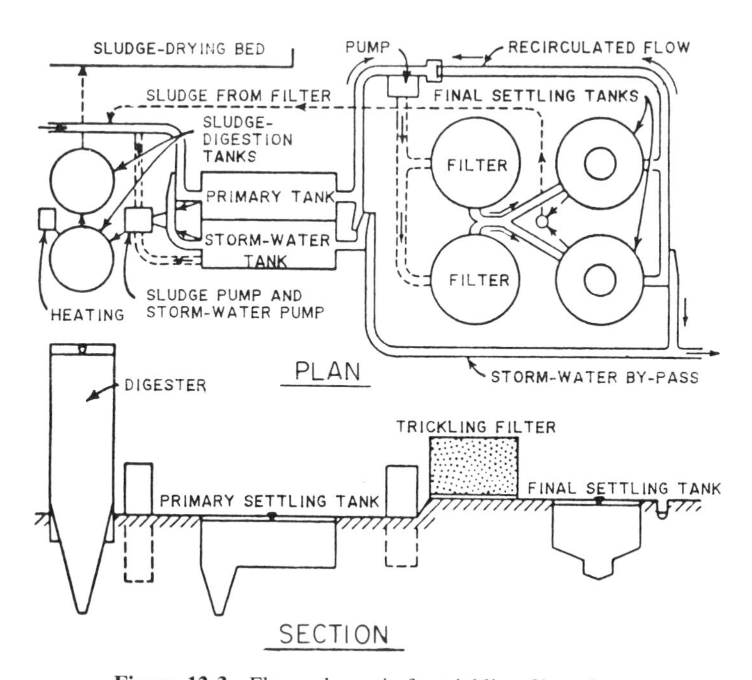

Figure 12.3 Flow schematic for trickling filter plant.

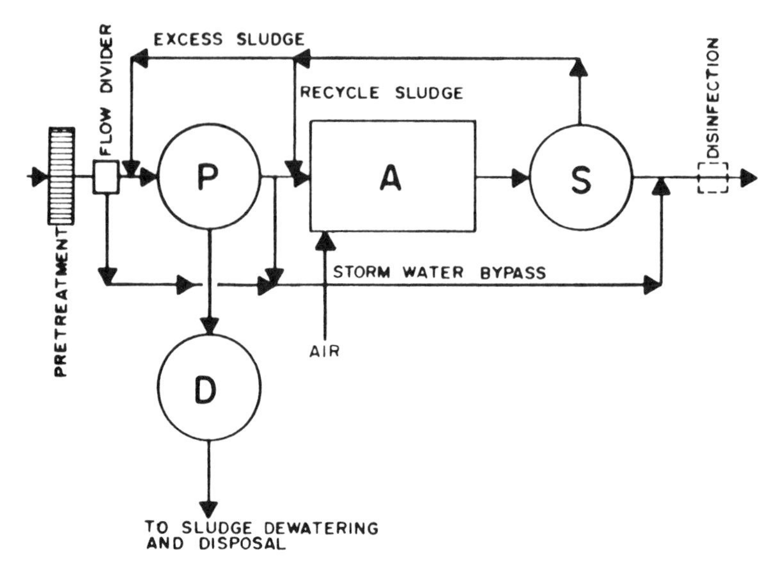

Figure 12.4 Flow schematics for activated sludge plant: P, primary clarification; A, aeration chamber; S, secondary clarifier; D, digester.

Final Clarifier. Detention time 3 hr. Useful volume 3 × 625 = 1 875 m^3. Select two circular clarifiers, each 23.3 m in diameter and 2.2 m deep, with a surface area of $3.14 \times 23.3^2/4 = 426.4$ m^2, which during dry weather flow conditions will have an overflow rate (including the recycle) of $2 \times 625/(2 \times 426.4) = 1.46$ $m^3/(m^2\text{-hr})$. During wet weather flow conditions reduce or stop the recycle. The maximum wet weather flow that can be treated (with the recycle discontinued), assuming a maximum acceptable overflow rate of 2.5 $m^3/(m^2\text{-hr})$, is $2.5/0.73 \times$ DWF = 3.4 × DWF.

Example 4. *Activated sludge unit, conventional one-stage process, with a primary clarifier, sludge digester, and secondary clarifier (for the flow schematic see Fig. 12.4).*

50 000 inhabitants, wastewater flow 300 liters/(cap-day), combined sewers with no stormwater retention, little industrial wastewater contributions. Expected BOD_5 removals 90–95%.

Wastewater flow per Example 1 625 m^3/hr or 173.6 liters/s.

Select the following:

1. *Bar screens* with 100-mm spacing.
2. *Automatically cleaned screens* with bar spacing of 25 mm. Volume of collected screenings (Table 6.1) is 3 liters/(cap-year) (higher for European cities), or 50 000 × 3 [liters/(cap-year)]/1 000 (liters/m^3) = 150 m^3/year.

3. *Aerated grit chamber* as shown in Figure 6.16. The amount of collected grit is about 5 liters/(cap-year) or 50 000 × 5/1 000 = 250 m^3/year.
4. *Primary clarifier* design according to Example 1.
5. *Pumps for primary sludge.* The volume of raw mixed sludge from the primary clarification (Table 9.1) is 2 liters/(cap-day) or 2 × 50 000/1 000 = 100 m^3/day, which is intermittently transferred throughout the day into the thickener.
6. *Thickener.* Select two thickeners each with a total volume equaling the daily volume of the raw sludge, which is 100 m^3. Depth 3 m, diameter 4.6 m.
7. *Sludge digester*, heated to 37°C. Volume 0.04 m^3/cap (Table 9.2) or 0.04 × 50 000 = 2 000 m^3.
8. *Gas volume* is about 20 liters/(cap-day) or 20 × 50 000/1 000 = 1 000 m^3/day.
9. *Sludge lagoon.* The amount of digested sludge (Table 9.1) is 2 liters/(cap-day). At 2.5% dry solids this would represent a volume of 2 [liters/(cap-day)] × 50 000/1 000 (liters/m^3) = 100 m^3/day or about 36 500 m^3/year. By storing the sludge in the lagoon for about 5 years the solids content will increase to about 20%. The annual stored volume of the sludge is then 36 500 × 2.5/20 = 4 562.5 m^3. The required area of the lagoon is 0.4 (m^2/cap) × 50 000 = 20 000 m^2. Select two sludge lagoons each with a surface area of 10 000 m^2 and a total volume of 60 000 m^3. At the annual sludge production of 4 562.5 m^3 and solids content of 20%, the lagoons will provide 60 000/4 562.5 = 13.2 years of storage.
10. *Filter press.* If storage lagoon space is not available the digested sludge can be mechanically dewatered and disposed onto a landfill. The amount of the digested sludge is 100 m^3/day with a solids content of 2.5%. By operating the filter press 5 days/week the daily volume of the sludge to be dewatered by the press is 100 × 7/5 = 140 m^3. For sludge conditioning use 8 kg of lime and 4 kg of 40% ferric chloride solution per 1 m^3 of sludge. As a result of chemicals addition, the solids content of the sludge will increase to about 3.5%. For thickening select two batch operated thickeners, each providing 140 m^3 of storage. After 1 day of thickening, the solids content in the thickened sludge will increase to about 7% and the sludge will reduce its volume to 140 × 3.5/7 = 70 m^3. Typical loading rates of filter presses are about 40 liters of sludge per 1 m^2 of the press plate per hour. The press will be operated about 7.5 hr each working day. Hence the required surface area of the press is 70 × 1 000/(7.5 × 40) = 233 m^2. Select a filter press with 1.5 m by 1.5 m plates with 2 m^2 effective surface area. The number of chambers is then 233/(2 + 2) = 59; select 65 chambers with 30-mm-filter cake thickness. At 45% solids content the volume of the filter cake to be disposed onto a landfill is 70 × 7/45 = 11 m^3/day.

11. *Aeration chamber.* Use a loading rate of 20 PE/m^3. The volume of the unit is then 50 000/20 = 2 500 m^3, which at a flow rate of 625 m^3/hr will provide about 4 hr of aeration time. Air requirements with fine bubble aeration in a 3 m deep tank is (Table 7.8) 1.7 m^3/(PE-day) or 50 000 × 1.7 = 85 000 m^3/day. The peak capacity of the blower is 1.5 × 85 000/(24 × 60) = 88.5 m^3/min. Organic loading is 20 PE/m^3 × 40 g BOD_5/(PE-day) = 800 g BOD_5/(m^3-day). Sludge concentration should be maintained at 3 000 mg/liter (same as g/m^3), which will yield a sludge load (F/M ratio) of 800/3 000 = 0.27 g BOD_5/(g of SS/day). At this F/M ratio, the expected BOD removal efficiency (Fig. 7.24) is 92% and the sludge age is 5 days. Select four chambers 60 m long, 3.6 m wide, and 3 m deep.
12. *Final clarifier* can be designed using either the solid flux concept or the limiting volumetric sludge loading. In the former procedure a relationship of the zone settling velocity versus the sludge concentration must be established by laboratory measurements.

    The limiting volumetric sludge loading of the final clarifier should be at or below 0.3 m^3/(m^2-hr). The volumetric sludge loading equals the Sludge Volume Index (SVI) expressed in m^3/kg times the sludge concentration (kg/m^3) times the hydraulic loading [m^3/(m^2-hr)].

    If the expected SVI is 150 liters/kg = 0.15 m^3/kg and the sludge concentration in the aeration basin is 3 000 mg/liters = 3.0 kg/m^3 then the limiting hydraulic loading is

$$\frac{0.3\ \text{m}^3/(\text{m}^2\text{-hr})}{0.15\ \text{m}^3/\text{kg} \times 3.0\ \text{kg}/\text{m}^3} = 0.66\ \text{m}^3/(\text{m}^2\text{-hr})$$

    For the dry weather flow of 625 m^3/hr the clarifier surface area should be 625/0.66 = 946 m^2. During wet weather conditions the flow will be doubled, therefore, the surface area should be increased to accommodate the increased flow. Select a clarifier 2.4 m deep with a diameter of 44 m and a surface area of 1520 m^2. The volume of the clarifier is 2.4 × 1 520 = 3 649 m^3 and the detention time during dry weather flow is 3 649/625 = 5.83 hr and 2.9 hr during wet weather flow.
13. *Wasted sludge.* According to the Table 9.1 the amount of wet secondary sludge (sludge yield) is about 5 liters/(cap-day) or 5 × 50 000/(24 × 60 × 60) = 2.9 liters/s. The sludge will be withdrawn from the clarifier underflow-recycle and returned to the inflow of the primary clarifier or into the aerated grit chamber.

Example 5. *Rapid infiltration system with primary clarification and sludge digestion.*

50 000 population connected to combined sewers, wastewater production 300 liters/(cap-day).

1. *Primary clarification* and sludge digestion according to Example 1.
2. *Rapid infiltration* on sandy soils, sand layer at least 1.0 m, groundwater level minimum 1.5–2 m below the surface. According to Table 7.2 select loading of 2 000 population equivalents per 1 ha. The required area is 50 000/2 000 = 25 ha.
3. *Hydraulic loading*, dry weather flow

$$50\ 000 \times 0.3\ \left[\mathrm{m^3/(cap\text{-}day)}\right] = 15\ 000\ \mathrm{m^3/day} \times 7\ \mathrm{days}$$

$$= 105\ 000\ \mathrm{m^3/week}$$

$$\text{Surface loading} = 105\ 000\ (\mathrm{m^3/week})/25\ (\mathrm{ha}) \times 10^{-4}\ (\mathrm{ha/m^2})$$

$$= 0.42\ \mathrm{m/week} = 42\ \mathrm{cm/week}$$

Referring to Figure 7.2, select 50 flooding basins each with 25 (ha) × 10 000 (m^2/ha)/50 = 5 000 m^2 area. Each basin may be, for example, 50 m wide and 100 m long.

4. *Application.* The system is operated daily during daytime hours (e.g., from 7 A.M. to 7 P.M.). During the operation hours the daily sewage volume of 15 000 m^3 is flooded onto the soil at a depth of about 0.12 m. The number of basins flooded each day is then 15 000 (m^3)/0.12 (m) = 125 000 m^2 = 25 basins. The basins are therefore operated in a 2-day cycle, flooding followed by 1 day drying and recovery.

   Each basin should be flooded in a period of about 15 min. The daily volume of wastewater applied to each basin is then 15 000/25 = 600 m^3.
5. *Storage.* Storage basin is needed for nighttime (12 hr) wastewater flow. From Figure 5.1 the average night flow is 0.65 times the daily average, or 0.5 × 0.65 × 15 000 = 4 875 m^3. Design a basin 2 m deep with a surface area of 2 500 m^2. The basin may require aeration to keep the wastewater from turning anaerobic.
6. *Conduits.* Conduits to each basin should be capable of delivering the total application within a 15-min period. The total application per basin is 600 m^3. Then the flow rate is 600 (m^3)/0.25 (hr) = 2 400 (m^3/hr)/3 600 (s/hr) = 0.66 m^3/s.

   Using Nomograph 2 or 5 from Appendix 1, the diameter of the pipeline from the storage basin to the infiltration basins should be 0.65 m if the slope is 10 m/km = 1%.
7. *Underdrains.* If the system is not designed for groundwater recharge the treated wastewater must be collected by drainage. Select 20-cm-diameter perforated drainage tiles located in the middle of each basin and laid in a gravel bed. The Manning roughness factor for PVC or clay tiles is about 0.012. If the drainage pipes are laid at a 1% slope (10 m/km), using

Nomograph 5 in the Appendix, the capacity flow can be found as $Q = 37$ liters/s and the velocity in the tile is 1.2 m/s.

With the flow of 600 m^3 into the basin the drainage time will be about $600/0.037 = 16\ 276$ s or about 4.5 hrs.

BIBLIOGRAPHY FOR PART II

R. M. Arthur (ed.), *Procedures and Practices in Activated Sludge Process Control.* Butterworth, Boston, MA and London, 1983.

H. E. Babitt, *Sewerage and Sewage Treatment.* Wiley, New York, 1958.

D. Barnes and F. Wilson, *The Design and Operation of Small Sewage Works.* E. & F. N. Spon, Ltd., London, 1976.

J. Brix, K. Imhoff, and R. Weldert, *Die Stadtentwässerung in Deutschland* (*Urban Drainage in Germany*). Fischer, Jena, 1934.

B. Böhnke et al., *Abwassertechnologie* (*Wastewater Technology*). Springer Verlag, Berlin, 1984.

P. G. Bourne (ed.), *Water and Sanitation.* Academic Press, Orlando, FL, 1984.

K. W. Brown, G. B. Evand, and B. D. Frentrup (eds.), *Hazardous Waste Land Treatment.* Butterworth, Boston, MA and London, 1983.

G. C. Cushnie, Jr., *Removal of Metals from Wastewaters, Neutralization and Precipitation.* Noyes Data Corp., Park Ridge, NJ, 1984.

G. C. Cushnie, Jr., *Electroplating Wastewater Pollution Control Technology.* Noyes Data Corp., Park Ridge, NJ, 1985.

G. W. Dawson and B. W. Merier, *Hazardous Waste Management.* Wiley, New York, 1986.

W. W. Eckenfelder, Jr. and D. J. O'Connor, *Biological Waste Treatment.* Pergamon Press, New York, 1961.

L. B. Escritt and W. D. Hawororth, *Sewerage and Sewage Treatment, International Practice.* Wiley, Chichester and New York, 1984.

G. M. Fair, J. C. Geyer, and D. A. Okun, *Water and Waste-water Engineering.* Wiley, New York, 1966.

K. Imhoff, W. J. Müller, and D. K. B. Thistlethwayte, *Disposal of Sewage and Other Waterborne Wastes.* Butterworth, London, 1971.

K. R. Imhoff, *GWF, das Gas-und Wasserfach: Wasse/Abwasser* **120,** 563 (1979).

IUPAC, *Reuse of Water in Industry.* Buterworth, London, 1963.

J. B. Johnston and S. G. Robinson, *Genetic Engineering and New Pollution Control Technology.* Noyes Data Corp, Park Ridge, NJ, 1984.

H. Liebmann, *Handbuch der Frischwasser-und Abwasserbiologie* (*Handbook of Freshwater and Wastewater Biology*). Oldenbourg Verlag, Munich, FRG, 1960.

G. F. Lingren, *Guide to Managing Industrial Hazardous Waste.* Butterworth, Boston, MA and London, 1983.

R. C. Loehr, *Pollution Control for Agriculture.* Academic Press, Orlando, FL, 1984.

K. Mudrack and S. Kunst, *Biology of Sewage Treatment and Water Pollution Control.* Halsted Press (Div. of Wiley), New York, 1986.

J. W. Patterson, *Wastewater Treatment Technology.* Ann Arbor Sci. Publ., Ann Arbor, MI, 1977.

J. W. Patterson, *Industrial Wastewater Treatment.* Butterworth, Boston, MA and London, 1985.

R. B. Pojasek, *Toxic and Hazardous Waste Disposal,* 4 Vols. Ann Arbor Sci. Publ., Ann Arbor, MI, 1980.

D. F. Soule and D. Walsh (eds.), *Waste Disposal in the Oceans, Minimizing Impact, Maximizing Benefits.* Westview Press, Boulder, CO, 1983.

W. J. Weber, Jr., *Physicochemical Processes for Water Quality Control.* Wiley (Interscience), New York, 1972.

PART III

PROTECTION OF RECEIVING WATERS

CHAPTER 13

Discharge of Wastewater into Surface Waters

13.1 USE OF SURFACE WATERS

Effects of Wastewater Discharges

The degree of treatment of wastewater discharges in many cases depends on the desired water quality of the receiving waters into which the effluent from the treatment plant is to be discharged. Earlier, engineers were interested in establishing a "safe dilution ratio" for treatment plant effluents at which the dry weather outflow from the plant would safely mix with an average low flow in the receiving stream. For example, in England such dilution ratios were mandated in 1915 by the Royal Commission on Sewage Disposal and ranged from 1:500 (1 part of sewage per 500 parts of stream flow) for effluents receiving only screening and grit removal to 1:50 for biologically treated effluents. In Germany, a rule-of-thumb type dilution ratio of 1:20 is sometimes used for a stream receiving discharge from a single biologically treated effluent.

The dilution ratio does not indicate the pollution content of the receiving waters, its loading, or whether there are other wastewater discharges located on the stream. To be more exact, an overview of the entire dissolved oxygen balance of the stream must be delineated. Later in the next chapter different procedures will be presented by which the dissolved oxygen content of receiving waters and its response to wastewater discharges can be estimated. Using such computations, it is possible to estimate the necessary degree of treatment that with self-purification of the stream will maintain a desirable minimum oxygen level, provide for a viable fish population, and assume a margin of safety against damages caused by inadequate oxygen concentrations.

For streams with multiple outfalls the dissolved oxygen computation will provide only basic insight into the problem. The dissolved oxygen balance is affected by a number of factors and also often depends on the photosynthesis of macrophytes (water plants) and microphytes (green microorganisms, also called phytoplankton). In such cases, monitoring is necessary. One of the first continuous dissolved oxygen monitoring systems was established in the 1960s on the Ohio River.[1] In this monitoring program, dissolved oxygen is continuously measured, transmitted, and statistically evaluated. The resulting frequency curves then provide a very useful source of information.

In many cases the dissolved oxygen computation alone is not sufficient. For example, streams from which water is used for water supply or for contact recreation require other parameters to be evaluated such as bacterial contamination or levels of various toxic substances.

All proposals for inclusion of treatment plants in sewage disposal systems must consider the quantity of sewage flow. It may often be more economical to employ on-site (septic tank) sewage disposal methods for individual houses or small subdivisions. In such cases, suitability of soils, groundwater conditions, and housing density are the most important factors to be considered.

Nutrient Problem.[2–9] In some instances, tertiary treatment may be required where nutrients (nitrogen and phosphorus) are removed from the treated wastewater effluent and prevented from entering the receiving water bodies.

In lakes and in streams with velocities below 0.3 m/s and sufficient light, it is thought that prolific algae growths may occur when more than 0.3 mg N/liter and 0.02 mg P/liter are available in the water. As a result of the life processes of plants and phytoplankton, the oxygen content of water bodies fluctuates with a high value in the late afternoon and a low value during the night. The magnitude of these fluctuations is proportional to the intensity of the algal life processes and their quantity. When the water plants and algae die, an oxygen demand is imposed on water by the decomposing organic matter. Most of the dead plant organic matter will settle into the benthic (bottom) deposits and, as a result, subsequent decomposition will recycle the phosphorus for new algal growth during the next growing season.

Stumm[10] stated that 1 g of phosphorus in the algal life cycle can bring about an ultimate oxygen demand of 160 g. This corresponds, according to the information given in Table 5.2 to an ultimate oxygen demand of raw wastewater of $3.0 \times 160 = 480$g/(cap-day).

Through these processes, the available oxygen can be completely depleted. Deep lakes and estuaries with a small inflow are particularly vulnerable. Algae can disrupt the water treatment process by clogging the filters and also affect the taste of the produced drinking water.

Domestic sewage is not the only source of nutrients. Agriculture also has a significant impact as well as urban runoff. German data attribute approximately 3.2 g of phosphorus/(cap-day) to sewage and 500 g of P/(ha-year) to agricultural operations. Data from experimental watersheds in the Great Lakes Region of the United States and Canada yielded unit loading of phosphorus from agricultural farm operations ranging from 0.1 to 10 kg P/(ha-year). Corresponding loadings of nitrogen ranged from 1 to 50 kg N/(ha-year).[11] In the Ruhr area of Germany, removal of 90% of the phosphorus from sewage would not bring about control of algal growths in the receiving lakes. However, Nusch and Koppe[12] have shown that at least massive and lasting algal blooms can be limited.

In some difficult to control cases, biologically treated sewage and runoff discharge into a water body must be completely eliminated. This may be accomplished by land disposal (irrigation) of treated sewage effluents and by building sewer bypasses around the lake with a discharge located downstream from the lake.[13,14] For various treatment methods used for carbon, nitrogen, and phosphorus removal from wastewaters see Chapter 8.

Water Quality Standards

The demands of receiving water bodies on treatment plants for municipalities and industries mostly depend on the loadings of biodegradable organics and pathogenic bacteria. In addition, it is necessary to evaluate and limit the loadings of toxic materials, acids, mineral oils, radioactive materials, pesticides, and other toxic organic chemicals and detergents.

In order to effectively control water quality, it is necessary to describe it in precise, technical quantitative terms. For the purpose of quantifying water quality, water is analyzed by various physical, chemical, and biological techniques that yield numerical values of concentrations of various substances present in the samples. One sample or analysis is not sufficient to classify the water quality of a stream or wastewater effluent. The stream or effluent classification should be based on a statistical evaluation of a number of samples taken at various locations, flows, and times. Presently, water quality models are used to interpret water quality monitoring data and to ascertain water quality at extreme low flow or future low flow conditions. The process of quantifying water quality then involves a comparison of the statistical or computed water quality characteristics obtained from appropriate water quality analyses and/or by models with a set of water quality standards.

In addition to effluent standards that prescribe allowable concentrations of various pollutants in the waste effluent or outfall, stream standards exist that are designed to protect the water quality of receiving waters for various uses including human health protection and propagation of aquatic life.

There are basically two types of standards used throughout the world.[14] Most of the European countries use a system of classifying streams into four "water quality classes" with few possible subclasses. The origin of this system can be traced to the Kolkwitz–Marson taxonomic saprobic classification, which classifies water quality according to the tolerance of various aquatic organisms to pollution. Such standards are in force in Germany[15] or in Great Britain[16] and most other European countries (including eastern Europe and the Soviet Union). Table 13.1 presents an example of the British standards. The present U.S. stream standards are based on the Water Pollution Control Act Amendments of 1972 and 1977, which gives authority to states to establish stream standards according to the Guidelines of the Environmental Protection Agency.[17] The standards are related to intended water uses of the state surface water. The beneficial uses that determine which set of standards is to be used include the following: (1) municipal, industrial, and domestic water supply, (2) water contact recreation, that is, swimming and water skiing, (3) noncontact recreation including boating, fishing, and aesthetics, (4) fish and wildlife protection and propagation, and (5) agricultural irrigation. A summary of U.S. stream standards is presented in Table 13.2.

TABLE 13.1 Water Quality Criteria for British Streams[a]

Quality Class	Limits (Observed 95% of the Time)	Remarks	Use of the Stream Water
1 A	Oxygen saturation >80%; ammonium (NH_4^+) ≤ 0.4 mg/liter, BOD_5 ≤ 3 mg/liter; not toxic to fish	Average BOD_5 ≤ 1.5 mg/liter, no visible pollution	High-quality water usable for drinking and other uses including contact recreation and fishing
1 B	Oxygen saturation >60%; BOD_5 ≤ 5 mg/liter; ammonium (NH_4^+) ≤ 0.9 mg/liter; not toxic to fish	Average BOD_5 ≤ 2 mg/liter; average ammonium ≤0.5 mg/liter; no visible pollution	Water of lesser quality than 1A but usable for the same uses as 1A
2	Oxygen saturation >40%; BOD_5 ≤ 9 mg/liter; not toxic to fish	Average BOD_5 ≤ 5 mg/liter; no visible pollution other than slight color because of humic compounds, little foam formation below dams	Usable for drinking after physical–chemical treatment. Fishing possible for rough fish, moderate amenity value
3	Oxygen saturation >20%; aerobic conditions; BOD_5 ≤ 17 mg/liter		Fish absent or only sporadically present; low level industrial use; other uses only after treatment
4	Oxygen saturation worse than Class 3, sometimes anaerobic		Waters that are grossly polluted and are likely to cause nuisance

[a] Source: National Water Council, United Kingdom.[16]

In the United States both effluent and stream standards are in force. The minimum treatment requirement, regardless of the size of the receiving water body and its ability to assimilate wastewater effluents, is mandated by the effluent standards. For example, all municipal sewage sources must receive biological treatment resulting in average monthly BOD_5 (excluding the effect of nitrification) and suspended solids concentrations of less than 30 mg/liter. Industrial sources must implement the best available, economically feasible treatment technology. In many situations this may not be sufficient and the discharge may still result in a water quality situation that would violate the stream standards. Then the state regulatory agency will allocate the permissible waste loads according to stream standards. These are called water quality (stream) limiting situations.

TABLE 13.2 U.S. Stream Water Quality Criteria[17]

Water Use: Aquatic Life Protection	
Alkalinity	>20 mg $CaCO_3$/liter
Ammonium	<0.02 mg NH_3/liter
Beryllium	
Soft water	<11 μg/liter
Hard water	<1,100 μg/liter
Cadmium	
Sensitive fish	
Soft water	<0.4 μg/liter
Hard water	<1.2 μg/liter
Less sensitive fish	
Soft water	<4.0 μg/liter
Hard water	<12 μg/liter
Marine life	<5 μg/liter
Chlorine	
Salmonid fish	<2 μg/liter
Other	<10 μg/liter
Chromium	<100 μg/liter
Copper	<0.1 × 96-hr LC_{50}[a]
Cyanide	<5 μg/liter
Dissolved gases	<110% of saturated concentration
Iron	<1 mg/liter
Lead	<0.01 × 96-hr LC_{50}[a]
Mercury	
Freshwater	<0.05 μg/liter
Marine	<0.10 μg/liter
Oil and grease	<0.01 × 96-hr LC_{50}[a]
Dissolved oxygen	>5 mg/liter
pH	
Freshwater	6.5–9
Marine	6.5–8.5
Phenol	<1 μg/liter
Elemental phosphorus marine water	<0.1 μg/liter
Phthalates	<3 μg/liter
PCBs	<0.001 μg/liter
Selenium	<10 μg/liter
Silver	<0.01 × 96-hr LC_{50}[a]
Hydrogen sulfide	<2 μg/liter
Temperature	Criteria based on geographical conditions, fish type, and their resistance to temperature changes
Zinc	<0.01 × 96-hr LC_{50}[a]
Organic chemicals	
Aldrin/Dieldrin	<0.003 μg/liter
Chlordane	
Freshwater	<0.01 μg/liter
Marine	<0.004 μg/liter
DDT	<0.001 μg/liter
Demeton	<0.1 μg/liter
Endosulfan	
Freshwater	<0.003 μg/liter
Marine	<0.001 μg/liter
Endrin	<0.004 μg/liter
Guthion	<0.01 μg/liter
Heptachlor	<0.001 μg/liter
Lindane	
Freshwater	<0.01 μg/liter
Marine	<0.04 μg/liter
Malathion	<0.1 μg/liter
Methoxychlor	<0.03 μg/liter
Mirex	<0.001 μg/liter
Parathion	<0.04 μg/liter
Toxaphene	<0.005 μg/liter

TABLE 13.2 (*Continued*)

Water Use: Domestic Water Supply (Health)

Arsenic	<50 μg/liter	Organic chemicals	
Barium	<1 mg/liter		
Cadmium	<10 μg/liter	2,4-D	<100 μg/liter
Chromium	<50 μg/liter	1,4,5-TP	<10 μg/liter
Color	<75 CU	Endrin	<0.2 μg/liter
Copper	<1 mg/liter	Lindane	<4 μg/liter
Iron	<0.3 mg/liter	Methoxychlor	<100 μg/liter
Lead	<50 μg/liter	Toxaphene	<5 μg/liter
Manganese	<50 μg/liter		
Mercury	<2 μg/liter		
Oil and grease	Not present		
pH	5–9		
Phenol	<1 μg/liter		
Selenium	<10 μg/liter		
Silver	<50 μg/liter		
Chlorides and sulfates	<250 mg/liter		
Tainting substance	Not detectable by organoleptic tests		
Zinc	<5 mg/liter		

Water Use: Contact Recreation

Fecal coliforms: Based on a minimum of five samples taken over a 30-day period the fecal coliform bacterial level should not exceed a log mean of 200/100 ml, nor should more than 10% of the total samples taken during any 30-day period exceed 400/100 ml.

Aesthetics:

All waters be free from substances attributable to wastewater or other discharges that:
(1) settle to form objectionable deposits;
(2) float as debris, scum, oil, or other matter to form nuisances;
(3) produce objectionable color, odor, taste, or turbidity;
(4) injure or are toxic or produce adverse physiological responses in humans, animals, or plants; and
(5) produce undesirable or nuisance aquatic life.

Water shall be virtually free from substances producing objectionable color.

[a]96-hour LC_{50}—the concentration of a toxic compound that is lethal (fatal) to 50% of the organisms tested under the test conditions in a 96-hr lasting assay.

The Federal Water Pollution Control Act Amendment of 1972 and Clean Water Act of 1977 require that all navigable U.S. waters be suitable for contact recreation and fish and wildlife propagation and protection. Table 13.2 contains water quality criteria suggested by the Environmental Protection Agency to the states for their consideration for incorporation in the state stream standards. In addition, EPA listed criteria for a number of toxic substances.

It should be noted that there is a difference between water quality criteria and water quality standards. Water quality criteria are based on scientific data and have no legal basis. On the other hand, water quality standards are legally enforceable but are not necessarily founded on science.

13.2 TYPES OF SEWAGE OUTFALLS AND WATER INTAKES

The decomposition and removal of pollutants in receiving waters will be easier if the wastewater is kept aerobic. Anaerobic sewage discharges are toxic to fish and they use more oxygen at the point of discharge because of their often significant immediate oxygen demand. That is why U.S. environmental regulations require that sewage effluents must be aerated before they are discharged into a receiving water.

If an effluent contains different wastewater streams these should be either mixed together and allowed to react before the discharge into the receiving waters or individually separated and treated.

It is important to consider the *spatial* distribution of outfall locations. For example, in quiescent waters the vicinity of the wastewater outfalls may be devoid of oxygen. Therefore, it is necessary to bring the sewage in contact with the largest possible volume of the receiving water flow to optimize mixing. This can be accomplished by multiple discharge points or by pumping or recycling clean or purified water and mixing it with the sewage from the outfall. Modern sewage outfall designs use mathematical models to determine the spatial extent of the mixing zones (a zone in the receiving water body between the outfall and a cross section in which the receiving water flow is fully mixed with the waste discharge). The determination of mixing zones and proper outfall design may be quite important, depending on the regulatory agency. Poorly evaluated and designed mixing zones may result in excessive pollution near river banks and lake and sea beaches, reducing or even negating their use for recreation. It may sometimes be necessary to relocate the outfall downstream from an important beach or water intake.

The problem of mixing zones and their engineering evaluation can be roughly divided into two categories: the design of the location and configuration of waste outfalls so that high initial dilution and spread of the waste are achieved and the evaluation of transverse and vertical mixing so that proper location and configuration of the outfall will result in the smallest mixing zone.[14]

The best sewage mixing with the receiving water flow is accomplished with perforated pipes (diffusers) laid on the bottom. Sewage is generally warmer, and hence lighter than the receiving water and tends to rise through it and mix with it. In salt or brackish water, this favorable difference in density is accentuated. Horizontal or inclined directions of the outlets (ports) of the diffuser are more effective than vertical discharge, because the longer path followed by the rising sewage offers more opportunity for mixing. Sewage outfall pipes of large coastal and Great Lakes communities are often several kilometers long. The hydraulic design of sewage effluent outfalls into inland and coastal waters has been covered by Fisher et al.[18]

The *temporal* distribution of the effluent discharge volumes and loads may be more important than the spatial outfall location when the discharges and receiving water flows are highly variable. Retention basins, which retain a portion of the effluent during a time of pollution emergency and release the accumulated sewage gradually afterward, can simplify the wastewater discharge problem. In Germany, proposals have been made for streams serving both for water supply and wastewater disposal purposes that assign alternate daily times of intake for water supply and wastewater discharges. This requires storage basins and coordination at both locations. Using the storage basins, the discharge of heavily polluted industrial wastes can be limited to times of high flows in the receiving stream.

The water management operation of an industry can also take advantage of storage ponds for process water. It is possible to use stream water during times when it has adequate quality and interrupt the withdrawals when the stream is polluted. The use of retention basins for treated wastewater and storage basins for process (clean) water can sometimes be coordinated so that wastewater discharges will not adversely affect the process water.[19]

The quality of wastewater discharges may also be adjusted to the changing requirements of the receiving water with a variable degree of treatment. This may be advantageous for combined sewer systems, when it may be more safe and beneficial to temporarily overload some treatment units than to let the sewer content overflow without any treatment.

13.3 IN-STREAM WATER QUALITY MANAGEMENT ALTERNATIVES

Every surface water body that receives wastewater has a certain waste assimilative capacity that can be quantitatively ascertained. The waste assimilative capacity represents a pollution load that can be discharged from a source or a number of sources without causing harm to the biota in the receiving water or to the downstream uses of the water body. As long as the pollution load from wastewater remains below this permissible limit there will be no damage done to the water body and it may even be enhanced since the discharged organic matter and other components may provide nourishment to fish (fish ponds in Europe have been receiving this kind of nourishment for centuries). When possible and economically feasible, this natural waste assimilative capacity can be increased by technical means analogous to wastewater treatment.[20] In the previous U.S. editions of this book, Imhoff and Fair[21] listed several suitable engineering water quality management operations that included impounding reservoirs, stream canalization, regulated dilution, in-stream aeration, scour, and dredging.

1. Impounding reservoirs can be constructed (1) to enlarge the water surface, (2) to lengthen the detention time, and (3) to induce deposition and reten-

tion of suspended solids as sludge. Ponds, lagoons, and lakes, natural or man-made, are, in all essentials, biological purification plants, which operate in a manner similar to wastewater treatment plants. However, loading by residual solids can be allowed only in very limited amounts. For this reason, sedimentation should precede the discharge of wastewater into streams that include impoundments.

The use of reservoirs as polishing treatment units is feasible only up to a certain pollution level, determined by a positive oxygen balance.[22] When the pollution input is too large compared to the flow and surface area of the reservoir, the entire life process in the impoundment can be inhibited. It may then need more oxygen than can be provided, the water will become anaerobic, oxygen-dependent biological life may die, and additional damage may be done to the impoundment as these processes continue.

Deep reservoirs that stratify in the summer are often unsuitable for the assimilation of large quantities of wastewater.[23]

2. Stream canalization and stream lining of heavily polluted streams (e.g., the Emscher River in Germany) have been used in the Ruhr district and elsewhere as a second management alternative. The conversion of excessively polluted streams into large open drainage ditches with steep hydraulic gradients and smooth lining to promote high velocities of flow will (1) constrict the water surface but increase the rate of reaeration, (2) shorten the time of flow, and (3) prevent the settling of suspended solids. As a result, raw or insufficiently treated wastewater can be transported over long distances without creating a nuisance. Oxygen is taken into solution at the water surface, and the pollution load is carried along until the water course to which the canalized stream is tributary becomes large enough to accept the imposed load.

   Malz and Müller-Neuglück[24] have shown by observations that the bacterial count in the air in the vicinity of open sewer channels does not increase.

3. Low flow augmentation or increase of low-water flows, that is, dilution with controlled volumes of clean water, normally has two objectives: (1) it increases the available amount of dissolved oxygen (DO), and (2) it increases the surface area and reaeration capacity of the receiving water. The diluting water may come from one of three sources: (1) an upstream reservoir on the watershed, (2) a neighboring catchment (note that inter-basin transfer of flows may not be legally feasible in some parts of the United States), and (3) by recycling from lower, cleaner, and more diluted reaches of the stream.[25] In addition to their flood control benefits, upstream storage reservoirs can reduce the cost of treatment by providing dilution water for low flow augmentation.

4. Flushing or the cleansing action of periodic floods is important. Streams not subjected to flood scour miss their seasonal house cleaning and are forced to cope with greater quantities of sludge deposits than flood scoured streams. Bottom scour is reduced or lost when streams overflow their banks, when impounding works are constructed in river valleys, or when flood

waters are diverted into other watersheds or separate floodways. Periodic scour can be induced by flow control measures and it can be very effective in stream reaches in which sludge can accumulate, such as lake and oceanic estuaries and harbors. The best time to flush-out the deposits is shortly before the beginning of the warm season so that this will not exert an additional demand upon the available oxygen supply during this critical period of the year.

By controlled release of water from dams and storage reservoirs, the water flow is increased to a point at which the shear force of the flow is sufficient to move the sludge deposits away. The accumulated organic (sludge) deposits in the slow reaches are mostly anaerobic and upon resuspension they exhibit an immediate oxygen demand. Sufficient oxygen must be available in the flush water to prevent an undesirable drop in the dissolved oxygen concentrations.

Flushing tunnels were built in the late 1800s in Milwaukee, Wisconsin, whereby clean Lake Michigan water was pumped several kilometers upstream into slow-moving harbor reaches to alleviate severe water quality problems. The tunnels are still in use and, after renovation, they have been incorporated in the most recent water quality management–pollution abatement program.

5. Dredging. Instead of flushing the sediments and accumulated sludges, dredging can be considered as a fifth management alternative that helps to increase self-purification capacity. Deposited sediments may adversely affect water quality as a result of their bottom oxygen demand and their release of accumulated nutrients that can accelerate eutrophication. However, the dredging process, if not performed properly, may resuspend large quantities of the sediment, cause an immediate oxygen demand, and release stored and inactivated nutrients into water.[26] 1.1 million m^3 of polluted sediments have been dredged from the Baldeney reservoir on the Ruhr river without severe environmental consequences.
6. Aeration. Whenever the surface of a body of water is disturbed, absorption of oxygen from the atmosphere (reaeration) is accelerated. Cascades, sprays, spillways, and agitation all increase reaeration. When the oxygen level in a fishing stream or lake is likely to drop below the limit for fish protection (4–6 mg/liter depending on the type of fish) during either winter ice periods or summer droughts fish kills can be avoided with artificial turbine or surface aeration. Alternatively, the water surface can be sprayed or air can be injected under pressure through diffusers laid on the bottom.

   The efficiency of aeration increases as the oxygen deficit increases. The Lippe River Association in the Ruhr area of Germany used aeration during the extremely dry summer of 1959 to keep an overloaded stretch of the river aerobic and odorless.[27] By spraying oxygen-devoid wastewater in the air it was found that in a single spray 3 mg/liter of oxygen was added over a spray distance of 1 m.[28] Similar oxygen enrichment was measured as a result of splashing. The splash effect of weirs or spillway dams can enrich water with oxygen, particularly during times when turbine aeration is not available.[29] Several articles have described the aeration induced by turbines in hydropower plants.[30,31] These systems result in a power output loss. On the average 0.8–2.5 kg of O_2 can be supplied per kilowatt of power loss.[32,33] Various aeration devices and their efficiencies were summarized by Thackston and Speece[34] and Hunter and Whipple.[35] For information on the aeration of wastewater lagoons and oxidation ditches see Sections 7.4 and 7.5.

Investigations into the aeration of the lake Baldeney and the lower Ruhr reaches examined the efficiency and economy of floating centrifugal aerators, diffuser pipes, turbine aeration, and aeration from weirs. It was concluded that the turbine and weir aeration schemes provided the cheapest alternatives.[36,37] The Tennessee Valley Authority experience with turbine aeration was described by Davis et al.[38]

Destratification of impoundments, intake of water from the hypolimnion, and hypolimnetic aeration can improve the dissolved oxygen conditions of deep (stratified) reservoirs. In additions, they can minimize reducing conditions in the lower layers of the reservoirs, prevent or reduce the release of nutrients (phosphorus, iron, and manganese) from benthic layers, and assist in the control of eutrophication.[14,39–41]

Design computation examples of artificial stream and lake aeration will be presented as Procedure VII in Section 14.3.

7. Nutrient inactivation by chemical additions can be considered the seventh management alternative. This alternative can accomplish the following: (1) change the form of the nutrients from available forms to less available or unavailable to algae, (2) remove the nutrients from the zone in which algae can grow, and (3) prevent the release or recycling of nutrients back into the upper layers.[13] Research results indicate that good removal of phosphorus can be achieved by adding metallic coagulant aids, namely aluminum sulfate or aluminum chloride, if pH decrease in the water body is to be minimal. The precipitated flocs that settle to the bottom may provide a protective cover reducing nutrient release from benthal layers.

   Other chemicals that have been used or suggested for control of temporary emergency water quality problems include nitrate salts or chlorine. Use of these chemicals should be considered as a last resort since they may have a detrimental impact on the biota.

   Liming of lakes affected by acid rainfall is commonly used if the watersheds do not possess enough buffering capacity to neutralize the increased acid inputs.

   The control of nuisance algal blooms can be achieved with a variety of chemicals. Copper sulfate is the most common, but other chemicals including organic herbicides have also been used. The dosage of copper sulfate varies from 0.05 to 4 mg/liter depending on the form and stage of the algal development.[14,42] Dosages greater than 1 mg/liter can be damaging.[43]

8. Lakes and streams that are overgrowing with aquatic plants (weeds) can be controlled temporarily by mechanical floating cutters, however, the cutting may cause the weeds to proliferate.
9. Water quality of lakes exhibiting the symptoms of accelerated eutrophication can also be controlled by fish management and stocking to enhance grazing of green aquatic organisms.
10. Finally, water cooling or reduction of heated discharges will be beneficial since the decomposition rates will be slowed down.

13.4 WATER REUSE (WATER–SEWAGE–WATER CYCLE)

Principles of Wastewater Reuse and Reclamation

Each technical utilization of surface waters must give a thorough consideration to the water reuse–recycle. A stream represents a physical link in this recycle concept whereby water intakes and wastewater outfalls are interconnected. The question of wastewater disposal cannot be solved without simultaneously considering water supply. In every urban area, the water–wastewater recycle is always present to some degree. With exception of the backwaters of mountain streams, every city that uses river water for its source of water supply cannot avoid residuals from upstream wastewater discharges. For example, the Mississippi River after receiving wastewater discharges from a large portion of the U.S. population, is the source of drinking water for New Orleans, Louisiana and other downstream cities. It is not therefore correct to assume that clean water without any contamination is always available and that wastewater discharges can be directed to waters that will not result in human contact. On the contrary, as groundwater originates from surface water it could become contaminated to some degree. In addition, all wastewater discharged into a receiving water body will ultimately be used again, including use as a drinking water source. Therefore, care must be exercised so that the time between reuse is not too short or that there are no "shortcuts."

This train of thought was first introduced in a classical treatise by K. Imhoff in 1931.[44] This article described a situation that occurred in the Ruhr area during the extremely dry summer of 1929. Without any detrimental effect, the water–sewage–water cycle was entered three times (300% recycle). Because the water demand was increasing, a series of recycling pumping stations was constructed on the Ruhr River that enabled backpumping of Rhine River water upstream to the Ruhr area. In the drought of 1959, the high water demand could be met only by pumping over 100 millions m^3 of the Rhine water upstream through the Ruhr dams. The recycled water was reused by artificial recharge of groundwater aquifers.[45] In the 4-month long emergency drought period, the quality of the drinking water was very poor and about 5–10% of the population of the city of Essen suffered from gastroenteritis. Although the bacterial contamination of drinking water was safe, interaction between the poor quality and excessive chemical content was suspected as a cause.[46] In Kansas during the 1948 drought, the cycle was repeated 17 times.[47] In the 5-month long drought period in Chanute, Kansas, after eight recycles, extremely poor water quality conditions persisted.

Wastewater Reclamation. Reuse can be planned or unplanned, controlled or uncontrolled. Every community water treatment plant that exists below the discharge of a wastewater treatment plant is in the mode of uncontrolled reuse. Planned and controlled reuse are now in existence in many parts of the world, including Israel, South Africa, the Netherlands, the United States, and the Ruhr area of Germany. In the United States, reuse projects have been underway in California, Arizona, Long Island, New York, and Fairfax County, Virginia, as well as other places. For example, since 1980, the Santa Barbara (California) regional Water Reclamation project has been reusing wastewater for landscape irrigation and, simultaneously, recharging the groundwater aquifer. The cost of the reclaimed water compares favorably with other alternative water sources. The state of California has consequently established wastewater reclamation criteria for irrigation of food crops, irrigation of fiber and seed crops, landscape irrigation, recreational impoundments, and groundwater recharge. The reclaimed water used for groundwater recharge by surface spray shall be at all times of a quality that fully protects public health. In most cases, this requires tertiary treatment.[48] Similar water reclamation–reuse projects have been underway in Arizona and other arid states of the western United States.

To meet increasing water demands by the municipal and industrial sections of Israel, between 200 and 350 million m^3 per year of reclaimed municipal and industrial wastewater is reused by the agricultural, industrial, and municipal water consumers. Agricultural crops were divided into several groups according to relative hazards to human health when consumed and standards were issued for each group. Potable reuse guidelines recommended a minimum physical dilution of the renovated water from sewage to freshwater of 1:3 to be maintained at all times. A minimum storage retention time must be maintained between the end of the wastewater treatment process and the beginning of the water supply network. This minimum retention requirement depends on the type of storage (reservoir, groundwater aquifer, etc.). Fine media filtration must be a part of the wastewater treatment process.

In South Africa full-scale reclamation plants were installed in Windhoek, Daspoort (Pretoria), and Springs.[49] The South Africa Department of Water Affairs recommended that reclaimed water should be diluted and thoroughly mixed with freshwater with not less than 1 part of reclaimed water to 4 parts of freshwater and the mixing ratio should be kept reasonably constant at all times to prevent quality and taste variations.

Recharge by reclaimed wastewater can also be used to minimize saltwater intrusion into coastal freshwater aquifers.[50,51]

In the reuse cycle, part of the water flow is lost from the system. According to

Schroeder[52] the following losses can be anticipated:

Cooling water recycle	2–4%
Municipal drainage losses	20%
Rural water supply relying on on-site (septic tank) sewage disposal	70%
Land application of sewage	40%
Balanced irrigation	up to 100%

Loss of water by evaporation (cooling water systems) or by evapotranspiration (land disposal and irrigation) from a recycle system will cause salt buildup in the soil that should be remedied by flushing (leaching) the salts from the system. This is accomplished by replenishing the lost water in amounts that exceed the loss by evaporation or evapotranspiration. Excess wastewater (blowdown in cooling systems or return flow in land application) is polluted and cannot be disposed of indiscriminantly.

Stream Water Use Designation. Wastewater disposal may be facilitated if water is taken from a stream that is assigned only for water supply and not loaded with wastewater discharges, and wastewater disposal is diverted to another stream that is not used for any other potable or contact uses. It may then be possible to utilize less expensive treatment. Such water intake–waste disposal schemes have been implemented in the Ruhr area in which the River Emscher has been assigned solely for wastewater disposal and the entire flow of the river is treated prior to the confluence with the Rhine. Such schemes may not be possible in the eastern parts of the United States in which water law doctrines prohibit interbasin transfer of water. In addition, the aesthetics of designating a river for untreated wastewater disposal would be unpopular in the United States.

When the same stream is used for both purposes, water supply and wastewater disposal, the required degree of treatment will depend directly on the requirements and standards for the quality of withdrawn water. In the United States and many other nations, such streams would be assigned for water supply and more stringent water quality standards would apply to waste effluent discharges.

Streams assigned for industrial water supply often have a water demand greater than the low flow of the stream. This is particularly true if thermal power plants are located there and use the stream water for cooling. In this case, multiple water recycle is needed and the individual users must return the water after sufficient treatment. In the case of municipal water supply from such flow-deficient streams, there should be no objections against such schemes provided that the retention time between the outfall and the next intake is sufficient. In addition, natural and technological treatment processes must be employed to treat the effluents, particularly when ground filtration by artificial recharge through the stream bottom and man-made recharge facilities are used. In densely populated and industrialized regions, maintaining good water quality by good wastewater treatment is the best guarantee that the population and the industries will be safely and economically supplied with good quality water.

Problems with Wastewater Reclamation. As stated previously, during the reuse process, salts and certain residuals of the treatment process remain in the water and accumulate with each recycle.

It should also be noted that problems may arise with the reused water itself. For example, attacks on building materials caused by increased corrosiveness (aggressiveness) or a massive build up of algae caused by excessive nutrients may increase. In areas with confined and unconfined cattle grazing operations and in urban and suburban areas with poor and malfunctioning septic tank sewage disposal, inadequate protection and short-cutting of human and animal urine into aquifers and subsequent contamination of wells for potable water are particularly dangerous. This can be recognized by a very high nitrogen content in wells.[53] High nitrogen content of groundwater can also be caused by excessive application of fertilizers for growing crops and maintaining urban lawns.[11,54] The nitrate–nitrogen in potable water can react with hemoglobin in infants under 6 months of age (methemoglobinemia or "blue baby" syndrome) and also cause some other adverse effects in humans.[17]

On-site sewage disposal systems using septic tanks and soil infiltration may often be a cause of ground and surface contamination. Septic tank effluents may contain 40–80 mg of nitrogen/liter, 11–31 mg phosphorus/liter, and 200–450 mg/liter of BOD_5.[55,56] There have been many failures of septic systems caused by inadequate subsoil conditions or, more commonly, by overloading. Surfaced effluent excess contributes to the unaccounted pollution of streams and can result in faulty and septic conditions of small local streams. The ammonia from well-functioning septic systems is rapidly oxidized in the soil to nitrate–nitrogen that is not adsorbed by soils and readily moves into groundwater. Bacteria and viruses during recharge or in septic tank disposal systems die quite rapidly as sewage passes through the soil material.[57]

REFERENCES

1. E. Cleary, *ORSANCO Story.* Johns Hopkins Univ. Press, Baltimore, MD, 1967.
2. K. Imhoff, *Sewage Ind. Wastes* **27,** 322 (1955).
3. C. N. Sawyer, *J. N. Engl. Water Works Assoc.* **61,** 109–127 (1947).
4. G. A. Rohlich (ed.), *Eutrophication, Causes, Consequences, Correctives.* Natl. Acad. Sci., Washington, DC, 1969.
5. A. D. Hasler, *Ecology* **55,** 383–395 (1974).
6. W. Stumm and I. I. Morgan, *Stream Pollution by Algal Nutrients,* Trans. 12th Annu. Conf., p. 16. University of Kansas, Lawrence, 1962.
7. H. Bernhardt *Vom Wasser* **44,** 53 (1972).
8. R. A. Vollenweider, *Scientific Fundamentals of the Eutrophication of Lakes and*

Flowing Water with Particular Reference to Nitrogen and Phosphorus as Factors in Eutrophication, Tech. Rep. No. DAS/CSI/68-27. Organ. Econ. Coop. Dev., Paris, 1968.

9. G. F. Lee, W. Rast, and R. A. Jones, *Environ. Sci. Technol.* **12,** 900–908 (1978).
10. W. Stumm, *Advances in Water Pollut. Res., Proc. Int. Conf. IAWPR* **2,** 216 (1964).
11. V. Novotny and G. Chesters, *Handbook of Nonpoint Pollution: Sources and Management.* Van Nostrand-Reinhold, New York, 1981.
12. E. A. Nusch and K. Koppe, *Proceedings of the Eleventh Meeting of Gesselschaft für Wasser und Abwasser* (*GWA*), Essen, FRG, 1978.
13. R. C. Dunst et al., *Survey of Lake Rehabilitation Techniques and Experiences,* Tech. Bull. No 75. Wisconsin Dept. of Natural Resources, Madison, 1974.
14. P. A. Krenkel and V. Novotny, *Water Quality Management.* Academic Press, New York, 1980.
15. LAWA, *Die Gewassergütekarte in der BRD,* (*Water Quality Maps in the FRG*). Jaeger Druck GmbH, Speyer, 1976.
16. National Water Council, *Review of Discharge Consent Conditions.* HM Stationery Office, London, 1978.
17. *Quality Criteria for Water.* U.S. Environ. Protec. Agency, Washington, DC, 1976.
18. H. B. Fisher, E. J. List, R. C. Y. Koh, J. Imberger, and N. H. Brooks, *Mixing in Inland and Coastal Waters.* Academic Press, New York, 1979.
19. G. Weicker, *Gesund. Ing.* **61,** 582 (1938).
20. W. Müller, *GWF, das Gas-und Wasserfach* **97,** 944 (1956).
21. K. Imhoff and G. M. Fair, *Sewage Treatment.* Wiley, New York, 1956.
22. K. R. Imhoff, *GWF, das Gas-und Wasserfach* **96,** 1264 (1965).
23. P. A. Krenkel, E. L. Thackston, and F. L. Parker, *J. Sanit. Eng. Div., Am. Soc. Civ. Eng.* **95,** 37–64 (1969).
24. F. Malz and M. Müller-Neuglück, *Städtehygiene* **12,** 249, (1966).
25. K. Imhoff, *Dtsch. Wasserwirtschaft* **20,** 124, (1943).
26. W. Stumm and J. O. Leckie, *Phosphate Exchange with Sediments: Its Role in the Productivity of Surface Waters,* Pap. III-26, Proc. 5th Int. Conf. IAWPR, San Francisco, CA. (1970).
27. E. Knop, *GWF, das Gas-und Wasserfach* **103,** 1373 (1962).
28. W. G. Anderson, *Sewage Ind. Wastes* **22,** 118, (1950).
29. Water Pollution Research Laboratory, *Aeration on Weirs.* Dept. Environ., Stevenage, Herts, Engl., 1973.
30. F. W. Kittrell, *Sewage Ind. Wastes* **31,** 1965, (1959).
31. M. Eckoldt, *Dtsch. Gewässerkl. Mitt.* **10,** 1 (1962); 14, 1 (1966).
33. A. T. Wiley and B. F. Lueck, *TAPPI* **43,** 24 (1960).
34. E. L. Thackston and R. E. Speece, *J. Water Pollut. Control Fed.* **38,** 1614 (1966).
35. J. V. Hunter and W. Whipple, *J. Water Pollut. Control Fed.* **42** (8), R249 (1970).
36. K. R. Imhoff, *Adv. Water Pollut. Res., Proc. Int. Conf.* Paper II-7. IAWPR, (1969).
37. K. R. Imhoff, F. Grabbe, and D. Albrecht, *Vom Wasser* **40,** 284 (1968).
38. J. L. Davis et al., in *Proc. Int. Conf. Waterpower 1983,* p. 1326–1335. University of Tennessee, Knoxville, TN, 1983.

39. J. H. Symons, *Water Quality Behavior in Reservoirs.* U.S. Dept. of Health, Education and Welfare, Public Health Service, Cincinnati, OH, 1969.
40. J. H. Symons, J. K. Carnwell, and C. G. Roberto, *J. Am. Water Works Assoc.* **62** (5), 322 (1970).
41. H. Bernhardt and J. Clasen, *Water Sci. Technol.* **14,** 397 (1982).
42. K. M. MacKenthun, *J. Water Pollut. Control Fed.* **54,** 1077 (1982).
43. R. Czesny, *Vom Wasser* **24,** 92, (1952).
44. K. Imhoff, *Eng. News Rec.* **106,** 883 (1931).
45. H. W. Koenig, G. Rincke, and K. R. Imhoff, *Adv. Water Pollut. Res., Proc. 4th Int. Conf.,* Pap. I-4. IAWPR, (1971).
46. C. L. P. Trüb, H. Althaus, and J. Posch, *Archiv Hygiene Bakteriol.* **145** (4), (1961).
47. N. I. Veatch, *Sewage Works J.* **20,** 3 (1948).
48. AWWA Research Foundation, *Munic. Wastewater Reuse News* No. 4, Denver, CO, (1981).
49. L. R. J. Van Vuren, W. R. Ross, and J. Prinsloo, *Prog. Water Technol.* **8,** 455–466, (1977).
50. AWWA Research Foundation, *Munic. Wastewater Reuse News* No. 39, Denver, CO, (1980).
51. *Water Reuse Highlights.* AWWA Research Foundation, Denver, CO, 1978.
52. W. Schroeder, *Dtsch. Wasserwirtschaft* **19,** 344 (1941).
53. W. Bucksteeg and H. Thiele, *GWF, das Gas-und Wasserfach* **98,** 26 (1957).
54. R. H. Harmerson, F. W. Sollo, and T. E. Larson, *J. Am. Water Works Assoc.* **63,** 303 (1971).
55. L. J. Sikora, *Ground Water* **14,** 304–314 (1976).
56. J. Harkin, B. Al-Hajjar, and D. G. Kroll, *Effects of Detergent Formulation on the Performance of Septic Systems,* Rep. No. WRC 83-09. Water Res. Cent., University of Wisconsin, Madison, 1983.
57. R. G. Bell and J. B. Bole, *J. Environ. Qual.* **7,** 193–196 (1978).

CHAPTER 14

Self-Purification of Surface Waters

14.1 MECHANISMS OF POLLUTION DEGRADATION IN RECEIVING WATERS

The introduction of treated or untreated wastewater into a receiving water body will result in a stress on the aquatic ecosystem. The stress can be reduced or removed from the surface waters by several processes including physical dilution, dispersion and sedimentation, chemical reaction, and adsorption and biological degradation and stabilization. This inherent ability of water to purge itself of contamination and to return sewage-laden waters back to reasonable purity is called self-purification. As stated by Gordon M. Fair in the previous U.S. edition of this book,[1] "self-purification is a formidable natural force, the proper utilization of which is a heavy engineering responsibility in the design and operation of both sewage treatment and sewage disposal works."

Biological and biochemical degradation of pollutants in natural waters can occur in two ways: anaerobically, in which there is a shortage of oxygen, or aerobically, in which sufficient oxygen levels are maintained. Although anaerobic and aerobic self-purification may potentially have the same result—cleaner water—they are as different as the treatment of sewage in septic tanks on the one hand and in activated sludge units on the other. Under aerobic conditions, reasonably clean appearance, freedom from odor, and normal aquatic animal and plant population are maintained; under anaerobic conditions, the receiving waters may become black, unsightly, and malodorous, and the normal water flora and fauna may be destroyed. Furthermore, anaerobic decomposition proceeds at a slower rate than aerobic decomposition, and recovery is delayed. Septic water usually creates a nuisance and is often an abomination. Under todays environment, anaerobic conditions in a receiving water body are unacceptable. Therefore, the term self-purification will be used only when considering aerobic conditions.

The total mass of floating living small organisms in water is called "plankton," which is responsible for the self-purification process. Knöpp[2] measured the biological oxygen demand exerted by plankton and has also shown how to biologically estimate self-purification capacity and its progress by measuring the incremental oxygen uptake by plankton. To an environmental engineer, plankton corresponds to the activated sludge microorganisms in wastewater treatment, which is recycled in a typical aeration basin. It can be concluded that when wastewater is added to plankton-deficient streams, the self-purification capacity is reduced. Conversely, Karl Imhoff[3] stated that the self-purification capacity is increased as wastewater addition allows the plankton to multiply and develop. Plankton nourished by sewage is also better adapted to decompose bacteriologically resistant pollutants such as organic acids, many toxic compounds, and oils. Phenols are known to be amenable to bacterial degradation in flowing waters after acclimation of the bacteria.

Biomagnification. Waste assimilative capacity studies should include special attention to toxic substances that may accumulate in the tissues of the aquatic organisms. Such materials may affect the entire food web and their discharge into receiving waters should be minimal or not at all. The most dramatic example of problems to aquatic life caused by relatively small quantities of toxic materials is the occurrence of the pesticide DDT and an industrial compound PCB (polychlorinated biphenyls) in the fish of the North American Great Lakes. The contamination of fish by PCBs, which is a carcinogenic compound, resulted in a ban on commercial fishing in the Great Lakes. Other cummulative toxicants with low assimilative capacity in surface waters include most of the persistent pesticides such as the organochlorine compounds chlordane, aldrin, dieldrin, and heptachlor (see Table 13.2 for maximal allowable concentrations of these compounds in surface water bodies).

14.2 DISSOLVED OXYGEN BALANCE

Organisms higher in trophic levels need more oxygen than required to maintain aerobic conditions. Fish and other aquatic biota can exist only when the dissolved oxygen is above a certain threshold minimum (3–4 mg/liter) and for fish propagation the DO concentrations must be even higher (the U.S. EPA water quality guidelines require minimum DO concentrations of 5 mg/liter or more, depending on the type of fish). To maintain this minimum oxygen level in emergency situations, water can be artificially aerated.

The oxygen economy of the receiving water is of primary concern in the analysis of self-purification of surface waters. In order to maintain the proper balance, the oxygen demand exerted by a number of biological, biochemical, and chemical oxygen demanding reactions must not exceed the available oxygen supply including a safety factor. The allowable amount of biodegradable organic wastes that can be discharged into the receiving water body without severely upsetting the oxygen balance is then called the waste assimilative capacity.[4] In practical computations and planning studies, the waste assimilative capacity is the maximum mass of pollutant discharge that will not result in violation of the pertinent water quality standards in the receiving water body under specified low-flow conditions. This concept can be extended to both conservative and nonconservative substances.

The concepts of oxygen balance and waste assimilative capacity determination for oxygen-demanding organic wastes were first introduced by Streeter and Phelps based on their study of the Ohio River[5] and were later summarized by Phelps.[6] Streeter and Phelps' approach was limited to only two phenomena, namely, deoxygenation of water caused by bacterial decomposition of carbonaceous organic matter, and reaeration caused by the oxygen deficit and introduction of oxygen into the water from the atmosphere. Several investigators have questioned Streeter and Phelps' theory because the self-purification process in small streams does not necessarily follow the model. A classical disagreement of field observations with the theory was published by Kittrell and Kochtitzky.[7] It is now known that a number of processes and reactions participate in the dissolved oxygen balance. However, it should also be noted that the basic fundamentals of the Streeter–Phelps concept are sound and have been accepted.

Oxygen Sinks

The sinks of oxygen, that is, the biological and biochemical processes that use oxygen, include

1. Biological oxidation of biodegradable carbonaceous organic matter by bacteria and fungi.
2. Nitrification in which oxygen is utilized during the oxidation of ammonia and organic nitrogen to nitrates.
3. Sediment oxygen demand, in which oxygen is required by the upper layers of organic bottom deposits.
4. Respiration by algae and aquatic plants that use oxygen during the night to sustain their living process.
5. Oxygen utilization by chemical oxidation, which is sometimes called immediate oxygen demand.

Biological Deoxygenation. The measure of organic pollution in waste assimilative studies is the same as that used in the design of biological treatment plants. It is commonly expressed as the 5-day biochemical oxygen demand (BOD_5), which can be related to the population per capita equivalent (average per capita BOD_5 load) as follows:

settleable suspended solids	20
nonsettleable and dissolved solids	40
Total: BOD_5 =	60 g/(cap-day)

These BOD_5 loadings are typical for both German and U.S. conditions.[1,8,9] The loadings of BOD and suspended solids from households having garbage grinders should be increased by 30 and 50%, respectively. It should be noted that these values represent data for raw sewage. Actual receiving water loads include the effect of treatment and are much lower. About 90% of the BOD_5 load is removed in a well-functioning biological treatment plant.

In the oxygen balance, the BOD loads are entered as their ultimate values (BOD_u). The effect of the decomposition rate and temperature on the relationship between BOD_5 and BOD_u has been explained in detail in Section 5.2. The deoxygenation rate coefficients that should be used for the conversion of BOD_5 to BOD_u are[10]

	$k_{20°C}$ (day^{-1}) (log base 10)
raw sewage, high-rate treatment effluents	0.15–0.25
high-degree treatment biological effluent	0.05–0.07

These deoxygenation rates are for 20°C warm effluents. A value of $k = 0.1$ day^{-1} has been used as a rough estimate of the deoxygenation rate. With this deoxygenation rate approximately 20% of the BOD load is biochemically decomposed daily and the ultimate BOD is about $1.4 \times BOD_5$.

Example: Estimate the ultimate BOD load from a municipal treatment plant serving 25 000 equivalent population. Assume 85% BOD removal in the treatment plant.

The BOD_5 load:

$$25\,000\ (\text{pop.}) \times 60\ [\text{g BOD}_5/(\text{cap-day})] \times (1 - 0.85) \times 0.001\ (\text{kg/g})$$
$$= 225\ \text{kg/day}$$

For treated sewage effluents select $k = 0.06\ day^{-1}$. Then using Equation (1) from Section 5.2 the ultimate BOD load is

$$BOD_u = BOD_5/(1 - 10^{-k \times 5}) = 225/(1 - 10^{-0.06 \times 5}) = 448.9\ \text{kg/day}$$

If the flow in the receiving water body was $Q = 10\ m^3/s$ then the BOD_u concentration increase in the stream would be

$$L_0 = 448\,900\ (\text{g/day}) \times (1/86\,400)(\text{day/s})/10\ (\text{m}^3/\text{s})$$
$$= 0.52\ (\text{g/m}^3)(\text{same as mg/liter})$$

These considerations apply only to the carbonaceous (first stage) BOD, which assumes that nitrification does not occur (Fig. 5.3). A complete degradation of the carbonaceous BOD at 20°C is completed in about 20 days. When the flow time in the stream is greater than 10 days or when the effluents require a high degree of biological treatment, the second stage of deoxygenation–nitrification should also be considered. Nitrification of treated effluents also occurs when the receiving water body is shallow and the nitrifying organisms can be attached to the bottom. Because of a long residence time, nitrification almost always occurs in estuaries.

Variability of the Stream Deoxygenation Rate. For shallow streams with depths of less than 0.5 m during the dry weather period, the deoxygenation proceeds at a faster rate. The organic pollutants can be removed and degraded both by suspended planktonic microorganisms and by attached bottom growths. Imhoff and Fair in the previous U.S. edition of this book[1] stated that "in shallow or rapid streams, the rate of BOD is no more like that in standard laboratory determinations than is the rate of BOD in trickling filters or activated-sludge units." Shallow and turbulent receiving waters offer increased opportunity for contact of sewage matters with the sides and bottoms of the stream channel, the surfaces of rocks and debris, and the leaves and stems of growing water plants, all of which may be covered with biological slimes. The rate of BOD is then increased accordingly.

BOD removal in receiving waters is not uniform, the rates having a tendency to decrease as the pollution load is reduced by self-purification.

It should be noted that even under the most favorable conditions, the removal times for organic pollution in nature are much larger and the rates of removal much slower compared to conventional biological treatment plants. A removal of 90% of the BOD, which in deeper streams or lakes at 20°C would take 10 or more days, is accomplished in an activated sludge plant in 1–10 hr and in a trickling filter in about 1 hr.

A large portion of the organic suspended solids remains suspended during the removal, partially because the particles are very small, and partially because the current prevents them from settling onto the bottom. The limiting flow velocities are known from sewage treatment as follows: From the theory of sedimentation basins, it is known that current velocities of less than 0.05 m/s do not hinder settling of small particles. Earlier experiments of Streeter[11] performed in a hydraulic flume also confirmed that at velocities of about 0.05 m/s, the flow changes from a quiescent flow into a more turbulent one accompanied by the formation of eddies. In grit chambers it was observed that at velocities of ≥ 0.3 m/s, only sand and gravel settled out and the organic particles remained suspended. Therefore, at flow velocities of less than 0.05 m/s most of the particles remain in suspension only for a limited time until their settling velocity brings them to the bottom. In the velocity range of 0.05–0.3 m/s only a portion of the organic suspended particles remains in suspension. Suspended organics are removed biochemically in the same way as dissolved solids with approximately the same oxygen demand. In the analysis both dissolved organic matter and suspended nonsettleable organic solids are treated as one load.

In the mathematical models that describe the self-purification process, the BOD removal is represented by the following equation:

$$L = L_0 e^{-K_r t} = L_0 e^{-K_r X/v}$$

where L is the concentration of organics present at time t, L_0 is the initial concentration of organics, K_r is a removal rate coefficient describing all mechanisms involved in the process including sedimentation and oxidation, X is the distance, and v is the flow velocity. In many studies, reaction rates are reported and related to the exponential base 10 rather than to base e. To convert K (base 10) to K (base e) multiply by 2.3.

The overall removal rate, K_r, is then a composite of the individual removal processes. It should be noted, however, that they may not occur simultaneously. Hence,

$$K_r = K_d + K_3 = B + k + K_3$$

In this expression, k is the BOD removal rate representing the removal by planktonic organisms (similar to the laboratory BOD removal coefficient), B is the removal rate caused by attached bottom microorganisms, and K_3 is the rate representing the effect of sedimentation of suspended particles when the flow velocity is less than 0.3 m/s. Although values of k (base e) are usually less than 0.5 day^{-1}, the values of K_r in some shallow streams can exceed 5 day^{-1}.[12-17] $K_d = B + k$ is the overall coefficient of deoxygenation.

Sediment Oxygen Demand. Sludge deposits in water bodies are composites of settleable solids that have accumulated on the bottom of receiving waters while the currents were too sluggish to prevent sedimentation of suspended solids or to remove deposited solids by bottom scour. The organic fraction of sludge deposits decomposes. If the overlying water contains dissolved oxygen, the surface of the deposits remains aerobic, however, oxygen will penetrate into the mud as far as its diffusion, on the one hand, and its consumption, on the other, permit. Generally, the aerobic layer is relatively thin (about 4 mm according to Fair), but its thickness varies depending on the oxygen concentration in the overlying water and other factors. The rest of the sludge layer is cut off from the oxygen supply. In the anaerobic lower sludge layers, the aerobic bacteria cannot survive and are replaced by facultative and anaerobic microorganisms. The decomposition of the sludge in the bottom deposits is similar to that occurring in the anaerobic sludge digesters, however, the evolving gases may contain more nitrogen. For example, according to Fair, they contain approximately 14% carbon dioxide, 17% methane, and 69% nitrogen. In digesters at 20°C, the decomposition of sludge to technical end-point lasts about 45 days if the sludge is inoculated and acclimated. In noninoculated digesters, the removal lasts over 6 months. The decomposition process in receiving waters is even slower as a result of a lack of inoculation and the lower organic content of the sludges (the organic content is approximately 30% for bottom deposits compared to 70% for raw municipal sludge). The anaerobic decomposition process of bottom sludges may last for many years with interruptions occurring during cold winter months.

The degree of sludge decomposition can be estimated by measuring the oxygen uptake of the sludge. This can be determined with an oxygen probe submerged in a hermetically closed bottle containing sludge and aerated water, similar to the procedure described in Section 5.2. The BOD_5 observed is then a measure of the portion of the organic matter that could have been decomposed by the bacteria in an aerobic environment. The test can be completed in 5 days. In situ measurements of the sediment oxygen demand are accomplished by placing a cylindrical bottomless container onto the bottom sediments and measuring the oxygen decrease inside the container by a probe. The sediments should not be disturbed during the test.

Magnitude of SOD. The first scientific measurements of the sediment oxygen demand (SOD) performed by Fair and his co-workers[18] focused on the processes occurring in the upper aerobic layer and related the benthic oxygen demand to the rate of deposition of the organic sludge solids on the bottom. It was assumed that for uniform sludge accumulation of bottom deposits, constant temperature, and rate of deposition independent of increasing depth, the rate of oxygen demand of deposited sludge is equal to the oxygen demand that would be exerted by a single daily load of settleable solids during a period of sludge accumulation.[1] Recent investigations have revealed that the sediment demand includes oxygen demand from several separate processes: (1) biological respiration of all living organisms in the aerobic upper zone, (2) chemical oxidation of reduced substances in the sediment such as reduced iron, manganese, and sulfide,[19–21] (3) oxidation of methane produced in the lower anaerobic sediment layers by bacteria residing in the aerobic sediment–water interface, and (4) simultaneous nitrification–denitrification of ammonia evolving from the sediments.

Holdren et al.[22] found that frequently more than one-half of the SOD can be attributed to chemical and biochemical oxidation of reduced compounds in the upper aerobic layer. Rudd and Taylor[23] described in detail the impact on the SOD by the methane evolution in the anaerobic portion of the sediments and its subsequent oxidation in the aerobic sediment–water interface. Only dissolved methane exerts an oxygen demand. The excess gaseous methane escapes in bubbles; however, a small part of it can dissolve and add an organic load to the overlying water.

Most of the in situ measured SOD range from 0.5 to about 5 $g/(m^2\text{-day})$.[21,31,32] Bowman and Delfino[21] measured the differences between the in situ and laboratory estimations of the SOD and noted that these differ by a significant margin. Table 14.1 shows some measured oxygen demands for various types of sludge deposits in receiving water bodies.

The factors that affect the magnitude of the benthic oxygen demand include temperature, mixing (turbulence), and oxygen concentration in the overlying water. Temperature increases respiration, chemical oxidation rates, and the diffusion rates of both oxygen and reduced substances in sediments.[30,31] Early work by Fair and his co-workers[1,18] indicated that sludge depth influenced SOD, however, later investigators reported that SOD was independent of sludge depth. Oxygen concentration in the overlying water affects SOD only when the dissolved oxygen is below 3 mg/liter. At higher concentrations, the effect is not noticeable. Other factors that may influence the SOD include photosynthetic activity of benthic algae, salinity, and pH.

Most of the values of the benthic oxygen demand report SOD for undisturbed core sediments. When the sediments are resuspended the oxygen demand can increase dramatically as a result of the immediate oxygen requirement of the

TABLE 14.1 In Situ Measured Sediment Oxygen Demand

Type of Deposits	Sediment Oxygen Demand g/(m^2-day)	Temperature (°C)
Simulated sludges[18]		
1.42 cm deep	1.05	20
10.2 cm	4.65	20
Rivers		
River sludge[24]	0.9	
River muds[25]		
2 cm	3.4	
25 cm	6.4	
Sandy bottom[26]	0.2–1.0	
Mineral soil[26]	0.05–0.1	
Slime covered shallow rivers[17]	3.4–14.9	4–19
Lower Milwaukee River[27]	2.8–6	20
Illinois rivers[28]	0.27–9.8	9–31
Impoundments		
Lake Michigan near Green Bay[29]	1.6–1.9	12
Lake Erken[30]	2.6	18
	0.43	4
Lake Ramsen[30]	2.3	17
Lake Ekoln[30]	2.6	18

reduced compounds from the anaerobic layers that must be satisfied by the dissolved oxygen in water. Resuspension of sediments can be caused by high flows, stormwater, and combined sewer overflows (local scour), by the bottom invertebrate activity, and by dredging. James[33] found a linear increase in oxygen demand with increasing water velocity.

When the bottom sludge is resuspended by flood water very little damage is done since the high flows provide usually enough dilution and oxygen to mitigate the oxygen demand of the resuspended sediments. Floods also flush the streams and for some time after the flood the benthic oxygen demand can be reduced.

However, a resuspension of the sediments during summer, when the temperature reaches 20°C or more, is dangerous. The gas production may be so strong that the solids that have accumulated on the bottom during the preceding winter and spring months can be resuspended. The damage done to the stream can be severe, since the stream usually carries less flow and less oxygen. Later, it will be shown in Procedure III how the effect of the benthic oxygen demand on the dissolved oxygen concentration in the overlying water can be estimated. However, as stated herein, because of a number of factors affecting the magnitude of SOD, such computations are often inaccurate and speculative.

The deposition of bottom sludges cannot be avoided in lakes and impoundments or slow moving estuaries, even with the best treatment of point sources of pollution. A large portion of the bottom sediment deposition is the result of nonpoint sources. As shown in Chapter 3, urban and agricultural runoff has suspended solids concentrations that may exceed those found even in raw sewage. In many lakes and reservoirs and even in some streams, additional large quantities of organic solids are formed by photosynthesis. The algal matter remaining after the death of the organisms will then settle to the bottom and decompose in the same way as sewage solids. In some cases, satisfactory dissolved oxygen levels can be maintained if the bottom deposits accumulated during the preceding winter months are flushed by water releases from upstream reservoirs or by dredging prior to the summer. This would have approximately the same affect as natural flushing by floods during the spring season.

The reducing anaerobic conditions of the lower layers of the bottom deposits are also responsible for the release of nutrients, particularly phosphorus. This was well documented by Mortimer.[34] It is thought that phosphorus release from sediments is hindered by the formation of an oxidized layer at the sediment–water interface.[35]

Nitrification. The process of nitrification, in which organic nitrogen and/or ammonia is oxidized to a higher oxidation state, may be of significant importance to the dissolved oxygen balance of receiving waters. The oxygen requirements are about 4.3 g of oxygen per 1 g of the Total Kjeldahl Nitrogen (TKN = sum of organic nitrogen and ammonia). The theoretical oxygen demand for ammonia is even higher (4.56 g of O_2/g of NH_4^+).

The effect of nitrification has been known to sanitary engineers for many years. Although the nitrogenous oxygen demand (NOD) of unnitrified effluents was well understood, sanitary engineers generally dismissed this potential stream problem on the basis of the following factors[36]:

1. Nitrification is caused by specific organisms (*Nitrosomonas* and *Nitrobacter*), the population of which is usually minimal in surface waters.
2. The reaction rate coefficient for nitrogenous oxidation is small in relation to the coefficient for carbonaceous matter.
3. Oxidation of ammonia to nitrates simply converts dissolved oxygen to a form in which it is still available to prevent development of anaerobic conditions.

In addition, the growth of nitrifying microorganisms is much slower than that of heterotrophic microorganisms oxidizing carbonaceous organic matter. Thus, the carbonaceous BOD is oxidized first and nitrification may occur only when most of the carbonaceous BOD is oxidized. Also, the nitrifying microorganisms are strict aerobes and the process will not occur if the oxygen concentrations are near zero. Because of these factors, nitrification has not been found in many water quality surveys.

However, a study of the Grand River by the Michigan Water Resources

Commission[37] demonstrated the impact of nitrification on stream oxygen resources and clearly indicated the necessity of including organic nitrogen oxidation in oxygen balance calculations. Other examples of nitrification processes taking place in surface waters have been reported for the English estuaries of the Thames and Tees rivers by the Water Pollution Research Laboratory,[38] the Delaware estuary by Tuffey, Hunter, and Matulewich,[39] the Holston River in Tennessee by Ruane and Krenkel,[40] and other water bodies. Summarizing such findings from the literature, it can be concluded that the probabilities for nitrification to occur are highest when (1) receiving water bodies provide a long detention time for completion of the carbonaceous BOD removal and development of the nitrifying organisms, (2) shallow water bodies maintain aerobic bottom–water interface at which nitrifying organisms can be attached to the bottom, and (3) the effluents receive a high degree of treatment or are partially nitrified and the organic carbon to TKN ratio is very low.

Nitrification is a two-stage process carried out in an aquatic environment by autotrophic bacteria belonging to the family Nitrobacteriaceae. These bacteria prefer solid surfaces for growth, that is, they grow better if suspended particles or bottom slimes are present. The first of the nitrification reactions is the oxidation of ammonia to nitrite by *Nitrosomonas*:

$$2NH_4^+ + O_2 \longrightarrow 2NO_2^- + 2H_2O + 4H$$

The second step is the oxidation of nitrite to nitrate by *Nitrobacter*:

$$2NO_2^- + O_2 \longrightarrow 2NO_3^-$$

It should be noted that the second step is quite rapid compared to the first, explaining why only small quantities of nitrite are found in natural waters.

Nitrifying organisms are sensitive to pH and function best over a pH range of 7.5–8.0. The rate of nitrification decreases rapidly at dissolved oxygen levels below 2.0–2.5 mg/liter. When anoxic conditions prevail, denitrification has been observed, whereby nitrates are reduced to nitrogen gas and ammonia.

Nitrification in streams is mathematically described in a fashion similar to the carbonaceous BOD.[40–42] However, it is very difficult to predict the nitrification rates. For example, Krenkel and Novotny,[4] after reviewing literature data, found that the nitrification rates have been reported in ranges from 0.1 to 15.8 day^{-1}.

Because of the complex nature of the nitrification process and the undefined effects of various environmental factors on this process, modeling nitrification is a descriptive rather than a predictive effort at the present time. Therefore, field data need to be collected under environmental conditions similar to those under which the waste assimilative capacity of the receiving water body is to be estimated. The rates can be determined by calculating the slope of the rate of the nitrate increase in the receiving water body with respect to time on a semilogarithmic (first-order reaction) or arithmetic (zero-order reaction) plots.

Oxygen Sources

There are three normal sources of dissolved oxygen for receiving waters: (1) the DO contained in the receiving water and in the wastewater itself at the point of discharge, (2) the oxygen absorbed from the atmosphere whenever the DO content of the water falls below its saturation value (reaeration), and (3) the oxygen released by green plants and algae by photosynthesis under the influence of sunlight.

Oxygen can be supplied from the atmosphere only up to the saturation concentration. The saturation values of dissolved oxygen vary with temperature and in brackish or seawater also with the salt concentration. On the other hand, water can be oversaturated by photosynthetic oxygen sources. Under intense sunlight conditions in a water body that has high density of green plants and algae, the oxygen concentrations can be four times higher than saturation. However, the effect of photosynthesis on the oxygen content in technical computations of the oxygen balance is not usually included since it is highly unpredictable. Furthermore, during the night hours or in the absence of light, the plants and algae consume oxygen by their respiration. Some simpler water quality models included the photosynthetic oxygen production as a sine wave function.

According to Knöpp,[2] in German geographical locations, the oxygen production by photosynthesis of green plants and algae occurs only from April to October, with the highest rates of photosynthesis in August and September. In the United States, the period of active photosynthesis depends on the geographical and climatic conditions. The photosynthetic process takes place only in the upper 2 m thick layer or euphotic zone, where sufficient light can penetrate. In deeper layers only respiration (oxygen consumption) can occur.

Measurements on the Rhine River in July 1959 showed a daily photosynthetic oxygen production of about 1.6 g/m^3. Other German measurements on the smaller river Lippe yielded values of photosynthetic oxygen production between 2 and 13 g of $O_2/(m^2\text{-day})$. The production of photosynthetic oxygen also depends on the biological state of the water body. Algae may not develop in sufficient numbers if water still contains higher concentrations of biodegradable organics and herbivorous zooplankton predate on them.

Barge traffic can be an appreciable source of oxygen on navigable rivers since the propellers of the ships agitate water and mix it with air. During the dry summers of 1959 and 1961, a decrease of oxygen concentrations that would otherwise be lethal to fish in the Rhine River was thought to be prevented by increased aeration caused by barge traffic.

Atmospheric Reaeration. In technical computations, the amount of oxygen absorbed by reaeration from the atmosphere is related to the water volume and to the oxygen deficit, which is the difference between the actual oxygen concentration and the saturation concentration. The rate of oxygen transfer across the water surface at given temperature depends on the turbulence parameters such as depth and velocity and the interaction of wind and waves. For rough estimates of reaeration, it is possible to use Table 14.2, presented earlier by Fair.

In order to convert the surface aeration to mg/liter of the volumetric oxygen

TABLE 14.2 Oxygen Absorbed from the Atmosphere through the Water Surface at Constant Oxygen Deficits, K_a, in g/(m^2-day) According to Fair.[1]

	K_a Percentage Oxygen Saturation					
Type of Receiving Water	100	80	60	40	20	0
1. Small ponds	0	0.3	0.6	0.9	1.2	1.5
2. Larger impoundments	0	1.0	1.9	2.9	3.8	4.8
3. Large streams with low velocity	0	1.3	2.7	4.0	5.4	6.7
4. Large streams	0	1.9	3.8	5.8	7.6	9.6
5. Swift streams	0	3.1	6.2	9.3	12.4	15.5
6. Rapid streams	0	9.6	19.2	28.6	38.4	48.0

gain, the values from Table 14.2 are divided by the average depth of the receiving water body. Use of the table is illustrated on the following example.

Example: A 2-m-deep stream carries a flow of 28.3 m^3/s with an initial (upstream) oxygen concentration of 5 mg/liter. What is the maximum BOD_5 loading to the stream that will not result in a decrease of the oxygen concentration below 5 mg/liter. The stream temperature is 20°C. From Table 5.3 the oxygen saturation concentration for 20°C is 9.2 mg/liter. Then the oxygen saturation expressed as a percentage is $100 \times 5.0/9.2 = 54\%$. The stream is in category 4. By interpolating in Table 14.2, the oxygen absorption through the water surface for 54% saturation is 4.3 g/(m^2-day). Dividing by the depth the volumetric oxygen gain is $4.3/2 = 2.15$ g/(m^3-day) [=mg/(liter-day)]. The maximum ultimate BOD input should not exceed the oxygen gain by reaeration. Hence, the maximum BOD_u input is 2.15 g/(m^3-day) × 28.3 m^3/s × 86 400 s/day × 0.001 kg/g = 5 257 kg/day of BOD_u = 5 257/1.46 = 3 600.7 kg/day of BOD_5. The conversion from BOD_u to BOD_5 is according to Table 5.4.

In mathematical computations, the capacity of a stream to gain oxygen by reaeration is expressed by a volumetric coefficient of reaeration, K_2. Note that the values in Fair's Table 14.2 can be represented by the following formula:

$$M = \frac{K_a}{H} = K_2 D$$

where M is the volumetric gain of oxygen by reaeration [mg/(liter-day)], K_a is the surface absorption rate of the atmospheric oxygen in g/(m^2-day) from Table 14.2, H is the flow depth in meters, $D = (C_s - C)$ is the oxygen deficit, and C_s and C are saturation and in-stream concentrations of the dissolved oxygen, respectively. The unit of K_2 is day^{-1}.

Many formulas have been proposed to estimate the reaeration coefficient, K_2. Their summary and critical appraisal have been presented by British researchers Wilson and Macleod[43] and in the United States by Rathbun.[44] Simple equations relate the magnitude of the reaeration coefficient to a few hydraulic parameters such as velocity, depth, roughness, or slope as

$$K_2 = CV^m H^n S_e^p$$

where V is the mean stream velocity in m/s, H is the depth of flow in meters, S_e is the energy (bottom) slope in m/m, and C, m, n, and p are coefficients. Empirically derived coefficients for some commonly used equations are given in Table 14.3.[45–49]

The reaeration rate coefficients are expressed for a water temperature of 20°C. For other temperatures between 5 and 30°C the reaeration coefficient increases or decreases geometrically with the temperature by 2.4% per each degree of difference from the reference temperature of 20°C[50] or

$$K_{2-T} = K_{2-20°C} \times 1.024^{(T-20)}$$

The 2.4% geometrical rate of change of the reaeration coefficient with temperature is smaller than that for the BOD bottle test or carbonaceous BOD removal in streams, which is 4.7%.

Example: A stream similar to the previous example (depth $H = 2$ m, flow $Q = 28.3\ m^3/s$, and velocity $V = 0.4$ m/s) has an initial upstream dissolved oxygen concentration of 5 mg/liter ($= g/m^3$). What is the maximum BOD_5 load that would not decrease the DO below 5 mg/liter if the stream temperature is 30°C. For 30°C the saturation concentration is 7.6 mg/liter (Table 5.3). Then the oxygen deficit is $D = 7.6 - 5 = 2.6$ mg/liter. To estimate the reaeration coefficient use the O'Connor–Dobbins formula from which (using Table 14.3)

$$K_{2-20°C} = 3.73 \times 0.4^{0.5} \times 2^{-1.5} = 0.84 \text{ day}^{-1}$$

TABLE 14.3 Coefficient for the Reaeration Formula[a]

Author	C	m	n	P
O'Connor and Dobbins[45]	3.73	0.5	−1.5	0
Krenkel and Orlob[46]	173.6	0.408	−0.66	0.408
Churchill et al.[47]	5.01	0.969	−1.67	0
Owens et al.[48]	5.32	0.67	−1.85	0
Tsivoglou and Neal[49]				
Flow $Q \le 0.3\ m^3/s$	31 000	1.0	0	1.0
$0.3 < Q \le 1.4\ m^3/s$	21 000			
$1.4 < Q \le 8\ m^3/s$	15 000			
$Q > 8\ m^3/s$	8 600			

[a] Logarithmic base e, temperature 20°C.

which corrected to 30°C gives

$$K_{2-30°C} = 0.84 \times 1.024^{(30-20)} = 1.05 \text{ day}^{-1}$$

The volumetric oxygen gain is then

$$M = K_2D = 1.05 \times 2.6 = 2.73 \text{ mg/(liter-day)}$$

and the maximum BOD_u load is

$$BOD_u = 2.73 \text{ mg/(liter-day)} \times 28.3 \text{ m}^3\text{/s} \times 86\,400 \text{ s/day} \times 0.001 \text{ kg/g}$$

$$= 7\,447 \text{ kg/day of } BOD_u = 7\,447/1.4 = 5\,320 \text{ kg/day of } BOD_5$$

Fair's surface absorption values of atmospheric oxygen are somewhat conservative if compared with more recent measurements.

14.3 COMPUTATION OF THE OXYGEN CONCENTRATION

In the absence of nitrification and neglecting photosynthetic oxygen contributions, the initial oxygen concentration, oxygen consumption due to pollution and bottom decomposition of sludges, and reaeration are the three basic components of the dissolved oxygen balance from which the DO concentration of receiving waters can be ascertained. The goal of such computations is commonly to predict the dissolved oxygen concentrations at predetermined points and to prove that the planned capacity of a treatment plant is sufficient. Larger projects involve determining the so-called "oxygen sag curve" in which the dissolved oxygen concentrations are plotted against the distance or travel time of the receiving water body reach. It can then be determined at which point the dissolved oxygen concentration has its lowest (sag) point and then the maximum allowable loading or the waste assimilative capacity can be found by back calculation.

A classical representation of the oxygen sag curve and its components is presented in Figure 14.1.

In the following section, several simplified procedures will be introduced by which the dissolved oxygen of the receiving waters can be estimated and the waste assimilative capacity for biodegradable organic matter can be determined. A number of mathematical computer models have been developed that can make the process of waste assimilative capacity determination more accurate, considering all or most of the factors that have an impact on the dissolved oxygen concentration (photosynthesis, nutrient inputs, effect of currents, etc.). It should be realized, however, that even the best models may give results that are different from those observed in nature. The magnitudes of many of the coefficients used in constructing water quality models are not known precisely and the processes may be described by equations that oversimplify reality. Therefore, the use of more sophisticated

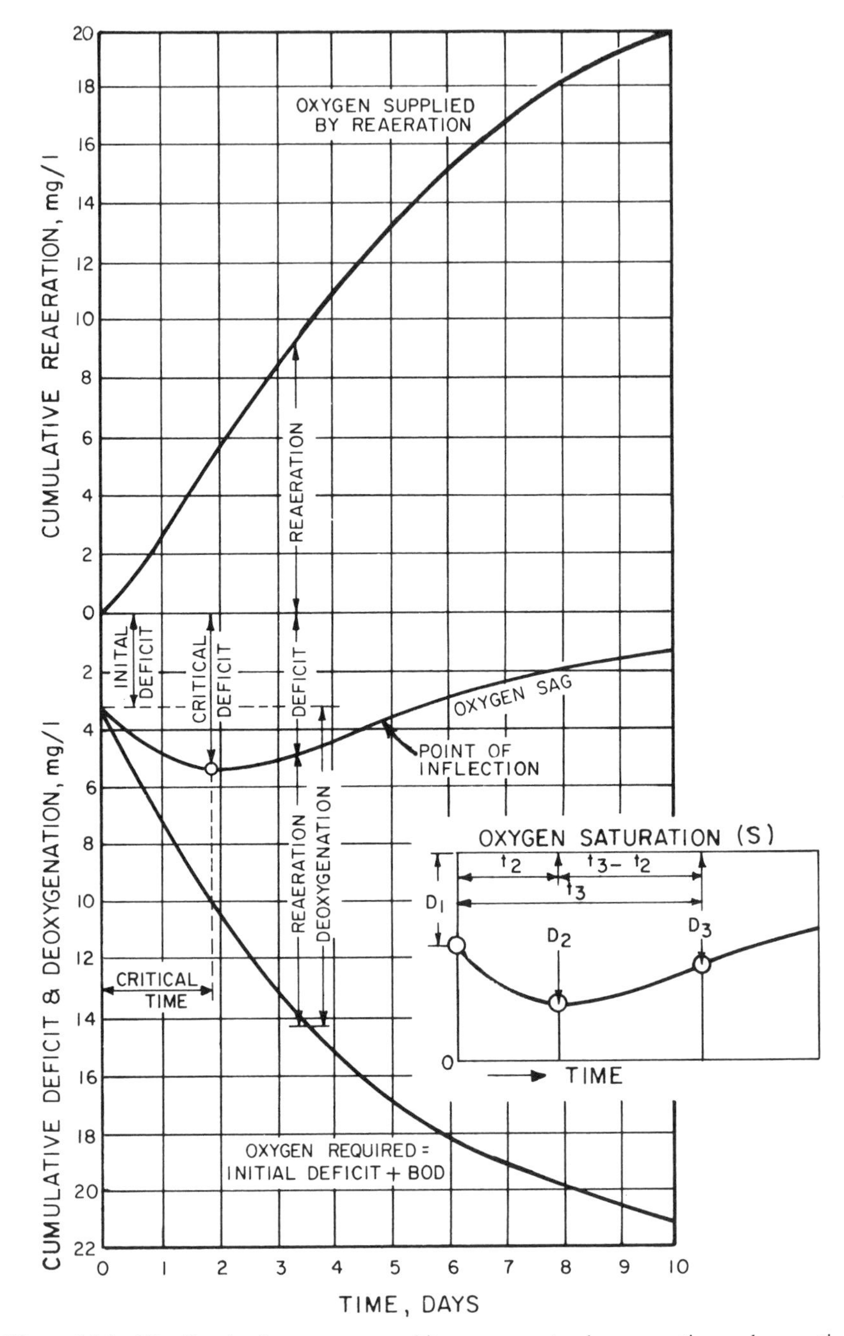

Figure 14.1 The dissolved oxygen sag and its components: deoxygenation and reaeration.

models should be preceded by water quality surveys, the data of which are used for calibration (model build-up) and verification of the model as described by Krenkel and Novotny.[4] Checking the results of the models or of the simplified procedures against observed data, even in the absence of special water quality surveys, is always advisable.

Before the models and procedures are used, the user or designer should be aware of possible differences between the dissolved oxygen balance processes taking place in small streams and large rivers. A model designed for large streams can give highly erroneous results if applied to a shallow stream and vice versa. The primary factors that will affect the selection of the model or procedure include the stream depth, velocity, bottom deposits and character, degree of pollution, topographical characteristics, presence of rapids, weirs, reservoirs, and dam outlets, navigation intensity, and stream currents. Krenkel and Novotny[4] suggested some generalizations that could aid in the selection of the model.

Large rivers with velocities exceeding 0.3 m/s and depths above 1.5 m (Fair's category 4 as defined in Table 14.4): These rivers will usually have sand or gravel bottoms with a very low potential for attached growths, minimal deposition of organic matter, and limited photosynthesis potential. In such rivers, the BOD removal rate is similar to that measured in the laboratory BOD bottle tests and most of the oxygen will be supplied by atmospheric reaeration. The potential for nitrification is low, and nitrification will occur only if sufficient detention time is available for nitrifiers to develop. Many major American and European Rivers (Mississippi, Missouri, Ohio, Danube, Rhine) can be modeled using this simplified concept).

Medium size and some large rivers with velocities below 0.3 m/s and impoundments (Fairs' category 2 and 3): Deposition of organic solids may result in benthic oxygen demand and increase the apparent BOD removal coefficient. After most of the carbonaceous BOD is removed, planktonic algae may develop in stagnant sections and photosynthetic oxygen production may be appreciable. The organic growths caused by primary (photosynthetic) production after dye-off will increase the amount of deposited organic solids and subsequent sediment oxygen demand and increase the BOD of the water. Although nitrification may occur in reaches that have sufficient detention times, some ammonia can be adsorbed on suspended particles and settle out without exerting an oxygen demand.

Shallow fast moving streams with velocity >0.3 m/s and depth <1.5 m (Fair's categories 5 and 6): These streams may show the greatest variability and fastest BOD removal under optimal conditions. Because of larger velocities, the bottom is mostly composed of rocks and gravel with optimal conditions for attached growths of biological slimes capable of decomposing organic matter. This results in a significant increase of the BOD removal rate. Also, most of the nitrifying organisms may be located in the attached growths, and nitrification in the immediate vicinity of secondary effluent outfalls is common. Very little sedimentation of organic suspended solids will occur and the oxygen demand of the attached growths is already included in the increased BOD removal rate.[14]

Smaller sluggish streams with velocity $<$ 0.3 m/s and depth $<$ 1.5 m and *shallow impoundments* (Fair's category 1 and 2): In these streams, the suspended organic matter may settle out and accumulate as bottom mud. Although the velocity may be low, the bottom layer is mainly composed of mud and fine sand that are constantly shifting and thus preventing the development of attached growths and algae. Rooted aquatic weeds and planktonic algae can develop in large densities if most of the carbonaceous BOD is decomposed and nutrients (nitrogen and phosphorus) are available. The increase in the overall BOD removal rate is only apparent and decreases as soon as the majority of the suspended organics settle out. Significant amounts of nutrients and other pollutants absorbed on suspended solids may also settle out. This reduction may be particularly significant for phosphorus and ammonia.

The modeling of small shallow streams is generally more difficult than that of large rivers. Neglecting processes occurring in the bottom layers and their participation in water quality changes may lead to gross errors and misinterpretations. Most of the subsequent "rule-of-thumb" simplified procedures (Procedures I and II) can be used only for the first group of streams (Fair's groups 3 and 4). For other streams and more complex wastewater discharges more detailed computerized water quality models are appropriate.

Simplified Procedure I. Balancing deoxygenation and reaeration in a fixed reach of a receiving water body.

Example 1: A city of 80 000 people is located on a small fish-abundant stream with a water surface width of 17 m, a depth of 2 m, and a flow of 10 m^3/s. The upstream oxygen content of the flow is near saturation. After 1 day of flow the stream becomes a tributary of a larger river. It is desired to show whether the stream can assimilate the dissolved organic solids without damage, and then determine the required treatment, if needed.

The 5-day municipal biochemical oxygen demand resulting from the dissolved (nonsettleable) solids is approximately 40 g/(cap-day). During the first day at 20°C, 30% or $0.3 \times 40 = 12$ g/cap must be satisfied by reaeration and the upstream oxygen content of the stream. In total $12 \times 80\,000 = 960\,000$ g of BOD must be satisfied. The flow velocity is $10/(2 \times 17) = 0.3$ m/s. In the reach the flowing water covers a surface area of $17 \times 0.3 \times 86\,400 = 440\,000$ m^2. The loading by dissolved organics per unit area is then $960\,000/440\,000 = 2.2$ g/(m^2-day). For reaeration, it can be assumed that the stream is in Category 4—a typical stream. Then, from Table 14.2, the reaeration rate of 2.2 g/(m^2-day) would require a saturation value of 77% (by interpolation). At 20°C, this corresponds (according to Table 5.3) to an oxygen content of $0.77 \times 9.2 = 7.1$ mg/liter in salt-free water.

In the above procedure, it was assumed that the upstream BOD is negligible and that the deoxygenation rate is similar to that of the BOD bottle test (i.e., there are no biological bottom or planktonic growths that would increase the BOD removal rate in the stream). It was also assumed that the effect of the bottom deposits on the dissolved oxygen is zero. However, even if the dissolved oxygen content of 7.1 mg/liter is reduced to account for the processes that have not been included in the BOD removal, the stream may still have enough oxygen resources to prevent damage to aquatic life. These conclusions will not be the same if the design temperature is higher, for example, 30°C.

Example 2: A larger reservoir with a surface area of 170 ha (1 700 000 m^2) and an average depth of 4 m impounds the water of a stream receiving the settled sewage from scattered communities. What is the maximum population that can discharge its sewage in the lake with and without biological wastewater treatment. The water temperature is 25°C.

Computation 2a considering no appreciable bottom deposits (no bottom sediment oxygen demand). From Table 5.3 the oxygen saturation for the water temperature of 25°C is 8.4 mg/liter. Assuming that the DO concentration in the lake should not drop below 5 mg/liter to preserve fish life, the minimum allowable oxygen saturation in the lake should be 100 × 5/8.4 = 60%. From Table 14.2, the oxygen absorption rate through the surface for an impounded stream (category 2) and 60% saturation is 1.9 g/(m^2-day) at 20°C, which, converted to 25°C, becomes $1.9 \times 1.024^{(25-20)} = 2.14$ g/(m^2-day). The lake could then assimilate the total BOD load of 2.14 g/(m^2-day) × 1 700 000 m^2 × 0.001 kg/g = 3 638 kg/day = 3 638 000 g/day.

Using the population equivalent values, the maximum population that can discharge its treated sewage in the lake can be quickly estimated. Assuming that the sewage is treated by sedimentation only (not viable in the United States), the population equivalent value for BOD_5 is 40 g/day at 20°C. The reaction rate for the BOD of the sewage can be estimated as 0.15 day^{-1}, which, using the BOD/BOD_5 ratio from Table 5.4, gives the BOD_u of the wastewater as 1.22 × 40 = 48.8 g/day. Using Table 5.5, the BOD removal rate in the lake at 25°C is 1.25 × 0.15 (day^{-1}) = 0.19 day^{-1}. If the lake has a detention time of 5 days then according to Table 5.4, 90% of the total BOD load is decomposed in the lake. Then the maximum allowable population equivalent is 3 683 000/(0.9 × 48.8) = 83 895. For biologically treated sewage effluents, assume 85% removal of the total BOD by treatment, hence 3 683 000/[(1 − 0.85) × 0.9 × 48.8] = 559 301 of allowed population.

If the detention time in the lake is shorter, less oxygen demand would have to be satisfied by reaeration of lake water. For example, for the detention time of 1 day and BOD lake water reaction rate of 0.19 day^{-1} only 36% of the BOD_5 load (Table 5.4) would be oxidized in the lake. Then the connected population number could be correspondingly larger. Use Tables 5.4 and 5.5 to estimate loading for other detention times and other design temperatures.

This simplified procedure assumes a well-mixed lake. To ensure such conditions, the lake should not be elongated and/or the sewage discharges should be evenly distributed.

Computation 2b considering packed sludge deposits: The oxygen demand of sludge deposits can only be estimated. From Table 14.1 it can be seen that the typical sediment oxygen demand of lake sludges is about 2 g/(m^2-day). Since this is greater than the estimated reaeration rate of the lake, the discharge of sludge-laden effluents into the lake should not be permitted.

By comparing the typical sediment oxygen demands for streams with the oxygen absorption rates listed in Table 14.2, one can arrive at a conclusion that for the first three categories of receiving waters the oxygen balance would be almost always endangered by the combination of the BOD from the wastewater and SOD from the sediments. Therefore, the sludge-laden effluents should not be discharged into the receiving waters in category 1 to 3. As stated before, SOD is not significant for the receiving waters in category 4 to 6.

Example 3. Sewage polishing (oxidation) ponds: According to Section 7.5 select an area of 1 ha = 10 000 m^2 per 2 000 connected population or 5 m^2 per capita. The sewage flow is 0.3 m^3/(cap-day)—biologically treated (BOD removal 75%). The effluent is aerated to 5 mg of O_2/liter and available dilution water (clean, saturated with oxygen) is 2 × 0.3 = 0.6 m^3/(cap-day). The mixed water flow is 0.3 + 0.6 = 0.9 m^3/(cap-day) and daily water loading is then 2 000 × 0.9 = 1800m^3/(ha-day). The detention time of the mixture is 10 days, which gives an effective depth of 10 (days) × 1 800 [m^3/(ha-day)]/10 000 (m^2/ha) = 1.8 m. The temperature is 25°C.

Oxygen Demand: The BOD_5 loading after treatment is about 60 (g of BOD_5/cap-day) × (1 − 0.75) = 15 g/(cap-day). Using the multiplier from Table 5.5, and assuming the BOD reaction rate at 20°C as being 0.1 day^{-1}, hence, the BOD_u is 1.46 × 15 = 21.9 g BOD_u/(cap-day). At 25°C, the BOD removal in the pond will progress at a rate 1.25 × 0.1 = 0.125 day^{-1}. According to Table 5.4, at 25°C and 10 days detention, the oxygen demand exerted by the wastewater will be about 95% of the BOD_u, or 0.95 × 21.9 = 20.8 g/(cap-day).

Oxygen Intake: The saturation oxygen concentration at 25°C is 8.4 mg/liter. (Table 5.3) If the oxygen concentration in the pond is allowed to drop to 2 g/m^3 (mg/liter) = 100 × 2/8.4 = 23.8% of saturation, then the oxygen uptake from the atmosphere is 1.15 g/(m^2-day) at 20°C (by interpolation in Table 14.2, category—small ponds). Converted to 25°C it becomes $1.024^{(25-20)}$ × 1.15 = 1.29 g/(m^2-day) or 1.29 × 5 = 6.45 g/(cap-day). The oxygen from sewage that can be utilized in the pond is (5 g/m^3 − 2 g/m^3) × 0.3 m^3/(cap-day) = 0.9 g/(cap-day), and that from the dilution water is (8.4 − 2) × 0.6 = 3.84 g/(cap-day). The total oxygen input is then 6.45 + 0.9 + 3.84 = 11.19 g/(cap-day), which is not sufficient to cover the oxygen demand of 20.8 g/(cap-day). During the summer, the benthal oxygen demand that would build up as a result of self-purification and algae growths in the pond should be added.

The results show that the pond does not have enough capacity of reaeration from the quiescent water surface for the given loading. Thus, such ponds must often rely on oxygen input by photosynthesis, which occurs only on sunny days. On cloudy days the pond may be in danger of becoming anoxic. It would then be necessary to reduce the load and/or to provide more dilution water and/or increase the surface area of the pond and/or to provide artificial aeration. If the pond is to rely on photosynthesis as a source of oxygen, the effective depth should be less than 1 m because of limited light penetration.

Procedure II. Evaluation of the permissible waste assimilative capacity and the oxygen sag curve according to the simplified procedure by Fair.[1] In working these examples, use Table 14.4, which was prepared by Fair according to the Streeter–Phelps equation.

TABLE 14.4 Estimation of the Allowable Loading of Receiving Waters and of the Critical Time of Flow[a,b]

Class of Receiving Water	*Lower Limiting Case.* Receiving water contains minimum allowable oxygen concentration (O_M) immediately below point of sewage discharge			*Upper Limiting Case.* Receiving water fully saturated with oxygen immediately below point of sewage discharge			Time of flow, t_c, to point of maximum deficit (minimum oxygen concentration), days (for the lower limiting case, the time of flow to the point of maximum deficit is zero)		
	15°C	20°C	25°C	15°C	20°C	25°C	15°C	20°C	25°C
1. Small ponds, and backwaters	$F = 0.6$	0.5	0.4	$F = 2.1$	1.6	1.3	$t_c = 5.9$	5.0	4.3
2. Sluggish streams, large lakes, and reservoirs	1.1	0.9	0.7	2.7	2.1	1.6	4.5	3.9	3.3
3. Large slow streams	1.6	1.2	0.9	3.2	2.5	2.0	3.8	3.2	2.8
4. Large normal streams	2.2	1.7	1.3	4.0	3.2	2.5	3.0	2.6	2.3
5. Swift streams	3.5	2.7	2.1	5.4	4.3	3.3	2.3	2.0	1.8
6. Rapids (variable)	22	17	13	25	20	15	0.6	0.6	0.5

[a] According to Fair.[1]

[b] Allowable 5-day, 20°C BOD of receiving water and added sewage at point of pollution (mg/liter) = load factor, $F \times$ critical DO deficit, D_c (mg/liter). Critical DO deficit (D_c) (mg/liter) = DO saturation value (10.2, 9.2, and 8.4 mg/liter at 15°, 20°, and 25°C, respectively) − minimum allowable DO concentration, O_M (mg/liter).

Example 4: 1. What is the maximum allowable 5-day, 20°C BOD of water in a large slow stream immediately below the point of sewage discharge if the water temperature is 25°C and the lowest allowable DO concentration of the stream is to be 5 mg/liter? Assume complete mixing of the effluent with the river water.

From Table 14.4, the load factor $F = 0.9$ (lower limit) to 2.0 (upper limit); also the critical deficit $D_c = 8.4$ (saturation value) −5.0 (lowest allowable value) = 3.4 mg/liter. Hence, the maximum allowable 5-day, 20°C BOD, equal to FD_c, lies between $0.9 \times 3.4 = 3.1$ mg/liter and $2.0 \times 3.4 = 6.8$ mg/liter.

2. If the 5-day, 20°C BOD of secondary treated sewage is 30 mg/liter, what is the required dilution in the river water?

The required dilution lies between $30/3.1 = 9.7$-fold and $30/6.8 = 4.5$-fold. If the river water itself has a BOD when it arrives at the point of sewage discharge, this value must be subtracted from the allowable values before making the division.

3. If the per capita flow of sewage is 300 liters/(cap-day) (0.3 m^3/day), what is the required stream flow in m^3/s per 1 000 connected population?

Since $0.3 \times 1\,000/86\,400 = 0.0035$ m^3/s per 1 000 population, the required stream flow lies between $(9.7 - 1.0) \times 0.0035 = 0.0305$ m^3/s and $(4.5 - 1.0) \times 0.0035 = 0.0123$ m^3/s per 1 000 population.

4. How far away from the point of sewage discharge will the lowest DO concentration be found?

From Table 14.4, the critical time, t_c, lies between 0 and 2.8 days. The distance in km (miles), therefore, equals the stream velocity in km/day (miles/day) times the critical time.

Example 5: The impoundment considered in Example 2 is to be evaluated using Fair's procedure. The impoundment has a surface area of 170 ha, an average depth of 4 m, and receives settled sewage. The dissolved oxygen concentration in the lake should not drop below 5 mg/liter at the lake water temperature of 25°C.

The oxygen saturation at 25°C is 8.4 mg/liter. (Table 5.3) If the minimum allowable DO to preserve fish life is 5.0 mg/liter then the required percentage of DO saturation is $100 \times 5/8.4 = 60\%$ and the permissible critical deficit is $D_c = 8.4 - 5.0 = 3.4$ mg/liter. The permissible loading of primary treated municipal sewage will lie between the following limits:

1. Lower limit. The inflow to the impoundment has a large but still permissible oxygen deficit, $D_c = 3.4$ mg/liter. From Table 14.4 the load factor for the impoundment (Category 2) is $F = 0.7$ and the permissible BOD loading is $L = FD_c = 0.7 \times 3.4 = 2.4$ mg/liter. Assuming per capita

sewage flow of 300 liters/day and a per capita BOD loading of sewage after primary treatment of 40 g/(cap-day), the BOD concentration of the settled sewage is about 40 (g/cap-day)/0.3 (m^3/day) = 120 g/m^3 (= mg/liter). The necessary dilution by the reservoir freshwater flow is 120/2.4 = 50-fold. Thus, the fresh water inflow into the reservoir must be (50 − 1) × 300/1 000 = 14.7 m^3/(cap-day). The volume of the reservoir with a depth of 4 m is 170 (ha) × 10 000 (m^2/ha) × 4 m = 6 800 000 m^3. Therefore, the permissible connected population is 6 800 000/(1 × 14.7) = 462 585 persons, if the detention time is 1 day. For detention times other than 1 day, divide the population number by the number of days.

2. Upper limit. The inflow is saturated with oxygen. Permissible BOD at 25°C according to Table 14.4 is $L = FD_c = 1.6 \times 3.4 = 5.44$ mg/liter. The flow time to the sag point of the oxygen curve (critical time) is x_c = 3.3 days (Table 14.4). The necessary dilution is 120/5.44 = 22-fold or (22 − 1) × 300/1 000 = 6.32 m^3/(cap-day). Permissible loading is then 6 800 000/(3.3 × 6.32) = 326 045 people when the detention time is 3.3 days or more.

The examples of this simple procedure are mostly used as a "rule-of-thumb" estimation of the waste assimilative capacity for a single sewage discharge. For multiple discharges, simple computer models become more convenient.

Procedure III. A procedure for the estimation of the critical oxygen concentration at the sag point that can be programmed on a programmable calculator or a microcomputer: Estimation of the critical deficit and sag point concentration.

1. Input and print: deoxygenation rate (K_r)[day^{-1} base e)], oxygen saturation (S) (mg/liters), sediment oxygen demand (SOD) [g/(m^2-day)], upstream flow (Q) (m^3/s), BOD_5 of the effluent (L_W) (mg/liter), DO of the effluent (D_W) (mg/liter), effluent flow (Q_W) (m^3/s), stream flow velocity (V) (m/s), flow depth (H) (m), temperature (T) (°C), upstream BOD_5 (L_U) (mg/liter), and upstream DO (D_U) (mg/liter).
2. Initialize:

Initial BOD

$$L_I = 1.46\left[\frac{(Q_W L_W + L_U Q)}{Q_W + Q}\right]$$

Initial oxygen deficit

$$D_I = \frac{(S - D_W)Q_W + (S - D_U)Q}{Q_W + Q}$$

Reaeration coefficient (O'Connor–Dobbins formula)

$$K_2 = 3.73 \frac{V^{1/2}}{H^{3/2}} 1.024^{(T-20)}$$

Convert coefficient of deoxygenation to temperature T

$$K_R = K_r \times 1.047^{(T-20)}$$

3. Compute time to the sag point (in days)

$$t_2 = \frac{1}{K_2 - K_R} \ln\left\{\frac{K_2}{K_R}\left[1 - \frac{K_2 - K_R}{K_R L_1}\left(D_1 - \frac{\text{SOD}}{HK_2}\right)\right]\right\}$$
$$\text{if } t_2 < 0 \text{ then } t_2 = 0$$

Compute distance to the sag point (in km)

$$X = 86.4 V t_2$$

Compute sag point (critical deficit)

$$D_2 = [K_R L_1 e^{(-K_R t_2)}]/K_2 + \text{SOD}/(HK_2) \qquad \text{If } t_2 = 0 \text{ then } D_2 = D_1$$

Convert to the critical dissolved oxygen concentration

$$C_c = S - D_2 \qquad \text{if } C_c < 0 \text{ then } C_c = 0$$

4. Print: Critical DO concentration C_c (mg/liter)
   Time to the sag point t_2 in days
   Distance to the sag point X in km

Example 6: Estimate the waste assimilative capacity of a medium-sized stream receiving treated sewage effluent from a single source. The dissolved oxygen in the stream should be maintained at or above 5 mg/liter to protect fish life. The stream flow is 3.00 m^3/s, the stream velocity is 0.25 m/s, and the depth is 1.3 m. The upstream DO = 7.0 mg/liter and BOD_5 = 3 mg/liter. The stream temperature 25°C and a stream survey yielded a value of the stream deoxygenation rate of $K_r = 0.4$ day^{-1} at 20°C. The waste effluent flow Q_W is 0.2 m^3/s.

The waste assimilative capacity cannot be directly computed since both t_c and D_2 depend on the initial BOD concentration, L_1. Therefore, the computation is of a trial-and-error type where two or more values of L_1 (or L_W) are selected and the waste assimilative capacity (allowable initial BOD concentration) is obtained by interpolation.

The oxygen saturation for 25°C is 8.4 mg/liter. Since the flow velocity is less than 0.3 m/s, deposition of organic solids will occur and the stream will exhibit sediment oxygen demand. From Table 14.1 select SOD of 2 g/(m^2-day).

1. In the first trial computation select the BOD_5 of the sewage L_W = 120 mg/liter (settled sewage). Then input the following values in the model:

$$
\begin{aligned}
K_r &= 0.4 \text{ day}^{-1} & S &= 8.4 \text{ mg/liter} \\
\text{SOD} &= 2.0 \text{ g/(m}^2\text{-day)} & Q &= 3 \text{ m}^3/\text{s} \\
L_W &= 120 \text{ mg/liter} & D_W &= 4 \text{ mg/liter} \\
Q_W &= 0.2 \text{ m}^3/\text{s} & V &= 0.25 \text{ m/s} \\
H &= 1.3 \text{ m} & T &= 25°\text{C} \\
L_U &= 3 \text{ mg/liter} & D_U &= 17.0 \text{ mg/liter}
\end{aligned}
$$

Then in the computation

$$L_1 = 1.46\left(\frac{0.2 \times 120 + 3 \times 3}{0.2 + 3}\right) = 15.05 \text{ mg/liter}$$

$$D_1 = \frac{(8.4 - 4) \times 0.2 + (8.4 - 7) \times 3}{0.2 + 3} = 1.59 \text{ mg/liter}$$

$$K_2 = 3.73 \frac{0.25^{0.5}}{1.3^{3/2}} 1.024^{(25-20)} = 1.41 \text{ day}^{-1}$$

$$K_R = 0.4 \times 1.047^{(25-20)} = 0.50 \text{ day}^{-1}$$

$$t_2 = \frac{1}{1.41 - 0.50} \ln\left\{\frac{1.41}{0.50}\left[1 - \frac{1.41 - 0.50}{0.50 \times 15.05}\left(1.59 - \frac{2}{1.41 \times 1.3}\right)\right]\right\}$$
$$= 1.07 \text{ days}$$

$$X = 86.4 \times 0.25 \times 1.07 = 23.11 \text{ km}$$

$$D_2 = [0.5 \times 15.05e^{(-0.5 \times 1.07)}]/1.41 + \frac{2}{1.41 \times 1.3} = 4.21 \text{ mg/liter}$$

$$C_c = 8.4 - 4.21 = 4.19 \text{ mg/liter}$$

2. Second trial. Since C_c from the first trial for L_W = 120 mg/liter is less than 5 mg/liter, in the second trial select L_W = 30 mg/liter (biologically treated sewage) and repeat the computation. The result of the second trial is

$$t_2 = 0.98 \text{ days} \qquad \text{and } C_c = 5.82 \text{ mg/liter}$$

3. By interpolating (as long as $t_2 > 0$ in both computation there is a straight line relationship between the BOD load and oxygen concentration of the sag point) the maximum BOD_5 concentration for the DO standard of S_t = 5 mg/liter is

$$L_W \text{ (allowable)} = \frac{L_W(1)\,[C_c(2) - S_t] - L_W(2)\,[C_c(1) - S_t]}{C_c(2) - C_c(1)}$$

$$= \frac{120 \times (5.82 - 5.0) - 30 \times (4.19 - 5)}{5.82 - 4.19}$$

$$= 75.2 \text{ mg of BOD}_5/\text{liter}$$

The waste assimilative capacity (maximum allowable BOD) can also be interpolated graphically as shown in Fig. 14.2).

The procedure for the computation of the waste assimilative capacity can be expanded to include the upper and lower limits of the allowable BOD loads similar to Fair's concept previously introduced.

1. The upper limit can be considered when the initial dissolved oxygen content is near the oxygen saturation. Then since $D_1 = 0$ and if SOD $= 0$ as well, the equations for the critical time and the critical (sag point) deficit can be reduced to

$$t_2 = \frac{1}{K_2 - K_r} \ln\left(\frac{K_2}{K_r}\right) = A$$

then

$$D_2 = \frac{K_r}{K_2} e^{-K_r A} L_1 = CL_1$$

and

$$C_c = S - D_2$$

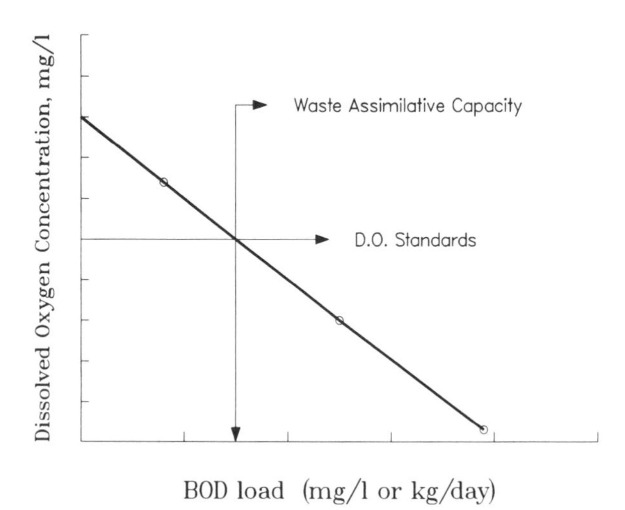

Figure 14.2 Waste assimilative capacity determination.

where A and C are constants. This yields a linear relationship between the sag point oxygen concentration, C_c, and the initial BOD load, B_l, as shown in Figure 14.2. Hence,

$$\text{maximum } L_l = \frac{K_2(S - S_r)}{K_r e^{-(K_r A)}}$$

In the above equation, S is the oxygen saturation value and S_r is the DO standard.

2. The lower limit can be considered when the initial dissolved oxygen concentration equals the DO standard, S_r. Then

$$\text{maximum } L_l = \frac{K_2(S - S_r) - \text{SOD}/H}{K_r}$$

Example 7: For the conditions of Example 6, estimate the upper and lower limit of the waste assimilative capacity.

1. The upper limit time is

$$A = \frac{1}{1.41 - 0.5} \ln\left(\frac{1.41}{0.5}\right) = 1.14 \text{ days}$$

and the concentration (neglecting the benthal oxygen demand) is

$$\text{maximum } L_l = \frac{1.41 \times (8.4 - 5)}{0.5e^{-0.5 \times 1.14}} = 16.95 \text{ mg/liter}$$

which gives the BOD_5 concentration in the waste effluent as

$$\text{maximal } L_W = \frac{16.95 \times (0.2 + 3)/1.46 - (3 \times 3)}{0.2} = 140.75 \text{ mg/liter}$$

2. The lower limit (including the benthal oxygen demand):

$$\text{maximum } L_l = \frac{1.41 \times (8.4 - 5) - 2.0/1.3}{0.5} = 6.51 \text{ mg/liter}$$

which gives the maximal BOD_5 concentration in the effluent of

$$\text{maximum } L_W = \frac{6.51 \times (3 + 0.2)/1.46 - (3 \times 3)}{0.2} = 26.34 \text{ mg/liter}$$

For the intermediate case with the initial dissolved oxygen concentration between the DO standard and DO saturation, Example 6 yielded the maximal effluent concentration of $BOD_5 = 75.2$ mg/liter.

Based on numerous computations of the waste assimilative capacity of streams in all categories, the following conclusions can be drawn:

1. If the sediment oxygen demand is negligible, then the maximum permissible initial BOD_u concentration (lower limit) equals $f = K_2/K_r$ times the permissible oxygen deficit $(S - S_r)$. For the majority of typical streams, f lies between 2 and 5. If the permissible oxygen deficit for typical U.S. conditions is around 3 mg/liter, this would give the maximum permissible BOD_u concentrations between 6 and 15 mg/liter, which would correspond to initial BOD_5 concentrations of 4–10 mg/liter, respectively.
2. Primary sedimentation (effluent BOD_5 about 120 mg/liter) may be sufficient if 12- to 25-fold dilution by clean stream flow is available. This can be related to the unit flow of the stream. Assuming a BOD_5 load of 40 g/(cap-day) by settled sewage, the number of population equivalent per 1 liter/s stream flow can be between 10 and 20 (280–560 population per 1 cfs). This dilution ratio depends on geographic location. For example, in Germany the accepted ''rule-of-thumb'' ratio of connected population to flow is 25 persons per 1 liter/s of flow. This is the result of the generally colder temperatures of German streams.

   It should again be noted that primary treatment is not satisfactory for discharge of effluents in U.S. rivers and lakes.
3. For biologically treated sewage effluents ($BOD_5 = 30$ mg/liter as daily average), the available dilution by clean stream flow should be 3- to 8-fold. In the United States, the present environmental regulations require installation of biological treatment in municipal treatment plants regardless of the stream waste assimilation capacity. For multiple sewage sources, the sewage outfalls in a reach with less than 6 days flow time interact and more complex models are used for determination of the waste assimilative capacity. In receiving waters with low stream velocities the effect of benthal oxygen demand must be included, which generally reduces the waste assimilative capacity.

Procedure IV: Simple dissolved oxygen model. The dissolved oxygen concentration curve can be computed using a modified Streeter–Phelps equation.[4] This will give a detailed description of the dissolved oxygen concentration curve that can be plotted against the time of flow and/or the distance. Such curves are useful in developing oxygen ''profiles'' for multiple effluent discharge and are also a necessary documentation for waste load allocation plans. A simple model will be presented first for a single uniform reach of a stream and later expanded for multiple reaches and waste effluents.

Assuming negligible runoff and photosynthesis within the reach, the BOD and DO variations can be described by the following equations:

BOD variations

$$L_t = L_1 e^{-K_r t} = L_1 e^{-K_r(x/v)}$$

Oxygen deficit variations

$$D_t = D_1 e^{-K_2(x/v)} + \frac{K_r L_1}{K_2 - K_r} [e^{-K_r(x/v)} - e^{-K_2(x/v)}]$$
$$+ [1 - e^{-K_2(x/v)}] \frac{\text{SOD}}{K_2 H}$$

In these equations:

L_t = BOD_u (mg/liter) in the stream after time, $t = x/v$
L_1 = initial BOD_u (mg/liter) in the reach
D_t = oxygen deficit (mg/liter) after time t
D_1 = initial oxygen deficit (mg/liter)
K_r = deoxygenation coefficient, 1/day
K_2 = reaeration coefficient, 1/day
H = depth of flow in the reach, m
SOD = benthal oxygen demand, g/(m^2-day)
x = length of the reach, km
v = flow velocity, km/day = 86.4 × m/s.

In a reach in which the travel time is less than the time to the sag point, t_2, as defined in the preceding procedure, the minimum oxygen concentration (maximal deficit) occurs at the end of the reach. In this case the modified Streeter–Phelps equation should be used.

Example 8: Determine dissolved oxygen concentrations in a reach 30 km long in 5 km increments. The following parameters and input variables describe the situation:

$K_r = 0.5 \text{ day}^{-1}$ $\qquad K_2 = 1.4 \text{ day}^{-1}$

$H = 1.3$ m $\qquad v = 25$ km/day

$L_1 = 15$ mg/liter $\qquad D_1 = 1.4$ mg/liter

SOD = 2 g/(m^2-day)
Temperature 25°C
Oxygen saturation $S = 8.4$ mg/liter
Flow $Q = 3.2 \text{ m}^3/\text{s}$
K_r and K_2 have already been converted to 25°C as shown in Example 6.

The time increment for the computation $\Delta t = 5\ (\text{km})/25\ (\text{km/day}) = 0.2$ days. The computation proceeds as shown in the following table:

Distance (km)	Time (days)	$e^{-K_r t}$	$e^{-K_2 t}$	BOD_u L (mg/liter)	Deficit D (mg/liter)	DO Concentration $C = S - D$ (mg/liter)
0	0	1.0	1.0	15.0	1.4	7.0
5	0.2	0.92	0.75	13.8	2.74	5.66
10	0.4	0.82	0.57	12.3	3.35	5.05
15	0.6	0.74	0.43	11.11	3.81	4.58
20	0.8	0.67	0.32	10.05	4.11	4.29
25	1.0	0.61	0.24	9.15	4.25	4.15
30	1.2	0.55	0.19	8.23	4.15	4.25

The results of Example 6 (critical time $t_2 = 1.07$ days and critical oxygen concentration $C_c = 4.19$ mg/liter) are almost identical to the results of the above computation.

To continue the example, assume that in river km 30 another effluent is discharging 0.1 m^3/s of settled sewage ($BOD_5 = 120$ mg/liter). Then river km 30 becomes the beginning of a new reach and the computation can continue after the initial B_1 and D_1 are ascertained. Hence,

$$L_1 = \frac{8.23 \times 3.2 + 1.46 \times 120 \times 0.1}{3.2 + 0.1} = 13.29 \text{ mg/liter}$$

and the initial deficit (assuming that the sewage is aerated to a DO of 4 mg/liter with the deficit = 8.4 − 4 = 4.4 mg/liter) is

$$D_1 = \frac{4.15 \times 3.2 + 4.4 \times 0.1}{3.2 + 0.1} = 4.16 \text{ mg/liter}$$

These values are then substituted in the first line of the computational table and the computation will proceed similarly to the end of this reach. A new reach must also be established at a confluence of the receiving stream with a larger tributary or when hydraulic conditions (velocity, and depth) change. The BOD load and the dissolved oxygen concentration can also be plotted versus distance. In Germany, the population equivalent of the BOD load in persons per unit of flow instead of BOD concentrations is used in water quality planning studies.

This concept of progressive computation of the oxygen profile is a foundation for several computer water quality models. Of note is DOSAG developed by the Texas Water Development Board in the late 1960s.[51] This model has gained wide popularity and has been used by a number of regulatory agencies and consulting companies in their water quality and waste allocation studies.

The DOSAG model, as any more sophisticated water quality model, should be calibrated and verified by data from field surveys. A technique of waste assimilative capacity determination using the DOSAG model was presented by Krenkel and Novotny.[4]

Procedure V. Water quality models for estuaries: Most of the important components of the equations for the oxygen balance of estuaries remain almost the same as for streams. However, the movement of water masses by tidal action makes the dispersion of pollutants an important factor. A simple version of an estuarial oxygen balance model was published by O'Connor.[52] The model is similar to the classical Streeter–Phelps equation with the exponents expanded to include the effect of tidal dispersion.

Procedure VI. Models for photosynthesis and algal growth: A number of sophisticated computer models evolved in the 1970s and 1980s that are capable of simulating algal growth in receiving water bodies. The QUAL model, presently maintained by and available from the U.S. EPA Environmental Laboratory in Athens, Georgia, was originally developed for the Texas Water Development Board.[53] The original model has been modified by various agencies and authors. It is capable of simulating a wide range of water quality constituents, including temperature, BOD, DO, nitrogen components, phosphorus, chlorophyll a, algal biomass, and a number of conservative substances. Bingham, Lin, and Hoag[54] and Van Benschoten and Walker[55] have recently published an example of the application of the QUAL-II (the most popular version) model in wasteload allocation.

QUAL-2E is a public domain version of the model that is run on IBM-PC (personal computer) types of computer hardware.[56]

A simpler analytical model of photosynthesis and its effect on the oxygen balance was published earlier by O'Connor and DiToro.[57] In this model, photosynthetic oxygen gain and loss by respiration are represented by a Fourier (cosine) series and the equations are then analytically solved.

These models require extensive calibration and verification by field water quality survey data that should be gathered under conditions similar to those under which the waste allocation plan is to be developed.

Procedure VII. Computation of artificial aeration of receiving flowing waters according to Imhoff[58]:

This procedure begins preferably with a measured critical dissolved oxygen concentration curve (profile), which is to be raised at one or more places. One can then estimate the oxygen profile that will result with the aeration devices if all other conditions remain the same.

Weirs and turbine aeration are the most convenient aeration devices, however, floating aerators and diffused air have also been used. The oxygen concentration increase at the point of aeration is computed according to the following equation (Fig. 14.3):

$$\Delta C_{\mathrm{O}} = \frac{B_{\mathrm{L}} D}{180 \times \mathrm{Q}} K' \qquad (1)$$

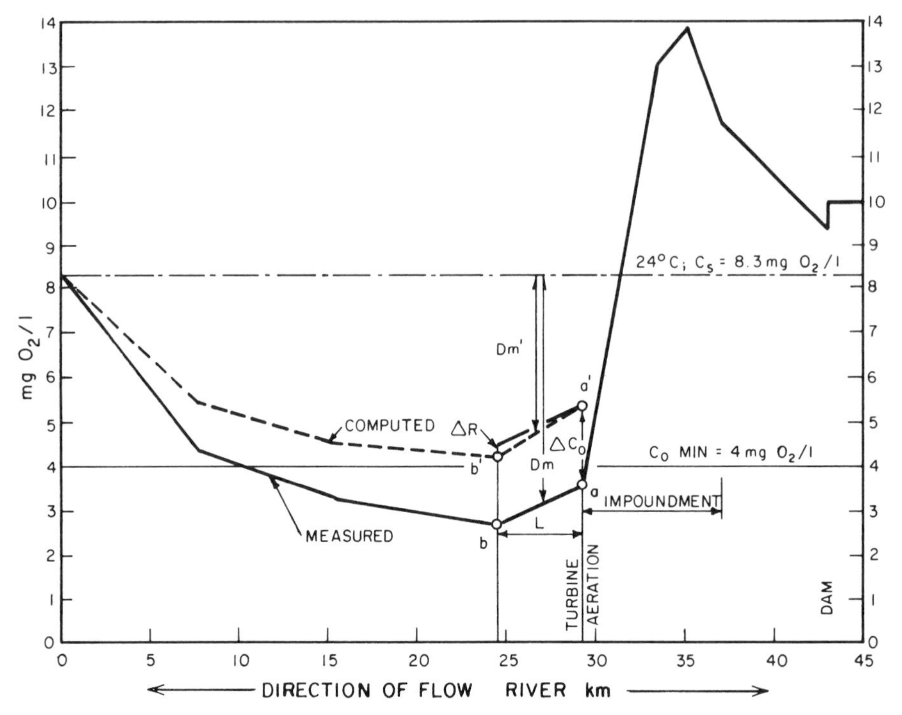

Figure 14.3 Computation of the oxygen profile below a place of in-stream aeration from measured (computed) DO concentration without aeration.

where ΔC_O = oxygen concentration increase at the point of aeration (mg/liter)
B_L = aeration power input in kW [for weir aeration substitute 9.81 × flow (m^3/s) × fall height (m)]
D = average oxygen deficit (%)
Q = flow (m^3/s)
K' = coefficient

The coefficient, K', represents the oxygen yield per kW-hr at 50% average oxygen deficit (mean of the upstream and downstream deficit at the point of aeration) at 20°C. When accurate data are not available, K' can be estimated from the following table:

	K' (kg O_2/kW-hr)
Cascades and rapids	1.5
Sharp-crested weir	0.6
Weirs and spillways	0.4
Turbine aeration	1.0
Surface aeration by floating aerators	0.5
Diffuse aeration	0.4

The oxygen supply by diffused aeration is independent of water temperature. For other aeration techniques the values of K' must be related to the corresponding oxygen saturation values, or $K'_T = K_{20°C} C_{sT} / C_{s20}$.[59]

The oxygen concentration profile after artificial aeration is then computed step by step from the measured critical oxygen curve as is illustrated for one section on Fig. 14.3.

After selecting B_L, the oxygen difference ΔC_O is known from Equation (1). a' lies on Fig. 14.3 the distance ΔC_O above a. b' can then be computed from the following equation:

$$b' = b + \Delta C_O - \Delta R \qquad (2)$$

Flow, temperature, organic decomposition, benthal oxygen demand, and the effect of algal growths for both curves remain the same. Thus, the different courses of the curves are due only to reaeration. As a result of the smaller average deficit after aeration, $D_{m'}$, there will be a reduction of natural reaeration, bringing the elevated (after aeration) line closer to the measured one at distance X by a value of ΔR. This can be computed from

$$\Delta R = (D_m - D_{m'}) \frac{XK_2}{v \times 24 \times 3600} \qquad (3)$$

where ΔR = drop in the natural reaeration (mg/liter)
$D_m = S - (a + b)/2$ = average oxygen deficit of the measured line (mg/liter)
$D_{m'} = S - (a' + b')/2$ = average oxygen deficit of the computed line (mg/liter)
X = length of the stretch (m)
K_2 = coefficient of natural reaeration for the stretch (1/day)
v = velocity (m/s)

With these values, the location of the point b' can be obtained by combining Equations (2) and (3) and rearranging them to

$$b' = \frac{v \times 48 \times 3\,600 \times (b + \Delta C_O) + XK_2(b - \Delta C_O)}{v \times 48 \times 3\,600 + XK_2} \qquad (4)$$

The next point of the aerated oxygen curve will be computed by analogy with Equation (4) by setting $\Delta C_O = b' - b$ and repeating the procedure.

Example 9: As a result of efficient removal of biodegradable organics in a reservoir, the oxygen concentration in the stream (Fig. 14.3) has dropped below the permissible minimum oxygen concentration of 4 mg/liter at river km 29.3. The minimum measured downstream oxygen concentration at km 24.5 was 2.2 mg/liter, hence the oxygen concentration at the critical point should be raised by

1.8 mg/liter, which is to be accomplished by turbine aeration in the power plant of the reservoir located at river km 29.3. The course of the elevated oxygen curve downstream from the aeration point is computed by changing the measured critical oxygen profile.

For this example the parameters are

$$Q = 20 \text{ m}^3/\text{s} \qquad b = 2.7 \text{ mg/liter}$$

$$\Delta C_O = 1.8 \text{ mg/liter} \qquad X = 4\,800 \text{ m}$$

$$H = 3 \text{ m} \qquad v = 0.1 \text{ m/s}$$

$$T = 25°\text{C} \qquad S = 8.4 \text{ mg/liter}$$

Using the Owens et al.[48] reaeration formula (Table 14.3) the reaeration coefficient for the reach is

$$K_2 = 5.84 \times 0.1^{0.67}/3^{1.85} \times 1.024^{(25-20)} = 0.184 \text{ day}^{-1}$$

From Equation (4), b' can be found as

$$b' = 4.32 \text{ mg/liter}$$

The next point of the aerated oxygen profile can be determined by analogy from Equation (4) for km 15.5. In this case $\Delta C_O = 4.32 - 2.7 = 1.62$ mg/liter, and for the measured $b = 3.3$ mg/liter and corresponding hydraulic parameters of the reach, X, v, and H, the next point b' can be determined. A second aeration point can be considered, however, this would not be economical for this example.

14.4 WASTE ASSIMILATIVE CAPACITY FOR NUTRIENTS

The problem of excessive loadings of surface waters by nutrients, primarily by nitrogen and phosphorus, is particularly important for lakes, reservoirs, and estuaries. In these water bodies, the classical dissolved oxygen concept may not work in evaluating the waste assimilative capacity, and the production of organic matter by algae and larger aquatic plants may be greater than the BOD contributions from municipal and industrial wastewater discharges. Oxygen levels are affected by photosynthesis and respiration, and BOD concentrations are affected by the planktonic organisms and their residues.[4,60,61]

The production of organic matter in a slow moving or stagnant water body is related to its trophic status. Eutrophication is a process by which a water body progresses from its origin to its extinction according to the level of nutrient and organic matter accumulation. The term oligotrophic is used to describe the youngest state of a surface water body—lake, estuary, or reservoir—and is characterized by water with very low mineral and organic contents. As the nutrient content is increased by wastewater disposal or runoff, algae increase in numbers. These organisms are called producers, and they initiate the entire cycle of production of organic matter in the water body. At a later state of the development process the water body becomes mesotrophic, and when the nutrient and organic matter productivity are high, the water becomes eutrophic. The final stages of the water body existence are pond, marsh, swamp, or wetland. The symptoms of eutrophication are shown in Figure 14.4 and the methods of evaluating the trophic status have been published by Carlson,[62] Uttormark and Wall,[63] and U.S. EPA.[64]

The related changes caused by advanced eutrophication (pH variations, hypolimnetic oxygen depletion, organic substances, increases in turbidity, and a shift of the fish population from game to rough fish) may interfere with the uses of the water. In addition, possible taste and odor problems caused by algae can make water less suitable or desirable for human consumption. When the concentration of algae exceeds a certain nuisance value, the situation is termed an algal bloom.

Limiting Nutrients. Although many nutrients contribute to algae development, only those that are in the shortest supply will limit the growth rate. From the array

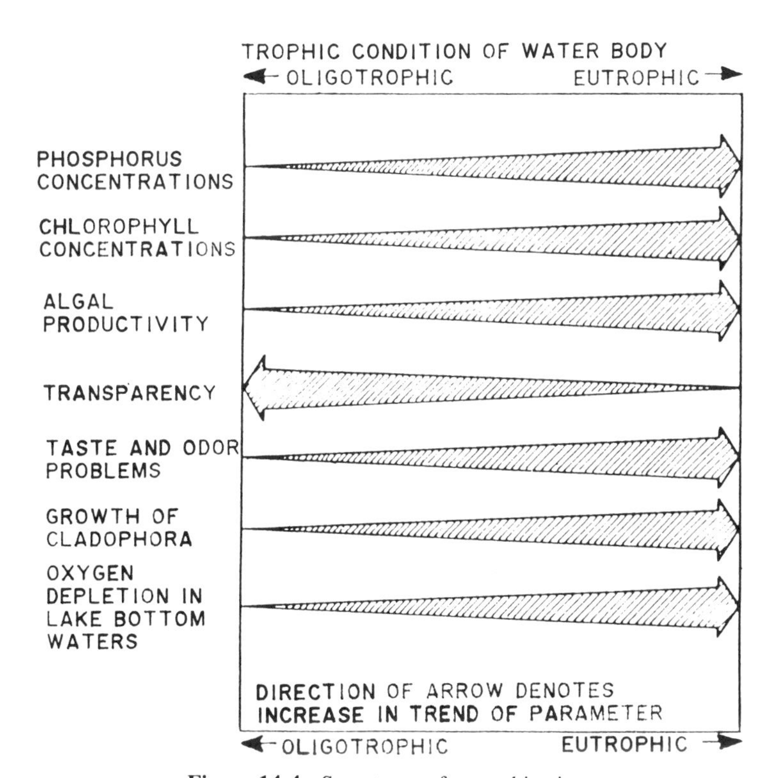

Figure 14.4 Symptoms of eutrophication.

of nutrients, nitrogen and phosphorus are usually considered to be controlling or limiting nutrients. In turbid waters, light penetration may also be a limiting factor.

From analyzing the algae growths it was found that when the nitrogen/phosphorus ratios were greater than 15:1, phosphorus was the possible limiting nutrient, the input of which should be curbed in order to limit the eutrophication process and its symptoms. When the ratio was less than 15:1, nitrogen was limiting. However, these ratios may vary from 5:1 to 20:1. Several laboratory and field techniques are available to estimate the limiting nutrient. The limiting nutrient can also be estimated by plotting phosphorus versus nitrogen concentrations during various seasons. If the line of the best fit intercepts the nitrogen axis (positive intercept on the nitrogen axis and negative one on the phosphorus axis) phosphorus may be considered as being the limiting nutrient and vice versa.

The majority of inland lakes in the United States, including the Great Lakes, are phosphorus limited whereas many estuaries and streams are nitrogen limited. Control of eutrophication can be achieved by curbing the inputs of the limiting nutrient to the water body.

Procedure VIII. Estimating permissible nutrient loadings. Sawyer[65] noted that algal blooms may occur when the concentrations of inorganic nitrogen (NH_4^+, NO_2^-, NO_3^-) and inorganic phosphorus exceed respective values of 0.3 mg/liter of N/liter and 0.02 mg P/liter. Vollenweider,[66] using a simple input-output model that included sedimentation and considering the limiting concentrations proposed by Sawyer, arrived at "admissible" and "dangerous" loadings of N and P that are related to the depth of the water body, H, and the annual flushing rate ρ = annual inflow/lake volume. A chart showing the Vollenweider relationship for phosphorus is shown in Figure 14.5. Corresponding admissible and dangerous loadings for nitrogen would be approximately 15 times higher.

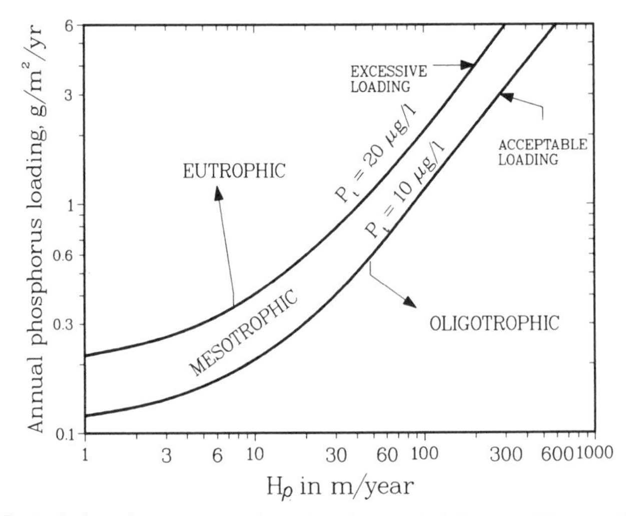

Figure 14.5 Relation between nutrient loading and lake trophic conditions (after Vollenweider[66]).

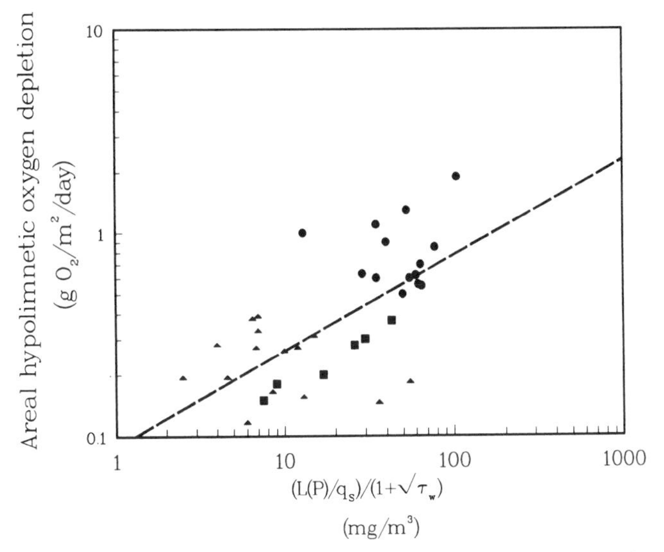

Figure 14.6 Phosphorus loading of lakes and hypolimnetic oxygen depletion (reported with permission from Lee et al.[67] Copyright American Chemical Society).

Assuming that phosphorus is the limiting nutrient Lee et al.[67] have expanded the Vollenweider model to include chlorophyll a, transparency measured by the depth of visible submergence of the Secchi disc, and the hypolimnetic oxygen depletion of the lake. Figure 14.6 shows the results of Lee et al. for dissolved oxygen. On the abscissa:

$L(\mathrm{P})$ = area of phosphorus loading $[\mathrm{mg\ P}/(\mathrm{m}^2\text{-year})]$

q_s = hydraulic loading (m/year) $= H\tau_w = H\rho$

H = mean depth (m) = water body volume/surface area

$\tau_w = 1/\rho$ = hydraulic residence time (years)

= water body volume/annual runoff volume

Example 10: A lake is receiving nutrient inputs from treated wastewater and runoff. Estimate the maximum loading of nitrogen and phosphorus to the lake that would keep the lake below eutrophication levels and maintain summer dissolved oxygen levels above 5 mg/liter.

Lake characteristics: surface area = $10\ \mathrm{km}^2$
average depth = 8 m
Annual inflow including
precipitation = $30 \times 10^6\ \mathrm{m}^3/\mathrm{year}$

Annual flushing rate

$$\rho = \frac{\text{annual inflow}}{\text{lake volume}} = \frac{30\ 000\ 000\ (\text{m}^3/\text{year})}{10\ (\text{km}^2) \times 8\ (\text{m}) \times 10^6\ (\text{m}^2/\text{km}^2)} = 0.38\ \text{year}^{-1}$$

From Figure 14.5 (Vollenweider's chart), the admissible and dangerous loadings of phosphorus for $H\rho = 8 \times 0.38 = 3.04$ are

$$\text{Dangerous P loading} = 0.28\ \text{g}/(\text{m}^2\text{-year})$$

$$\text{Acceptable P loading} = 0.13\ \text{g}/(\text{m}^2\text{-year})$$

If 0.20 g of P/(m^2-year) is selected the loading will be

$$\begin{aligned}\text{max P load} &= 0.2\ [\text{g}/(\text{m}^2\text{-year})] \times 10\ (\text{km}^2) \times 10^6\ (\text{m}^2/\text{km}^2) \times 0.001\ (\text{kg}/\text{g} \\ &= 2\ 000\ \text{kg}/\text{year} = 5.48\ \text{kg of P}/\text{day}\end{aligned}$$

The allowable N load using the Vollenweider model would be about 15 times greater, or

$$\text{max N load} = 15 \times 5.48 = 82\ \text{kg of N}/\text{day}$$

To estimate the maximal loading using the Lee et al. chart (Fig. 14.6) several assumptions must be made:

1. The oxygen is used in the hypolimnion (approximately 5 m depth from the bottom) and there is no oxygen interchange between the upper (epilimnion) and lower (hypolimnion) layers.
2. The average temperature in the hypolimnion is 15°C and the initial oxygen concentration at the beginning of the summer growing season is near saturation.

For $S = 10.2$ mg/liter and the available oxygen in the hypolimnion is

$$\Delta C = S - 5.0 = 10.2 - 5.00 = 5.2\ \text{mg}/\text{liter}$$

The allowable oxygen depletion rate (3-month period)

$$\frac{\Delta C}{\Delta t} H_{\text{h}} = \frac{5.2\ (\text{g}/\text{m}^3) \times 5\ (\text{m})}{90\ (\text{days})} = 0.29\ \text{g}/(\text{m}^2\text{-day})$$

From the Lee et al. chart (Fig. 14.6) the loading parameter for 0.29 g/(m^2-day) of oxygen depletion rate is

$$[L(\text{P})/q_{\text{s}}]/(1 + \sqrt{\tau_{\text{w}}}) = 16\ \text{mg}/\text{m}^3$$

Herein

$$q_s = H\rho = 3.04 \qquad \text{and} \qquad \tau_w = 1/\rho = 1/0.38 = 2.63$$

Then

$$L(\mathrm{P}) = 16q_s(1 + \sqrt{\tau_w}) = 16 \times 3.04 \times (1 + \sqrt{2.63})$$
$$= 127 \text{ mg}/(\text{m}^2\text{-year}) = 0.13 \text{ g}/(\text{m}^2\text{-year})$$

The corresponding P loading is then

$$\text{max P load} = \frac{0.13\ [\text{g}/(\text{m}^2\text{-year})] \times 10\ (\text{km}^2) \times 10^6\ (\text{m}^2/\text{km}^2)}{365\ (\text{days/year}) \times 1\ 000\ (\text{g/kg})}$$
$$= 3.56 \text{ kg of P per day}$$

14.5 USE OF THE MODELS

The waste assimilative capacity of receiving water bodies and waste load allocation are usually determined for extreme low-flow conditions and temperatures. In the United States such conditions are typically represented by a low-flow characteristic that for 7 days would not be exceeded with a probabilistic expectancy of once in ten years (so called Q_{7-10}). In many other countries the low-flow characteristics are based on the cumulative probability curve of a long-term flow series. In such cases, a flow magnitude with 98–99% probability of exceedence is selected. The temperature selected for the waste assimilative capacity determination usually corresponds to the highest average monthly temperature. To select a temperature with a very low probabilistic expectancy such as that for the flows would lead to unrealistically low expectancy of the event and overdesigning of the treatment facilities. This is the result of the fact that a joint probability of two independent phenomena (that is low flow and high temperature) equals a product of multiplication of the individual probabilities.

The model that is selected must be calibrated and verified and agreed upon by all participants in the waste assimilative capacity process, including the regulatory agency and the wastewater dischargers.

The stream or estuary water quality surveys that collect data for calibration and verification should be conducted during a period when dissolved oxygen is available at all locations over the survey course. A low-flow–high-temperature period (late summer) is desirable. Conditions reflecting the effects of highly variable storms or flood flow should be avoided.

The nine waste assimilative capacity determination procedures presented herein are recommended for the following tasks:

- Procedure I for a quick rule-of-thumb determination for a single waste source located on a limited stretch of a stream.

- Simplified procedures II only as a rule-of-thumb procedure for quick answers and when there is a need for fast overview results.
- Procedures III and IV for simple desktop calculations using programable microcomputers or minicomputers. The DOSAG model is a simple public model available from the U.S. Environmental Protection Agency, Athens, GA.
- Procedure V for waste assimilative capacity determination for estuaries.
- Procedure VI for complex conditions involving multiple sources and when modeling the effects of photosynthesis are needed. The QUAL model that is now also available for personal computers (IBM PC or compatibles) is a public model and can be obtained from the U.S. Environmental Protection Agency, Athens, GA.
- Procedure VII for computation of the effects of artificial aeration.
- Procedure VIII for estimation of waste assimilative capacity of quiescent water bodies for nutrients (nitrogen and phosphorus).

An article by Hines et al.[68] provides a good overview on the uses and limitations of models in the waste assimilative capacity determination process. A number of technical handbooks and textbooks have been published recently that deal extensively with the design of models of water quality and their applications from which treatises by James,[69] Shen,[70] Orlob,[71] Thomann and Mueller,[72] and Tchobanoglous and Schroeder[73] are recommended herein.

All these procedures and computations of the self-purification capacity of surface waters are related to the predominant kinds of pollution, namely decomposition of biodegradable organic matter and, with it, they address the potential of communicable diseases. With advances in industrial development in recent times, other pollutants have been introduced against which the self-purification capacity is almost helpless. Examples are industrial salts, radioactive materials, detergents, mineral oils, organic chemicals including many commercial pesticides and herbicides, PCBs (polychlorinated biphenyls), and other toxic materials. Because the waste assimilative capacity for these materials is very low at best, or, more commonly, nonexistent, these materials should be prevented from entering the receiving waters and their production, distribution, and use should be curtailed, strictly controlled, or, as a last resort, banned. Such components can also damage the biota in the treatment process.

REFERENCES

1. K. Imhoff and G. M. Fair, *Sewage Treatment.* Wiley, New York, 1940 and 1956.
2. E. Knöpp, *Schweiz. Z. Hydrol.* **22**(1), 152 (1960).
3. K. Imhoff, *Kommunalwirtschaft,* p. 35 (1955).
4. P. A. Krenkel and V. Novotny, *Water Quality Management.* Academic Press, New York, 1980.

5. H. W. Streeter and E. B. Phelps, *Public Health Bull.* 146, (1925).
6. E. B. Phelps, *Stream Sanitation.* Wiley, New York, 1944.
7. F. M. Kittrell and O. W. Kochtitzky, Jr., *Sewage Works J.* **19,** 1032–1049 (1947).
8. G. M. Fair, J. C. Geyer, and D. A. Okun, *Water and Waste-water Engineering.* Wiley, New York, 1966.
9. A. E. Zanoni and R. J. Rutkowski, *J. Water Pollut. Control Fed.* **44,** 1756–1762 (1972).
10. W. W. Eckenfelder, Jr., *Water Quality Engineering for Practicing Engineers.* Barnes & Noble, New York, 1970.
11. H. W. Streeter, *Sewage Works J.* **9,** 315 (1936).
12. K. Bosko, *Adv. Water Pollut. Res., Proc. 3rd Int. Conf.* IAWPR, Vol. 1, p. 40 (1967).
13. V. Novotny, *Adv. Water Pollut. Res., Proc. 4th Int. Conf. IAWPR,* Pap. I-3, (1969).
14. V. Novotny and P. A. Krenkel, *Water Res.* **9,** 233–241 (1975).
15. R. M. Wright and A. J. McDonnel, *J. Environ. Eng. Div., Am. Soc. Civ. Eng.* **105,** 323–355 (1979).
16. S. O. Ajayi and O. Osibanjo, *Water Res.* **18,** 505–506 (1984).
17. J. D. Boyle and J. A. Scott, *Water Res.* **18,** 1089–1099 (1984).
18. G. M. Fair, E. W. Moore and H. A. Thomas, Jr., *Sewage Works J.* **14,** 270, 756, and 1200 (1941).
19. W. Wang, *Water Res.* **14,** 603–612 (1980).
20. M. J. Barcelona, *Water Res.* **17,** 1081–1093 (1983).
21. G. T. Bowman and J. J. Delfino, *Water Res.* **14,** 491–499 (1980).
22. G. S. Holdren, D. E. Armstrong, and R. F. Harris, *Relation between SOD and Phosphorus Release in Lake Sediments,* Pap. 56th Water Pollution Control Federation Conf., Sec. 39, Atlanta, GA, 1983.
23. J. W. M. Rudd and C. D. Taylor, *Advances in Aquatic Microbiology* (M. R. Drop and H. W. Jannach, eds.), Vol. 2, pp. 77–150. Academic Press, New York, 1980.
24. W. H. Oldaker, A. A. Burghum, and H. R. Parhen, *J. Water Pollut. Control Fed.* **40,** 1688–1701 (1968).
25. A. S. McDonnel and S. D. Hall, *Proc. Ind. Wastes Conf. (Purdue Univ.)* **22,** 414 (1967).
26. R. V. Thomann, *System Analysis and Water Quality Management.* ERA, New York, 1972.
27. W. A. Kreutsberger, T. L. Meinholz, M. Harper, and J. Ibach, *J. Water Pollut. Control Fed.* **52,** 192–201 (1980).
28. T. A. Butts and R. L. Evans, *Sediment Oxygen Demand Studies of Selected Northeastern Illinois Streams.* Illinois State Water Surv. Circ. No 129, Urbana, (1978).
29. D. J. Patterson, E. Epstein, and J. McEvoy, *Water Pollution Investigation: Lower Green Bay and Lower Fox River,* EPA Rep. No. 905/9-74-017. U.S. Environ. Prot. Agency, Chicago, IL, 1974.
30. N. Edberg and B. V. Hofsten, *Water Res.* **7,** 1265–1294 (1973).
31. A. J. McDonnel and S. D. Hall, *J. Water Pollut. Control Fed.* **41,** R353–R363 (1969).
32. J. Fillos and A. N. Molof, *J. Water Pollut. Control Fed.* **44,** 644–662 (1972).
33. A. James, *Water Res.* **8,** 955–959 (1974).
34. C. H. Mortimer, *J. Ecol.* **30,** 280–329 (1942).

35. C. H. Mortimer, *Limnol. Oceanogr.* **16,** 387–404 (1971).
36. H. G. Wild, C. N. Sawyer, and T. C. McMahon, *J. Water Pollut. Control Fed.* **43,** 1845–1854 (1971).
37. Michigan Water Resources Commission, *Oxygen Relationship of Grand River-Lansing to Grand Ledge-1960 Survey.* MWRC, Lansing, MI, 1962.
38. Water Pollution Research Laboratory, *Nitrification in the BOD Test,* Notes Water Pollut., No. 52. Stevenage, Herts, England, 1971.
39. T. J. Tuffey, J. V. Hunter, and V. A. Matulewich, *Water Res. Bull.* **10,** 555–565 (1974).
40. R. R. Ruane and P. A. Krenkel, *J. Water Pollut. Control Fed.*, **50,** 2015–2028 (1978).
41. C. T. Wezernak and J. J. Gannon, *J. Sanit. Eng. Div., Am. Soc. Civ. Eng.*, **94,** 883 (1968).
42. F. Lopez-Bernal, P. A. Krenkel, and R. J. Ruane, *Prog. Water Technol.* **7,** 821–832, (1976).
43. G. T. Wilson and N. MacLeod, *Water Res.* **8,** 341–366 (1974).
44. R. E. Rathbun, *J. Hydraul. Div., Am. Soc. Civ. Eng.* **103,** 409–424 (1977).
45. D. J. O'Connor and W. E. Dobbins, *Trans. Am. Soc. Civ. Eng.* **123,** 641 (1958).
46. P. A. Krenkel and G. T. Orlob, *J. Sanit. Eng. Div., Am. Soc. Civ. Eng.* **88** (SA2), 53 (1962).
47. M. A. Churchill, H. L. Elmore, and R. A. Buckingham, *Adv. Water Pollut. Res., Proc. 1st Int. Conf. IAWPR 1962,* Vol. 1, p. 126–136, (1964).
48. M. Owens, R. W. Edwards, and J. W. Gibbs, *Air Water Pollut.* **8,** 469 (1964).
49. E. L. Tsivoglou and L. A. Neal, *J. Water Pollut. Control Fed.* **48,** 2669 (1976).
50. Anonymous, *J. Sanit. Eng. Div., Am. Soc. Civ. Eng.* **87** (SA6), 59 (1961).
51. Texas Water Development Board, *DOSAG 1, Simulation of Water Quality in Streams and Canals,* TWDB Rep. No. PB-202-974. Austin, TX, 1970.
52. D. J. O'Connor, *Trans. Am. Soc. Civ. Eng.* **105,** 641–684 (1960).
53. Texas Water Development Board, *QUAL-1-Simulation of Water Quality in Stream and Canals,* TWDB Rep. No. PB-202-973. Austin, TX, 1971.
54. D. R. Bingham, C. H. Lin, and R. S. Hoag, *J. Water Pollut. Control Fed.* **56,** 1118–1122 (1984).
55. J. Van Benschoten and W. W. Walker, Jr., *Water Resour. Bull.* **20,** 109 (1984).
56. *Computer Program Documentation for the Enhanced Stream Water Quality Model QUAL-2E.* National Council of the Paper Industry for Air and Stream Improvement, New York, 1985.
57. D. J. O'Connor and D. M. DiToro, *J. Sanit. Eng. Div., Am. Soc. Civ. Eng.* **96,** 547–571 (1970).
58. K. R. Imhoff, *GWF, das Gas-und Wasserfach* **110,** 543–545 (1969).
59. K. R. Imhoff and D. Albrecht, *Adv. Water Pollut. Res., Proc. 6th Int. Conf. IAWPR 1972,* Pap. A/11/22 (1972).
60. G. A. Rohlich (ed.), *Eutrophication: Causes, Consequences, Correctives.* Nat. Acad. Sci., Washington, DC, 1969.
61. P. E. Greeson, *Water Resour. Bull.* **5,** 16–30 (1969).
62. R. E. Carlson, *Limnol Oceanogr.* **22,** 361–369 (1977).
63. P. D. Uttormark and J. P. Wall, *Lake Classification—A Trophic Characterization of*

Wisconsin Lakes, EPA Rep. No. 600/3-75-033. U.S. Environ. Prot. Agency, Washington, DC, 1975.

64. National Eutrophication Survey, *An Approach to Relative Trophic Index System for Classifying Lakes and Reservoirs, Survey,* U.S. EPA, Work Paper No. 24. Pacific Northwest Lab., Corvallis, OR, 1973.
65. C. N. Sawyer, *J. N. Engl. Water Works Assoc.* **61,** 109–127 (1947).
66. R. A. Vollenweider, *Schweiz. Z. Hydrol.* **36,** 53–83 (1975).
67. G. F. Lee, W. Rast, and R. A. Jones, *Environ. Sci. Technol.* **12,** 900–908 (1978).
68. W. G. Hines, D. A. Ricket, G. W. McKenzie, and J. P. Bennett, *J. Water Pollut. Control Fed.* **47,** 2357–2370 (1975).
69. A. James, *An Introduction to Water Quality Modeling.* Wiley, New York, 1984.
70. H. W. Shen (ed.), *Modeling of Rivers.* Wiley, New York, 1979.
71. G. T. Orlob (ed.), *Mathematical Modeling of Water Quality.* Wiley, New York, 1983.
72. R. V. Thomann and J. A. Mueller, *Principles of Surface Water Modeling and Control.* Harper & Row, New York, 1987.
73. G. Tchobanoglous and E. D. Schroeder, *Water Quality: Characteristics, Modeling, Modifications.* Addison-Wesley, Reading, MA, 1985.

BIBLIOGRAPHY FOR PART III

A. K. Biswas (ed.), *Models for Water Quality Management.* McGraw-Hill, New York, 1981.

B. T. Bower, R. Barrù, J. Küchner, and C. S. Russel, *Incentives in Water Quality Management. France and Ruhr Area.* Resources for the Future, Johns Hopkins Univ. Press, Baltimore, MD, 1981.

L. W. Canter, *River Water Quality Monitoring.* Lewis Publ., Chelsea, MI, 1983.

G. D. Cooke, E. B. Welch, S. A. Peterson, and P. R. Neuroth, *Lake and Reservoir Restoration.* Butterworth, Boston, MA and London, 1986.

A. James (ed.), *The Use of Water Quality Models in Water Pollution Control.* Wiley, Chichester, England, 1978.

J. C. Lamb, *Water Quality and its Control.* Wiley, New York, 1979.

J. P. O'Kane, *Estuarine Water Quality Management.* Pitman, Boston, MA, 1980.

K. H. Reckhow and S. C. Chapra, *Engineering Approaches for Lake Management,* 2 vols. Butterworth, Boston, MA and London, 1983.

S. Rinaldi, R. Soncini-Sessa, H. Stehfest, and H. Tamura, *Modeling and Control of River Quality.* McGraw-Hill, New York, 1979.

T. G. Sanders, R. C. Ward, J. C. Loftis, T. D. Steele, D. D. Adrian, and V. Yevjevich, *Designing of Networks for Monitoring Water Quality.* Water Resour. Publ., Littleton, CO, 1983.

H. I. Shuval, *Water Quality Management under Condition of Scarcity.* Academic Press, New York, 1980.

T. H. Y. Tebbutt, *Principles of Water Quality Control.* Pergamon Press, Oxford, 1983.

T. D. Waite, *Principles of Water Quality.* Academic Press, Orlando, FL, 1984.

CHAPTER 15

Conversion of SI units to U.S. (English) Customary Units[a]

To convert from (SI)	To (U.S. English)	Conversion factor
Meters (m)	Feet (ft)	Multiply by 3.28
Meters	Yards	Multiply by 1.094
Meters per second (m/s)	Miles per hour (mph)	Multiply by 2.237
Centimeters (cm)	Inches (in.)	Multiply by 0.39
Milimeter (mm)	Inches	Divide by 25.4
Kilometer (km)	Mile (mi)	Divide by 1.608
cm^2	Square inch	Divide by 6.45
m^2	Square foot (ft^2)	Multiply by 10.76
Hectare (ha = 10 000 m^2)	Acre	Multiply by 2.46
km^2 (= 100 ha = 10^6 m^2)	Square mile	Multiply by 0.387
Liters (= 1 dm^3)	Cubic foot (ft^3)	Divide by 28.3
m^3	ft^3	Multiply by 35.4
Liters	U.S. gal	Divide by 3.78
Liters	Imp. gal	Divide by 4.54
Liters/(s-ha)	Inches/hour	Multiply by 0.014
m^3	Million U.S. gal (mg)	Divide by 3 780
m^3	Barrel	Multiply by 8.54
m^3	Acre-ft	Divide by 1 230
Liters/s	Million gal/day (mgd)	Divide by 43.75
Liters/s	gal/min (gpm)	Multiply by 15.87
Gram (g)	Pounds (lb)	Divide by 454
Gram	Grain	Multiply by 15.43
Kilogram[b] (kg = 1 000 g)	Pounds	Multiply by 2.2
Newton (= 0.1 kg[b])	Pounds	Multiply by 0.225
Metric ton (= 1 000 kg)	U.S. ton	Multiply by 1.1
Metric ton	English ton	Multiply by 0.98
Milligram per liter (mg/liter = g/m^3)	Parts per million (ppm)	Multiply by 1.0
mg/liter	Grain/ft^3	Divide by 2.29
Microgram per liter (μg/liter = 10^{-3} g/m^3)	Parts per billion (ppb)	Multiply by 1.0
g/m^2	lb/ft^2	Divide by 4 885

To convert from (SI)	To (U.S. English)	Conversion factor
g/m^3	lb/ft^3	Multiply by 6.24×10^{-5}
m^3/kg	ft^3/lb	Multiply by 16.03
$m^3/(m^2\text{-day}) = m/day$	gpd/ft^2	Multiply by 24.6
Bar (= 10^5 N/m^2)	psi (= $lb/in.^2$)	Multiply by 14.2
kg/m^2	lb/ft^2	Divide by 4.89
kg/cm^2	psi	Multiply by 14.49
kg/m^3	lb/ft^3	Divide by 16
Liters/m^2	gal/ft^2	Divide by 40.76
Watt (W = N × m/s)	Btu/hr	Multiply by 3.41
Kilowatts (kW = 1 000 W)	Horsepower (hp)	Multiply by 1.34
Kilowatt-hours (kW-hr)	Btu	Multiply by 3 409.5
W/m^3	hp/mg	Multiply by 5
kW-hr/(m^2 × °C)	Btu/ft^2/°F	Multiply by 176
kW-hr/(m^3 × °C)	Btu/ft^3/°F	Multiply by 53.6
Calories (gram) 1 calorie = 1.16×10^{-6} kW-hr	Btu	Divide by 252
Degree celsius (°C)	Degrees Fahrenheit (°F)	Convert as 1.8 × °C + 32

[a]1 m^3 of water weights 1 000 kg.
1 ft^3 of water weights 62.4 lb.
1 U.S. gal of water weights 8.34 lb.
1 Imp. (English) gal of water weights 10 lb.
1 day has 1 440 minutes and 86 400 seconds.

[b]Metric kilograms in this table are weight kilograms which equal 9.81 (m/s^2) × kg (mass) = 9.81 Newtons.

CHAPTER 16

Information Retrieval

16.1 REPORTS AND TECHNICAL JOURNALS

Most of the reports published by the U.S. Environmental Protection Agency can be obtained from the National Technical Information Service, U.S. Department of Commerce, Springfield, VA 22161.

The reports can be also obtained from major libraries that are depositories of U.S. Government documents.

British governmental documents can be obtained from Her Majesty's Stationary Office.

German Reports

Berichte der Abwassertechnischen Vereinigung (ATV). ATV Publ. Co., Markt 1, D-5205, St. Augustin, FRG.

Berichte der Abwasser- und Abfall Technik. ATV Publ. Co., Markt 1, D-5205, St. Augustin, FRG

Major North American Journals

Journal of the Environmental Engineering Division. Proceedings of the American Society of Civil Engineers, 345 East 57th St., New York, NY 10017

Journal of Water Pollution Control Federation (JWPCF), formerly *Sewage Works Journal* and *Sewage and Industrial Wastes.* WPCF, 601 Wythe Street, Alexandria, VA 22314

Environmental Science and Technology. American Chemical Society, 1155 16th St. NW, Washington, D.C. 20036

Journal of Environmental Quality. American Society of Agronomy, 677 S. Segoe Rd., Madison, WI 53711

Water Resources Bulletin. American Water Resources Association, 5410 Grosvenor Lane, Bethesda, MD 20814

Water-Engineering Management. Scranton-Gillete Communications, 380 NW Highway, Des Plains, IL 60016

Civil Engineering. American Society of Civil Engineers, 345 East 47th Street, New York, NY 10017

Canadian Journal of Civil Engineering. Water Research Council of Canada, Ottawa, Ontario K1A-OR6, Canada

European Journals

Water Research. International Association for Water Pollution Research and Control, Pergamon Press, Headington Hill Hall, Oxford OX3 OBW, England

Water Science and Technology formerly *Progress in Water Technology.* International Association for Water Pollution Research and Control, Pergamon Press, Headington Hill Hall, Oxford, OX3 OBW, England

Journal of the Institute of Water Pollution Control (JIWPC), formerly *Institute of Sewage Purification, Journal and Proceedings.* Ledson House 53, London Rd., Maidstone, Kent, England

Effluent + Water Treatment Journal. Thunderbird Enterprises Ltd., Omega Lodge, Troutstream Way, Rickmansworth, Herts. WD3 4JN, England

Water and Waste Treatment. DR Publications Ltd., Faversham House, 111 St James Rd., Croydon CR9 2TH, England

Phoenix International (English and German languages), Sonneggstrasse 21, CH-8006, Zurich, Switzerland

GWF, *Das Gas- und Wasserfach: Wasser/Abwasser.* R. Oldenbourg Publ. Co., Rosenheimer Strasse 145, D-8000 Munich, FRG

Gesundheits-Ingenieur (GI). R. Oldenbourg Publ. Co., Rosenheimer Strasse 145, D-8000 Munich, FRG

Industrieabwässer. Deutscher Kommunal-Verlag, Roseggerstrasse 5a, D-4000 Düsseldorf, FRG

Kommunalwirtschaft. Deutscher Kommunal-Verlag GmbH, Roseggerstrasse 5a, D-4000 Düsseldorf, FRG

Korrespondenz Abwasser (KA). Publ. Geselschaft zur Förderung der Abwassertechnik, Markt 1, D-5205 St. Augustin, FRG

Wasser und Boden. Paul Parey Publ., Spitalerstrasse 12, D-2000 Hamburg, FDG

Wasser, Luft und Betrieb. Publ. Krauskopf Verlag für Wirtschaft, Lessingstrasse 12–14, D-6500 Mainz, FRG

Schweizerische Zeitschrift für Hydrologie. Birkhäuser-Verlag, Basel, Switzerland

16.2 COMPUTERIZED INFORMATION RETRIEVAL

Most of the recent information stored in libraries, technical journals, government reports, and patent offices can be retrieved from the data base vendors using computer terminals. In the United States and worldwide, DIALOG and BRS are the most widely used data base retrieval systems in engineering applications. The needed information can be retrieved telemetrically as abstracts or summaries, using a computer terminal or a personal computer with a phone modem and a printer.

DIALOG, BRS, or other information data base systems can be accessed through a local telephone call to one of the data communication networks such as TYMNET, TELENET, or UNINET in North America or DIALNET. The following data bases contain information related to topics discussed in this book:

AGRICOLA (National Agricultural Library, Belsville, MD)—a worldwide journal literature and monographs related to agriculture, including use of effluents for irrigation and wastewater disposal on land.

DOE Energy (Department of Energy, Washington, D.C.)—all aspects of energy and related fields, including nuclear energy and radioactive waste disposal.

COMPENDEX (Engineering Information, Inc., New York, NY)—machine readable version of Engineering Index, covers approximately 4 500 journals and selected government reports and books.

ENVIRONMENTAL BIBLIOGRAPHY (Environmental Studies Institutes)—general human ecology, atmospheric studies, energy and land resources, water resources, over 500 periodicals covered.

ENVIROLINE (EIC/Inteligence Inc., New York, NY) covers worldwide environmental information, including fields such as management, technology, planning, law, and economics, 5 000 primary publications including Federal Register and Official Gazette.

NTIS (Department of Commerce, Washington, D.C.) is a catalog of government-sponsored research, development, and engineering plus analyses prepared by federal agencies, their contractors or grantees—identifying research, sponsors, and reports. Includes all reports published by the U.S. Environmental Protection Agency, National Science Foundation, and other governmental agencies.

POLLUTION ABSTRACTS (Cambridge Scientific Abstracts)—leading data base on pollution, including air pollution, environmental quality, pesticides, radiation, solid wastes, and water pollution and treatment.

WATER RESOURCES ABSTRACTS (U.S. Department of Interior)—materials covered by over 50 Water Resources Centers and Institutes in the United States, including water resources economics, hydrology, planning and management, water quantity and quality, and pollution.

APPENDIX

Nomographs For Hydraulic Computations

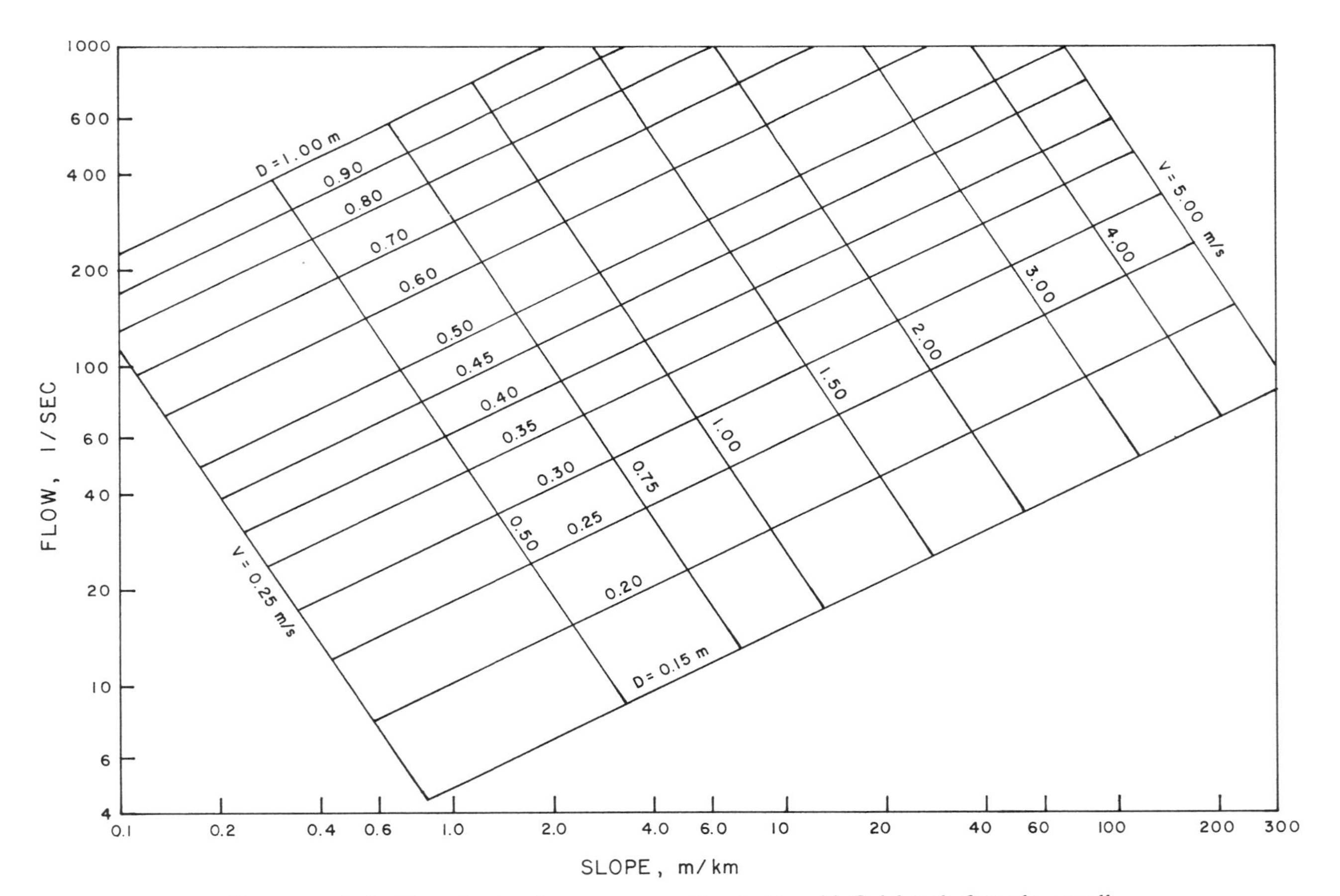

Nomograph 1 Flow in circular pipes according to Prandtl–Colebrook formula—small flows: $k = 1.5$ mm, $\nu = 1.31 \times 10^{-6}$ m^2/s.

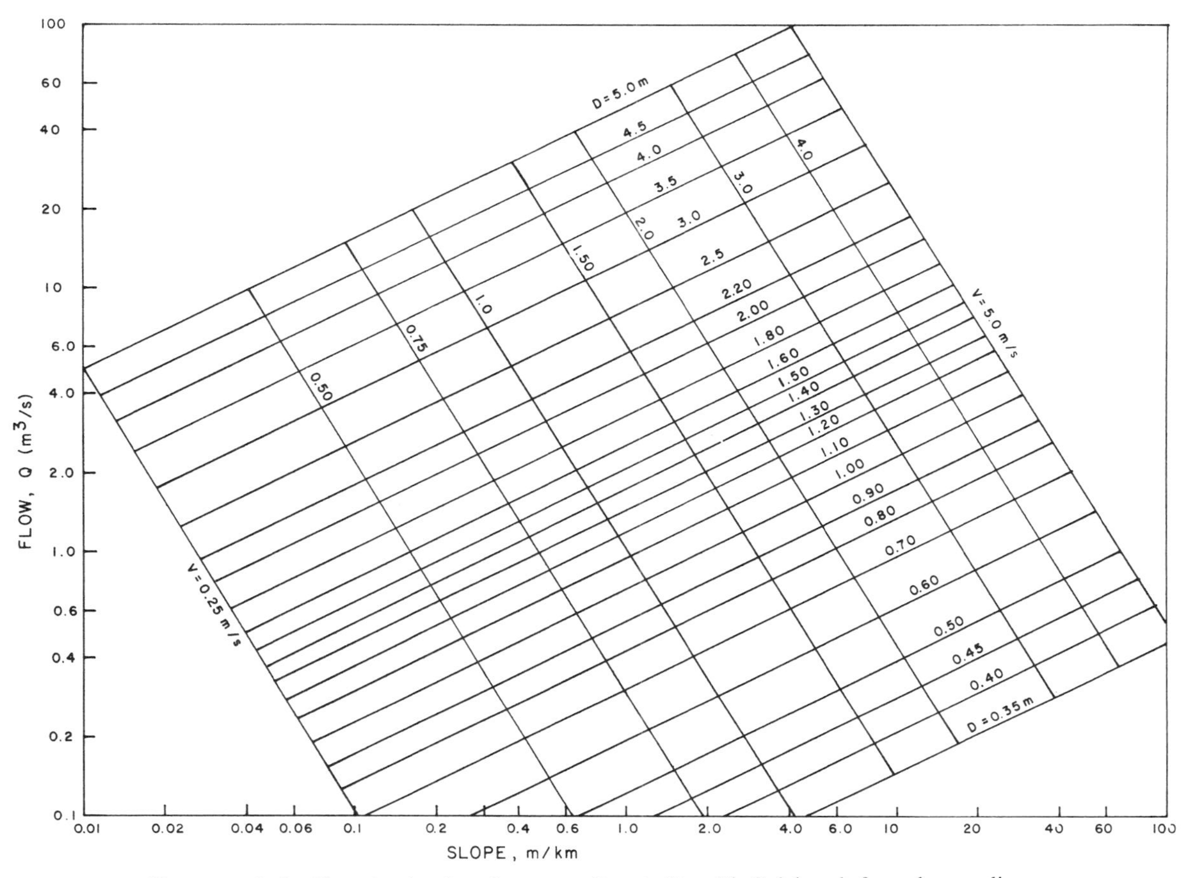

Nomograph 2 Flow in circular pipes according to Prandtl–Colebrook formula—medium and large flows: $k = 1.5$ mm, $\nu = 1.31 \times 10^{-6}$ m^2/s.

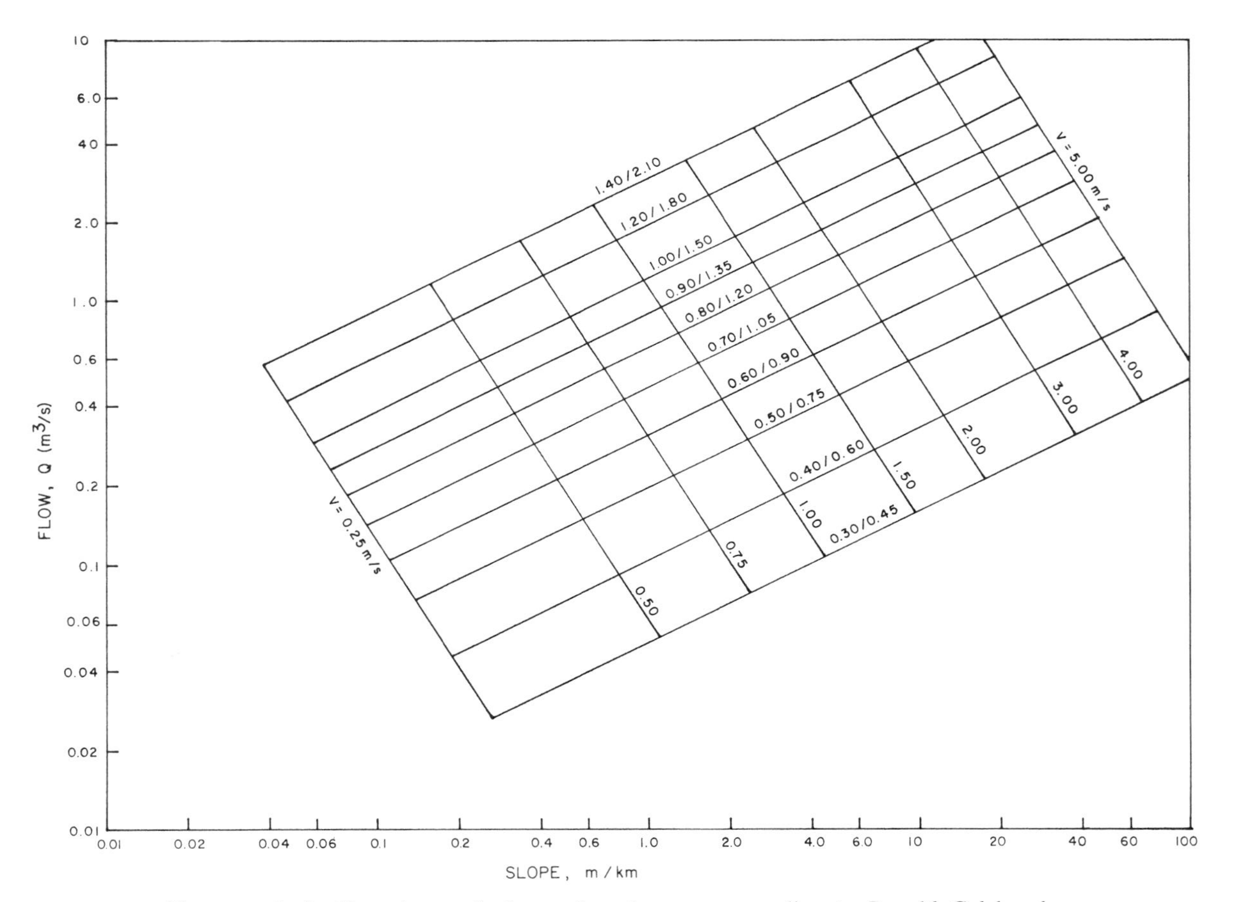

Nomograph 3 Flow in standard egg-shaped sewers according to Prandtl–Colebrook formula: $k = 1.5$ mm, $\nu = 1.31 \times 10^{-6}$ m^2/s.

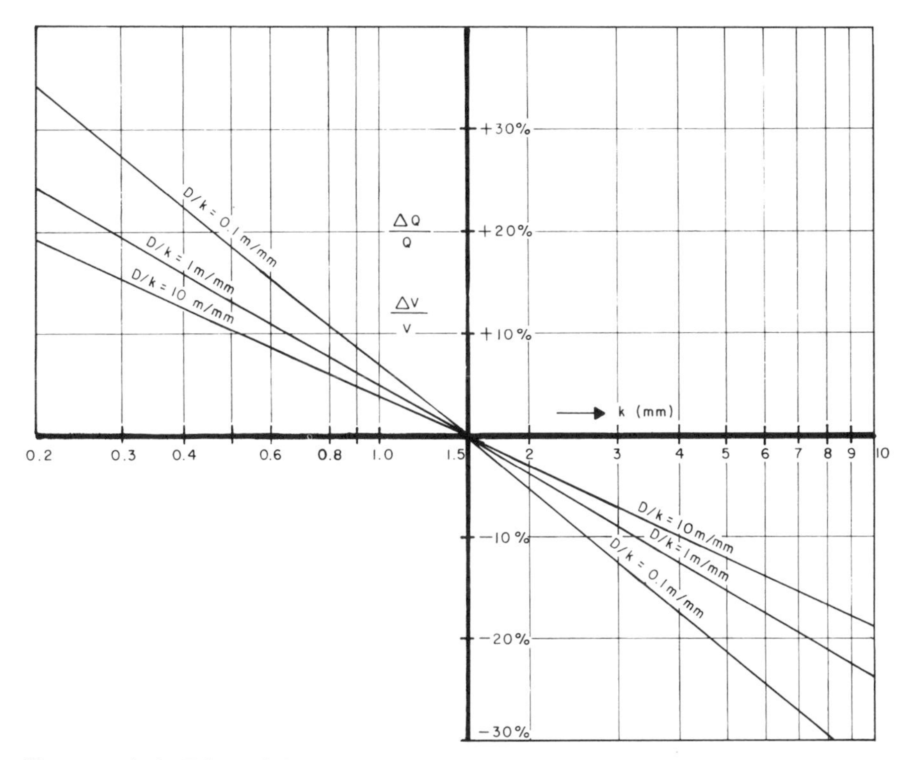

Nomograph 4 Effect of change of roughness on flow and velocity according to Prandtl–Colebrook.

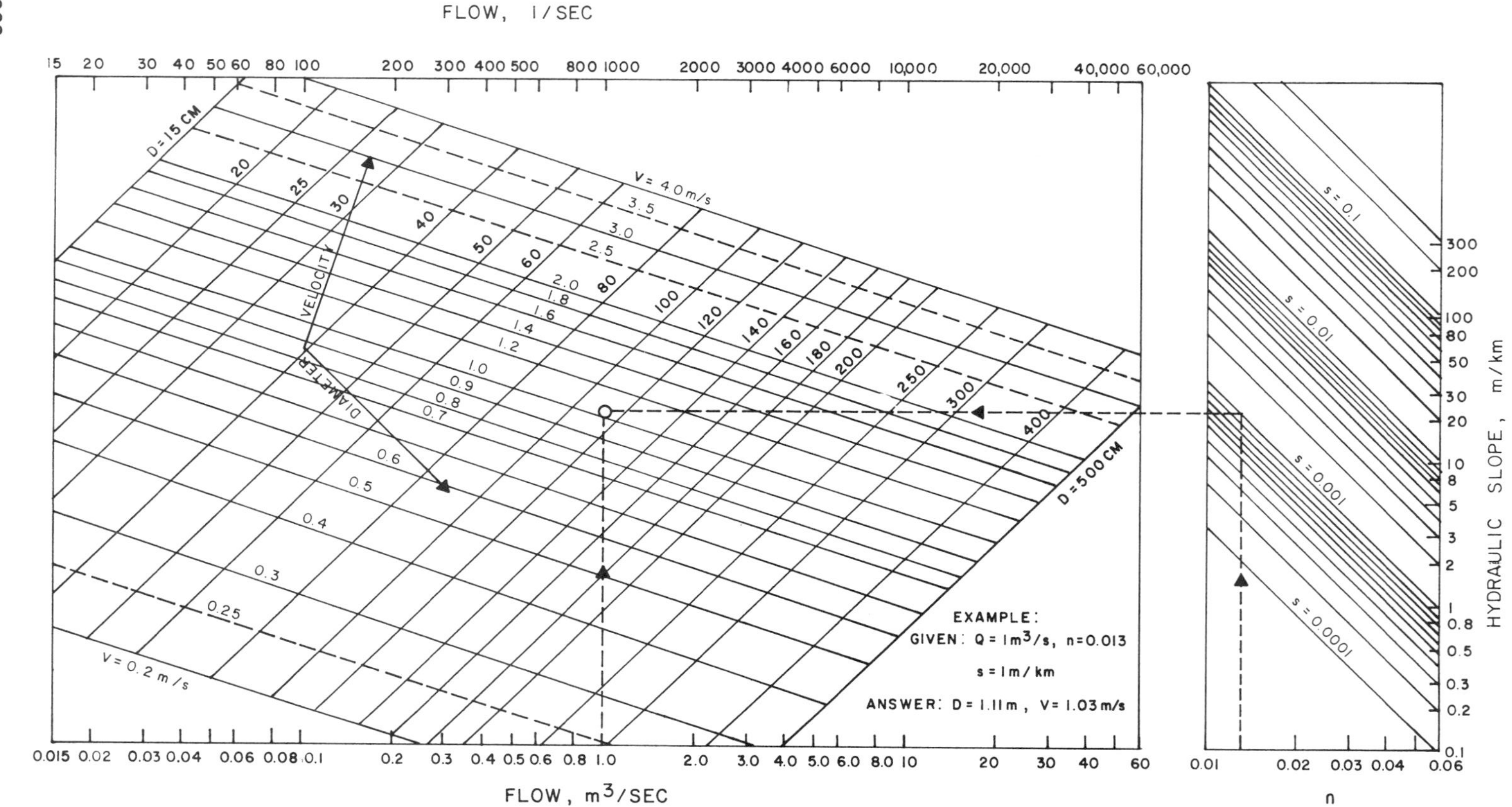

Nomograph 5 Manning equation for full flow in circular pipes.

Author Index

Subject Index